THE NORTHERN
GREAT BARRIER REEF

THE NORTHERN
GREAT BARRIER REEF

A ROYAL SOCIETY DISCUSSION
ORGANIZED BY
D. R. STODDART AND SIR MAURICE YONGE, F.R.S.

HELD ON 28 AND 29 JANUARY 1976

LONDON
THE ROYAL SOCIETY
1978

Printed in Great Britain for the Royal Society
at the
University Press, Cambridge

ISBN 0 85403 102 2

First published in *Philosophical Transactions of the Royal Society of London*:
Part A in series A, volume 291, pages 1–194;
Part B in series B, volume 284, pages 1–162.

Published by the Royal Society
6 Carlton House Terrace, London SW1Y 5AG

PREFACE

This book consists of papers originally presented at a meeting of the Royal Society held to discuss the results of the Royal Society and Universities of Queensland Expedition to the northern Great Barrier Reef, which took place over six months in 1973. The papers originally appeared in two groups in the *Philosophical Transactions of the Royal Society*, in series A (vol. 291, pp. 1–194) and series B (vol. 284, pp. 1–162). The first group comprised mainly geophysical and sedimentological papers, the second mainly morphological and biological papers. Because of this division, the order of the papers in each part and in this book departs from that of the original meeting. In its present form, papers on sea level change appear in each part; each part also includes papers which summarize preceding discussion.

Nevertheless, the structure (and the argument) of the book follows that of the Discussion Meeting. Introductory papers describing the Expedition and the reefs are followed by detailed consideration of the internal structure of the reefs determined by geophysical methods and coring, and of the carbonate sediments and rocks associated with them. Radiocarbon dating is an integral component of the interpretation of rock and sediment units, and radiocarbon dates are reported in detail. This leads naturally to a general consideration of the problem of sea level change in the light of the geological results, and sets the stage for the geomorphological studies in the second part. These are followed by a series of summary papers which bring together geophysical, geological, geomorphological and biological evidence to focus on the evolution of reef features over the last several thousand years, and to relate this history to the still contentious problem of Holocene sea level history.

The various indexes, place name, taxonomic, personal name and general, have been designed to help the reader trace common themes through these distinct yet related papers.

August 1978

D. R. STODDART
C. M. YONGE

CONTENTS

[Thirty-one plates; three pullouts]

† Parts A and B are paginated separately, each beginning at 1.

PART B

THE NORTHERN
GREAT BARRIER REEF

PART A

Phil. Trans. R. Soc. Lond. A. **291**, 3–4 (1978)

Printed in Great Britain

[3]

Introductory remarks

By C. M. Yonge, F.R.S.

Chairman of the Southern Zone Research Committee of the Royal Society

What we are here to discuss concerns the Great Barrier Reef of Australia. It is very fitting that we should do so in this place, because the Royal Society was intimately concerned with events that led to its discovery in 1770. We go back to 1716, to a communication printed in Latin in the *Philosophical Transactions* by Edmond Halley, then Savilean Professor of Geometry at Oxford and Secretary of this Society. There, and for no less an objective than the more accurate determination of the dimensions of the Universe, he drew attention to the unique opportunities to that end to be presented by observing the transits of Venus across the face of the Sun due on 6 June 1761 and 3 June 1769. In the event international observations in the former year were largely fruitless, giving added reason for adequate observations in 1769. One of the conclusions of the specially appointed Transit Committee of the Society was that one site for observation should be in the South Seas.

A direct appeal to George III produced one of the earliest grants of money for purely scientific purposes, and even more to the point the Admiralty was in almost enthusiastic agreement. Sweeping aside the Committee's proposal that Dalrymple should head the expedition, the Admiralty selected Mr James Cook, previously surveyor of the lower reaches of the St Lawrence and of the coasts of Newfoundland. He was now commissioned Lieutenant of H.M.S. *Endeavour*, and the transit was to be observed from the island of Tahiti recently discovered by Wallis on H.M.S. *Dolphin*.

Our concern with this voyage takes us to the later discovery of the east coast of Australia and to Cook's progress north surveying the coast and naming its geographical features. By 10 June 1770 he had sailed for some 600 miles within the complex of offshore reefs that constitute the southern half of the Great Barrier. At 10 o'clock on that day, Trinity Sunday, he 'hauld off north in order to get within a small low Island which lay 2 Leagues from the Main'. This is the first mention of the coral formation occupied by the Expedition of 1928–9. Disaster came the following day when 'the Ship Struck and stuck fast' on what is now Endeavour Reef. After repair of the vessel Cook's subsequent movements become of particular interest. Viewed from the summit of Grassy Hill, he noted with concern the 'number of Sand Banks and Shoals laying along the coast, the innermost lay about 3 or 4 Miles from the Shore, and the outermost extending off the Sea as far as I could see without my glass, some just appearing above water'. With the Southeast Trades blowing, his only hope of escape lay to the north. He proceeded cautiously in that direction, landing on Lizard Island, where from the 1100 ft high summit, he followed the here dramatically sharp line of the outer barrier reefs, perceiving an opening through which he later entered the open Pacific.

Matthew Flinders, who charted the more southern Cumberland Islands and then passed through Flinder's Passage in 1802, was impressed by the beauty of coral growths, writing of 'wheat sheaves, mushrooms, stagshorns, cabbage leaves, and a variety of other forms, glowing under water with vivid tints of every shade betwixt green, purple, brown and white . . .'.

The first comprehensive survey of the Barrier starting in the Capricorns and finishing at Cape York, and including the outer islands of Torres Strait, was carried out by H.M.S. *Fly* in 1842–3, engaged, in the words of its naturalist, Beete Jukes, in 'marking out a more secure road through some of these reefs and shoals'. Impressed with Darwin's recently published work on the origin of reefs, he postulated that the Barrier had grown up on a sloping, slowly subsiding platform.

T. H. Huxley was to come this way on H.M.S. *Rattlesnake* in 1848–9. Strangely oblivious to the supreme zoological interest of what surrounded him, the only comment in his diary when at anchor off Lizard Island concerns a first reading in Italian of Dante's *Divina commedia*! It was another matter with Saville-Kent. His massive *Great Barrier Reef of Australia*, published in 1893, contains a series of magnificently photographed and no less admirably reproduced views of exposed coral which must largely have been taken during low spring tides in winter.

It may have been Saville-Kent's success in viewing exposed corals in winter that led him to advise Alexander Agassiz to visit the Barrier during April and May 1896, when in the small *Croydon* he encountered nothing but persistent trade winds and spent most of his time in shelter. He abandoned the cruise at Cooktown, his major conclusion that the reefs had grown up on a submarine plateau formed by erosion and denudation. He was accompanied by A. G. Mayor, later, among so many other reef activities, to survey the fringing reef around Mer Island.

Recent interest in the Barrier springs largely from the activities of the Queensland Branch of the Royal Geographical Society of Australasia, which in 1922 established a Barrier Reef Committee to investigate 'the origin, growth and natural resources' of the reefs. Under the energetic chairmanship of H. C. Richards much significant geological work was carried out, but the need for biological work led to the invitation to the British Association which resulted in the Expedition of 1928–9 when we established a marine station on Low Isles and studied the cycle of events over 13 months. An associated geographical party led by J. A. Steers was financed by the Royal Geographical Society; in 1936 he paid a further visit, this time surveying southern reefs as well as northern ones.

The Barrier Reef Committee later established a field station on Heron Island in the Capricorn Group, while the Australian Museum, Sydney, has also been active, first at One Tree Island in the south, and, very recently, at Lizard Island. Transcending all came the establishment by the Commonwealth Government in 1972 of the Australian Institute of Marine Science at Townsville, now building most impressive accommodation 30 miles to the east near Cape Cumberland.

To retrace our steps, however, after the experience of his two expeditions, J. A. Steers realized the need for further information about reef structure, sedimentation and related geomorphological problems, and five years ago, in 1971, formulated proposals which came to the Southern Zone Research Committee of this Society. The interests of the University of Queensland at Brisbane and of the James Cook University of North Queensland at Townsville were obtained, with their agreement that David Stoddart should be the leader of a largely geomorphological expedition to the northern regions of the Great Barrier Reef in 1973. In this case the majority of the expedition members came from Australia, many of whom we are happy to see here today. It is the results of this expedition that are now our concern.

Phil. Trans. R. Soc. Lond. A. **291**, 5–22 (1978)   [5]
Printed in Great Britain

The Great Barrier Reef and the Great Barrier Reef Expedition 1973

By D. R. Stoddart

Department of Geography, Cambridge University, Downing Place, Cambridge, U.K.

The Great Barrier Reef, the world's longest barrier reef, extends for 2000 km along the northeast coast of Australia. It is a complex feature, with outer ribbon reefs and inner platform and patch reefs of widely differing forms. Most previous work has concentrated in the extreme south (at Heron and One Tree Islands) or in the central sector (at Low Isles). The 1973 Expedition worked in the northern sector, from Cairns to the latitude of Cape Grafton (11° 30′ S). The main aim of the Expedition was to elucidate the recent history of the reefs, especially in response to Holocene sea level change. Evidence was sought from shallow coring, geophysical surveys, studies of reef and inter-reef sediments, observations on modern reef communities, and the analysis of the geology and geomorphology of reef islands. The work of the field parties in each of these areas is briefly reviewed and related to the questions to be discussed in the following papers.

Introduction

The Great Barrier Reef off the northeastern coast of Australia (figure 1) is the longest in the world. It extends for 2000 km from the latitude of New Guinea to nearly 24° S, and the shelf it encloses, which carries many small reefs, covers some 250000 km². Yet since Cook's first exploration in 1770 it has received less scientific attention than might have been expected, with the conspicuous exception of the work of the Great Barrier Reef Expedition 1928–29. More recently, Maxwell (1968) and his associates have done much to extend our knowledge of the Reef and especially its sediments, and work at research stations at Heron Island and One Tree Reef in the extreme south and at Lizard Island in the north is making a fundamental contribution to our understanding of coral reefs. In this introductory paper I sketch some aspects of the general nature and environment of the reefs, indicate the problems which the 1973 Royal Society and Universities of Queensland Expedition to the northern Great Barrier Reef sought to solve, and describe briefly the organization and conduct of the expedition.

General geomorphology

Many workers have stressed the variability in form of reef structures along the length of the Barrier. Both Yonge (1930) and Steers (1929) draw attention to the presence of long linear ribbon reefs forming an almost continuous barrier along the shelf edge north of Trinity Opening, off Cairns, in latitude 17° S, in contrast with the more scattered and less regular edge reefs south of that point. More recently Maxwell (1968) has defined three main reef provinces:

(1) The Northern Province, between 9° and 16° S, where the shelf is generally less than 40 m deep and where ribbon reefs extend along 71 % of the shelf edge. The width of the shelf varies from 25 to 60 km, but it narrows to a minimum of 13 km at Cape Melville.

(2) The Central Province, between 16° and 20° S, characterized by extensive platform reefs

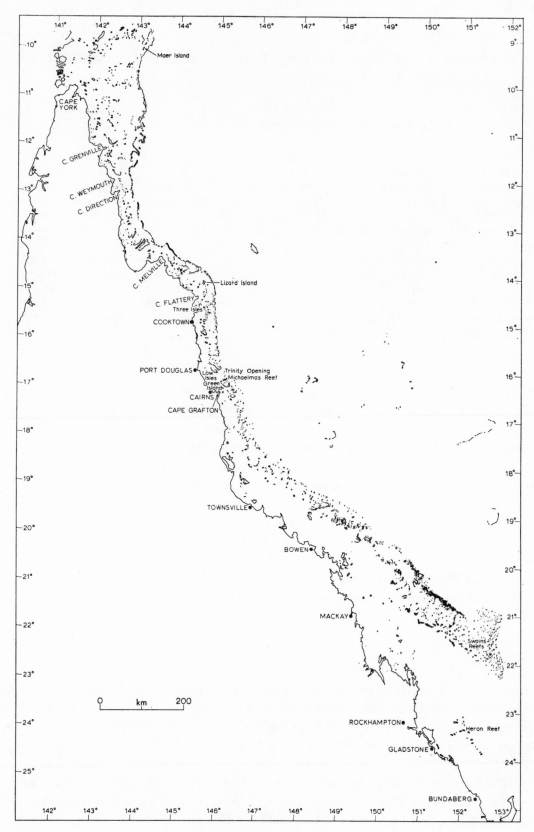

FIGURE 1. The Great Barrier Reef.

on the shelf itself; shelf edge reefs are less continuous and extend along 45% of the edge; and the shelf itself is both wider and deeper (40–70 m) than in the north.

(3) The Southern Province, where the shelf reaches an extreme width of over 300 km and is more than 80 m deep; complex dissected reefs extend along only 36% of the shelf edge in the Swain Reefs, and within the barrier there are scattered platform reefs, some with well developed lagoons (in the Bunker and Capricorn Groups).

We are here concerned only with the reefs of the Northern Province, north of 16° S. Here the mainland coast consists of bold headlands of granite (Capes Weymouth, Direction, Melville and Grafton), with Palaeozoic and Mesozoic graywackes and conglomerates also forming promontories (Capes Bathurst, Tribulation, Flattery and Bedford, and Lookout Point) (Keyser & Lucas 1968). Between Cairns and Cape Melville much of the coast is bold (Bird 1970), but north of Cape Melville there are extensive progradational sequences of beach ridges and mangroves, with high Pleistocene siliceous dunes. 'High islands' of non-reef rocks are common across the coastal shelf. Lizard Island, the highest (359 m) is formed of Permian granite and stands only 18 km from the shelf edge; other granite islands include Howick, Forbes, Haggerstone, Sir Charles Hardy's, and the Cockburn Islands. Less rugged islands in the great indentation of Princess Charlotte's Bay (Flinder's, Clack, King Islands) consist of flat-lying Mesozoic sandstones. Further north, towards New Guinea, shelf edge islands such as Maer and Darnley are volcanic. The granite islands suggest a highly irregular basement on which the reef sequences have formed, while the presence of gently dipping sandstones hints that some at least of the extensive platform reefs might be built on such foundations.

The coral reefs of the Queensland shelf have been classified by Steers (1929), Spender (1930) and Fairbridge (1950) on the basis of size, form, location, and the nature of sedimentary accumulations on them. Maxwell (1968), using air photographs, has shown for the first time the true complexity and variability of the reefs, and proposes a new descriptive–genetic terminology and classification, much of which, in the present state of knowledge, must be inferential. For present purposes in the Northern Province we need only distinguish:

(1) Fringing reefs. These are not extensively developed in the Northern Region.

(2) Platform reefs. These are extensive reefs, located on the shelf itself, with rather featureless surfaces, generally with maximum dimensions of 5–10 km though some reach more than 20 km in length. Some carry leeward sand cays (such as Combe and Stapleton), but most have no permanent dry land.

(3) Ribbon reefs. These are linear reefs 5–25 km long and 300–500 m wide, extending along the shelf edge in the Northern Province, and interrupted by gaps or entrances with depths similar to those of the adjacent shelf. These reefs are little known, though Yonge Reef was briefly described by Stephenson, Stephenson, Tandy & Spender (1931, pp. 82–86).

(4) Small inner-shelf reefs. These include the 'low wooded islands' of Steers (1929, 1937) or 'island reefs' of Spender (1930): small reefs, usually only 1–2 km in greatest dimension, with complex assemblages of shingle ridges, exposed limestone platforms, mangrove swamps and sand cays, which, because of their diversity, offer the greatest possibilities of interpreting the recent record of reef growth and sedimentation. The headquarters of the 1928–29 Expedition was located on one of these (Low Isles); Steers (1937, 1938) studied many more in 1936; and, with their surrounding areas, they formed the main focus of the 1973 Expedition.

D. R. STODDART

Environment

Four main environmental factors are significant, both in interpreting the reefs themselves and in planning investigations of them. These factors are tides, winds, rainfall and cyclones.

Tides in the Great Barrier Reef area (Easton 1970) are variable but generally large in comparison with most other reef provinces. In the south the range at springs varies from 5 to 10 m, while in the Central Province it averages 3 m. In the north, it varies from 1.7 m at Cairns to 2.5 m at Cape York, but little is known of tides between these two locations (table 1). The tides are mixed semidiurnal, with the diurnal component increasing northwards. One fact of great ecological interest is that there is a seasonal reversal in the diurnal incidence of low water: in summer (December–January) tides low enough to give reef-top emersion occur at night, while in the winter (May–August) they occur during the day. All elevations in this study are referred to low water datum at Cairns, i.e. the mean height of the lower low waters at springs.

TABLE 1. MEAN TIDAL RANGE AT SPRINGS IN THE NORTHERN PROVINCE OF THE GREAT BARRIER REEF

Data from Queensland Department of Harbours and Marine (1973).

locality	latitude S	range/m
Cairncross I.	11° 14½′	2.50
Hannibal I.	11° 35½′	2.44
Sir Charles Hardy's I.	11° 55′	1.83
Cape Grenville	11° 58′	2.01
Piper I.	12° 14½′	2.01
Restoration Rock	12° 37½′	1.77
Night I.	13° 10½′	1.77
Morris I.	13° 30′	1.80
Burkitt I.	13° 56½′	2.07
Pipon I.	14° 07½′	1.59
Flinders Is	14° 10′	1.89
Howick I.	14° 30′	1.77
Lizard I.	14° 40′	1.53
Low Wooded I.	15° 05′	1.77
Cooktown	15° 28′	1.77
Hope Is	15° 44′	1.71
Bailay Creek	16° 12′	1.28
Low I.	16° 23′	1.71
Port Douglas	16° 29′	1.77
Green I.	16° 45½′	1.71
Cairns	16° 55′	1.89
Fitzroy I.	16° 56′	1.77

Tidal levels in metres at Cairns are as follows (Queensland Department of Harbours and Marine 1973):

highest astronomical tide	2.9	mean low water neaps	1.2
mean high water springs	2.3	mean low water springs	0.5
mean high water neaps	1.6	lowest astronomical tide	−0.1

The Southeast Trades blow with great constancy and force in the Great Barrier Reef area for nine months of the year (March–November). Mean velocities are given as 18–29 km/h (10–15 knots), but north of Cairns these are frequently exceeded. As Steers (1929) noted, the Trades blow almost parallel to the trend of the coast, producing short, steep, difficult seas inside the Barrier; their effect on the morphology and ecology of the reefs has long been recognized (Hedley & Taylor 1908; Stephenson *et al.* 1931, p. 26). Cook sailed along the Barrier lagoon north of Cairns in August 1770, the time of year when the 1973 Expedition was in the field, and his journals are full of references to rough conditions: 'strong gales all this day' on the 8th, 'a prodigious great sea with breakers all around us' on the 9th, 'strong gales',

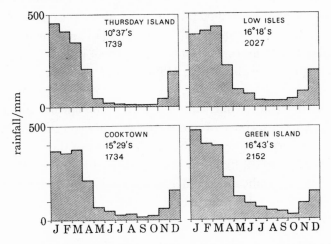

FIGURE 2. Monthly distribution of rainfall at stations on the northern Great Barrier Reef; the totals for each station are annual means in millimetres. Data from Brandon (1973).

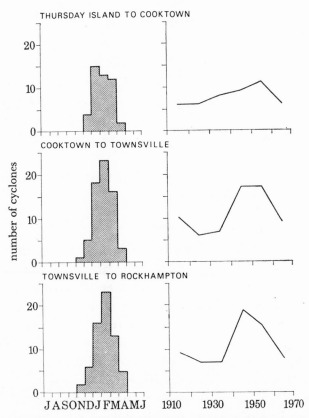

FIGURE 3. Mean monthly distribution of cyclones (left) and variation in frequency per decade (right) over the period of record (1910–69). Data from Coleman (1971).

'hardly any intermission', 'fresh gales', 'such boisterous weather' (Beaglehole (ed.) 1955). When the Trades cease, for the three months December–February, the winds are northwesterly and less regular, with days of calm.

Rainfall varies considerably with latitude on the mainland coast of Queensland, even over short distances (Brandon 1973), but there are few records from islands on the coastal shelf. The Trades are dry, and rainfall is concentrated in the summer months December–April (figure 2). Annual totals average about 2000 mm in the southern part of the Northern Province (Green Island 2152 mm, Low Isles 2027 mm), and less than this further north (Cooktown 1734 mm, Thursday Island 1739 mm). It is possible that reef islands far from the mainland, such as Raine Island, are considerably drier; Willis Island in the Coral Sea has a 48 year mean of 1047 mm.

Finally, cyclones have important ecological and geomorphic effects, and also form a severe constraint on reef work. They are concentrated in the summer wet season. Records appear to show (figure 3) that they are less frequent in the northern part of the Northern Province than in the south, but this could reflect differences in recording rather than occurrence (Coleman 1971). The available data also indicate considerable variations in frequency over time, with highest frequencies about 1950 and lowest between 1920 and 1930.

PROBLEMS

Three sets of interrelated problems were investigated during the 1973 expedition.

Morphology

The gross form of the reefs clearly reflects both the history of growth and the effects of environmental constraints. Many workers have discussed the presumed effects of wave refraction on reef outline in plan, but generally without detailed consideration of the time scales involved or of the linkages between form and process. Maxwell (1968, pp. 139–142) has in addition shown how the orientation of reefs changes with latitude, possibly in response to environmental controls. Morphological adjustments take place in more complex ways than simply in plan-form, moreover, and we were particularly concerned with the detailed distribution of such sedimentary bodies as boulder tracts, shingle ramparts, sand and gravel flats, and sand cays on the reef tops.

Structure

Form is but the external expression of structures which themselves reflect growth histories constrained by both environmental influence and substrate topography. Earlier work on the development of individual reefs on the Queensland coast emphasized the importance of vertical growth followed by lateral extension and differentiation as the reef top approached sea level and thus began to modify its own environment (Fairbridge 1950). Studies elsewhere have suggested that many reef structures may thinly veneer topography developed on older reef limestones exposed during Pleistocene low sea levels (Stoddart 1969, 1973), and this interpretation has been documented for Caribbean reefs by Purdy (1974a, b) and in the Indian Ocean by Braithwaite, Taylor & Kennedy (1973), using both seismic techniques and drilling in the first case and stratigraphic analysis of exposed reefs in the second. Maxwell's studies (1968) of Queensland reef development emphasized simple growth processes rather than karst-inheritance, although on the basis of external morphology rather than internal structure, but in a later paper (Maxwell 1970) he did draw attention to the apparent inheritance of reef forms from those of older sedimentary structures. While there have been several deep bores on the Great Barrier Reef – at Michaelmas Cay (1926), Heron Island (1937), Anchor Cay (1969), Wreck Island (1960), and several recent deep holes in the south – the study of these has been concerned with the deeper rather than the shallower parts of the record, and there is considerable disagreement over dating and interpretation (Richards & Hill 1942; Lloyd 1973). The 1973 studies were therefore concerned not only with surficial structures, both of sediments and limestone bodies, but also with the upper few tens of metres of the reef record, to determine the sequence of phases of accretion and erosion in late Pleistocene and Holocene times. To interpret these records, some attention was also given to the distribution and composition of modern growth communities.

History

It is impossible to consider problems of reef development, structure and form without being concerned with the effects of Pleistocene changes of sea level on the reefs. There is broad agreement that the last major fall in sea level was to more than 100 m below the present level, and that between 10000 and 6000 a B.P. the sea rose at a mean rate of 1 m/100 a, but there is controversy over (1) the lowest level to which the sea last fell, (2) the occurrence, level and duration of still-stands during the subsequent transgression (and during earlier transgressions and regressions), and (3) the date at which sea level reached its present level, and its subsequent behaviour. With respect to this last period, arguments have been adduced for sea level stability over the last few thousand years; for continuing but decelerating transgression; and for oscillation both above and below the present level. Since corals are marine animals which can grow up only to specific tidal levels, they are clearly sensitive to such fluctuations, and it should therefore be possible to determine the interrelations of changing sea level and of reef form and structure. Hopefully the details of the reefs can be used to construct an unambiguous record of changing sea level.

Maxwell (1968, 1973) has erected a relative chronology of sea level change based on evidence of now-submerged terrestrial drainage patterns on the shelf, on the distribution of relict and contemporary shelf sediments, and on shelf and reef topography. This calls for still-stands at -102 m, -88 m, -66 m, an extensive stand at -29 m, a fall to another extensive stand at -59 m, a rise to a further important level at -37 m, a brief stand at -18 m, a transgression to $+3$ m, and a fall to present sea level. Except in so far as these stages can be related by elevation and sequence to the general eustatic curve of Fairbridge (1961), it is fair to note that Maxwell's interpretation is largely inferential, and is not yet supported either by radiometric evidence or by stratigraphic information on reef structures.

Figure 4 gives part of Milliman & Emery's (1968) curve of Holocene sea level, with, superimposed on it, the levels of Maxwell's inferred still-stands; it is not apparent how the formation of extensive terraces and ridges can have taken place during the time intervals of still-stands which might be accommodated within the general curve. It is, of course, possible that some of the features identified by Maxwell are older than the last major transgression. Figure 4 also shows two low-level ^{14}C dates on Australian material: one of $13\,860 \pm 220$ a B.P. at -175 m on the southern Great Barrier Reef (Veeh & Veevers 1970) and one of $16\,910 \pm 500$ a B.P. at -132 m in the Timor Sea (Van Andel & Veevers 1967, p. 100): these may define the trough of the curve in the Australian region.

The evidence of very recent oscillations is controversial (Thom, Hails & Martin 1969; Gill & Hopley 1972; Cook & Polach 1973; Thom & Chappell 1975), but nevertheless crucial to an interpretation of recent reef history. Much of the work of the expedition was concerned with resolving ambiguities in the evidence for positive and negative sea level changes in the northern Great Barrier Reef area, ambiguities increased because of the macrotidal conditions and also because of the complex relation between growing marine organisms and tidal levels.

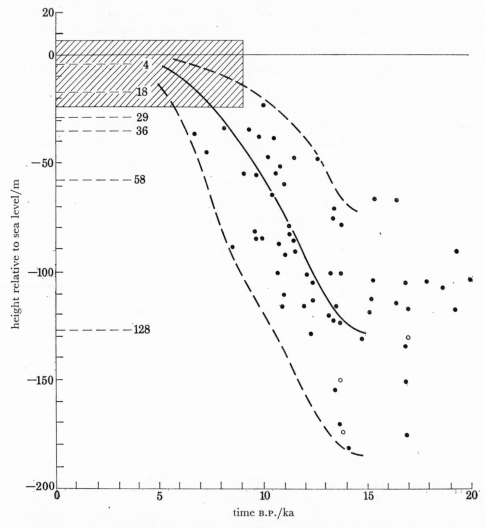

FIGURE 4. Holocene rise in sea level, after Milliman & Emery (1968). The open circles are additional data point for Australian samples published by Van Andel & Veevers (1967) and Veeh & Veevers (1970). The levels on the left of the graph are those of sea level still-stands inferred by Maxwell (1968, 1973). The shaded area represents the controversial recent period of sea level history with which the Expedition was concerned.

ORGANIZATION OF THE EXPEDITION

The Expedition, which began in mid-July and lasted until mid-November 1973, was divided into three phases, each concentrating on different themes in different areas. Table 2 lists the members of the Expedition with their main research interests. The Expedition used several vessels for varying periods: *Odyssey* was the main vessel for transporting personnel and also for the geophysical programme; *Calypso II* was used for the sediment sampling programme; *Privateer*, a barge, was used for transporting the drilling rig during Phase I; and *James Kirby*, the James Cook University research vessel, was used north of Cape Melville during Phase III. The Expedition also worked closely with the Royal Australian Navy during Phase II.

TABLE 2. MEMBERS OF THE EXPEDITION

J. Beaton, archaeologist, Department of Prehistory, Australian National University, Canberra (Phase II, part).

A. L. Bloom, geomorphologist, Department of Geology, Cornell University, New York (Phase II, part).

J. Collins, marine biologist, School of Biological Sciences, James Cook University, Townsville (Phase III).

P. Flood, sedimentologist, Department of Geology, University of Queensland, Brisbane (Phase I).

P. E. Gibbs, marine biologist, Marine Biological Association of the United Kingdom, Plymouth (Phases I–III).

D. Hopley, geomorphologist, Department of Geography, James Cook University of North Queensland, Townsville (Phases II, part, and III).

R. C. L. Hudson, marine biologist, School of Biological Sciences, James Cook University, Townsville (Phase III).

N. Kelland, geophysicist, Coastal Sedimentation Unit, Institute of Oceanographic Sciences, Taunton (Phase II, part).

E. Laundon, geophysics technician, Department of Geology, University of Queensland, Brisbane (Phase II, part).

R. F. McLean, geomorphologist, Department of Biogeography and Geomorphology, Australian National University, Canberra (Phases I and II).

M. Morton, sedimentology technician, Department of Geology, University of Queensland, Brisbane (Phase II, part).

G. R. Orme, deputy leader, sedimentologist, Department of Geology, University of Queensland, Brisbane (Phases I, part, and II).

H. A. Polach, radiometric dating, Radiocarbon Dating Laboratory, Australian National University, Canberra (Phase I, part).

I. R. Price, marine biologist, School of Biological Sciences, James Cook University, Townsville (Phase III, part).

G. E. G. Sargent, geophysicist, Department of Geology, University of Queensland, Brisbane (Phase II, part).

T. P. Scoffin, sedimentologist, Grant Institute of Geology, University of Edinburgh, Edinburgh (Phases I and II).

K. Shaw, drilling technician, Department of Biogeography and Geomorphology, Australian National University, Canberra (Phase I).

A. Smith, sedimentology technician, Department of Geology, University of Queensland, Brisbane (Phases I and II, part).

D. R. Stoddart, Expedition Leader, geomorphologist, Department of Geography, University of Cambridge, Cambridge (Phases I–III).

B. G. Thom, drilling operation, Department of Biogeography and Geomorphology, Australian National University, Canberra (Phase I).

J. E. N. Veron, marine biologist, School of Biological Sciences, James Cook University, Townsville (Phase I, part, and leader, Phase III).

J. Webb, geophysicist, Department of Geology, University of Queensland, Brisbane (Phase II, part).

L. P. Zann, marine biologist, School of Biological Sciences, James Cook University, Townsville (Phase III, part).

L. Zell, marine biology technician, School of Biological Sciences, James Cook University, Townsville (Phases I, II–part, III).

Phase I (figures 8 and 9)

Phase I was an initial phase of one month (19 July–19 August) in the Howick Group, centred at Howick and Bewick Islands. A team from the Australian National University under B. G. Thom carried out shallow drilling on Bewick and Stapleton Islands. A sedimentological team comprising G. R. Orme, T. P. Scoffin and P. G. Flood, working from *Calypso II*, sampled shelf and reef-top sediments. A geomorphological team of D. R. Stoddart and R. F. McLean studied island geomorphology and sediments, mainly from *Odyssey*. In addition, J. E. N. Veron collected corals, H. Polach supervised the collection of samples for radiometric dating, and P. E. Gibbs collected benthic fauna. This phase ended in mid-August when the Expedition returned to Cairns; it was followed by a charter flight over the areas already visited, for photographic purposes.

Phase II (figures 5–9)

Phase II extended from 20 August to 14 October and was logistically complex, partly because of the need to provide ship time for the geophysical party, partly to coordinate the programme with the movements of the Royal Australian Navy Hi-fix survey teams. The geomorphological and sedimentological parties therefore separated from the *Odyssey* and occupied a series of

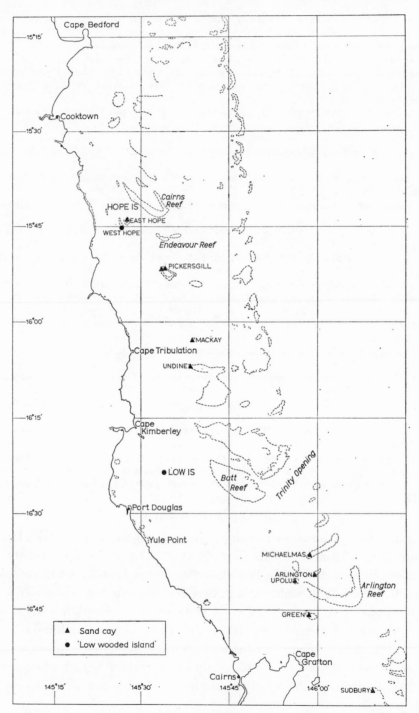

FIGURE 5. Southern sector of the Northern Province: reef islands mapped.

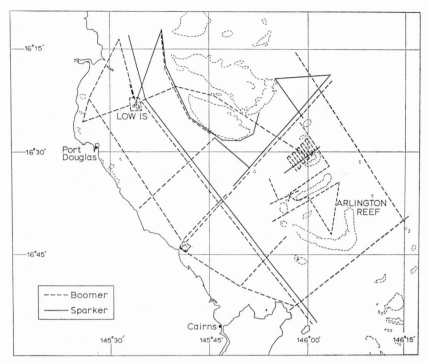

FIGURE 6. Southern sector of the Northern Province: seismic profiles.

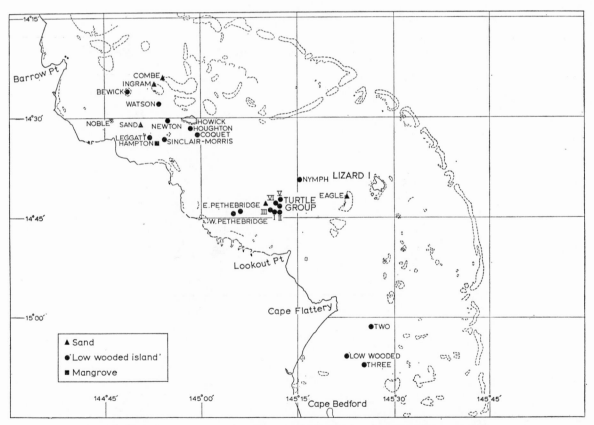

FIGURE 7. Central sector of the Northern Province: reef islands mapped.

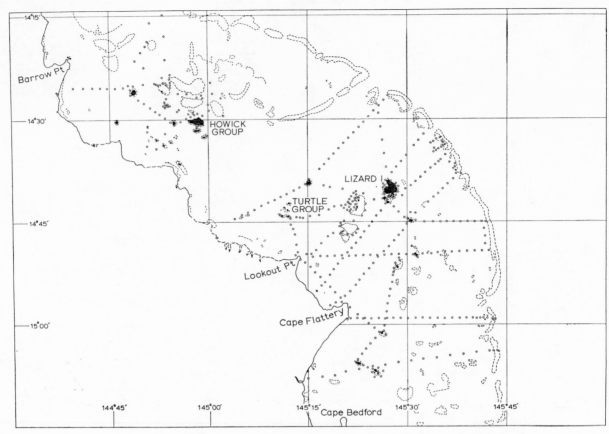

FIGURE 8. Central sector of the Northern Province: shelf sediment samples.

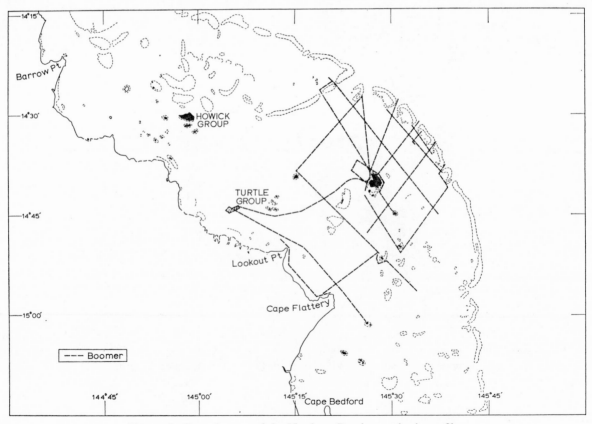

FIGURE 9. Central sector of the Northern Province: seismic profiles.

shore camps, serviced by *Calypso II*. The geophysical work was carried out with *Odyssey*, fitted with Hi-fix equipment operated by R.A.N. personnel.

The shore party comprised D. R. Stoddart, P. E. Gibbs, T. P. Scoffin and R. F. McLean. It occupied camps at Low Isles (headquarters of the 1928–29 Expedition) from 23–29 August, East Hope Island from 29 August to 5 September, and Three Isles from 6 to 26 September. D. Hopley joined the Expedition during this last period. Each of these camps was made the

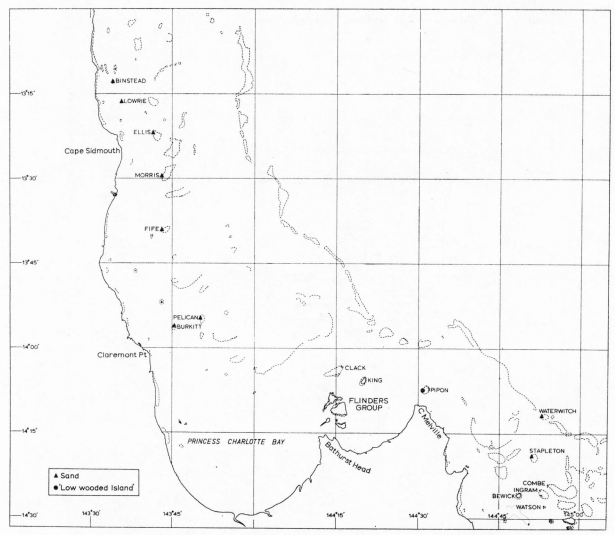

FIGURE 10. Northern sector of the Northern Province: reef islands mapped.

basis for studies of reefs and islands in the vicinity. Meanwhile the geophysical party with *Odyssey* worked its first area between Cairns, Arlington Reef and Low Isles (figure 6). The Royal Australian Navy then moved the Hi-fix navigation system to the area between Lizard Island and Cape Flattery. While *Odyssey* carried out geophysical surveys here (figure 9), the shore party, now joined by A. L. Bloom, moved north to the Turtle Islands, where it worked with *Calypso II* from 27 September to 14 October. At the end of this period the Expedition regrouped in Cairns after seven weeks in the field.

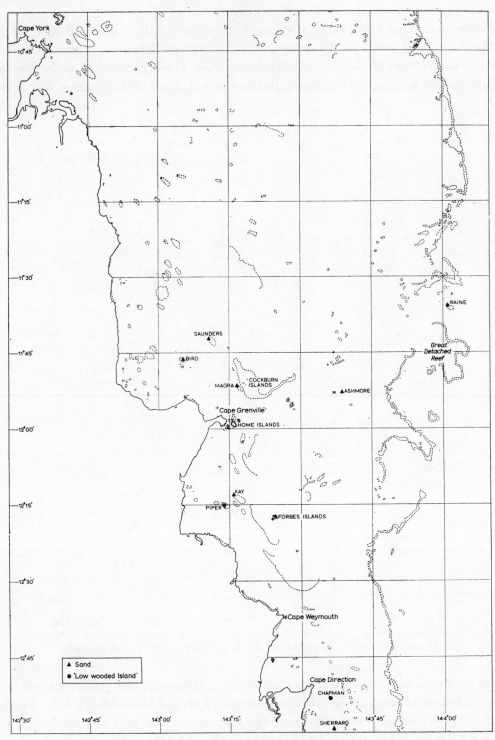

FIGURE 11. Far northern sector of the Northern Province: reef islands mapped.

Phase III (figures 10 and 11)

The Expedition again moved north on 20 October to Lizard, the Howick Group and Pipon Island, to rendezvous with James Cook University's research vessel *James Kirby* with a party from Townsville under J. E. N. Veron. *Odyssey* then returned to Cairns at the end of its charter. *Calypso II* also returned to Cooktown, but its charter was extended so that Orme and Flood could complete the sediment sampling programme of Phases I and II (figure 8). *James Kirby* with Stoddart, Gibbs, Hopley and Veron's biological party then moved north from Cape Melville to Cape Grenville, finally reaching Raine Island and the Great Detached Reef in early November. This was the furthest north point that the Expedition worked (11° 36′ N). Stoddart, Gibbs and Hopley were then landed at Iron Range and flew to Cairns, and *James Kirby* worked south back to Townsville. The Expedition effectively ended its field work in mid-November.

Summary of scientific projects

Geomorphology

Geomorphological studies were concentrated on reef islands, including simple sand cays, shingle islands, mangrove islands, and 'low wooded islands'. Approximately 60 islands were mapped at scales varying from 1 : 500 to 1 : 2500 by Stoddart. Profiles were levelled on many of these and related to the Cairns datum by Stoddart and McLean during Phases I and II, and by Hopley and Bloom during the latter part of Phase II. The geomorphological work was in the hands of Stoddart and Hopley during Phase III. The aim of this work was to identify and describe the sediment bodies which have formed islands during Holocene time, and to map the major vegetation units on them. These studies also provided basic data for some of the following projects, and provided data for comparison with maps made in previous years, notably by Steers (1938) in 1936.

Island sediments and rocks

McLean and Scoffin concentrated on the surficial sediments of the islands (sand and shingle of ridges, ramparts and cays) and cemented materials (upper and lower platforms and beach-rocks). They were concerned with the composition and origin of the sediments; McLean also worked on dating and on post-depositional changes in clast morphology, and Scoffin on the diagenetic fabrics and histories of the cemented rocks. Hopley studied beach-rock morphology, and Bloom the mangrove sediments.

Drilling programme

The Australian National University Gemco drill was used for coring on Bewick and Stapleton Islands during Phase I. This work was supervised by Thom with the assistance of Orme: the drilling itself was in the hands of Karl Shaw. The rig weighed 1900 kg and was not easy to handle on steep beaches with a high tidal range. Because of tide and weather conditions it was not possible to relocate the rig at Noble Island, the third site selected, and bad weather in fact prevented the barge *Privateer* returning to Cooktown with the rig until long after the charter period had ended. Drilling was carried out to 19.5 m on Stapleton and 29 m on Bewick.

Shelf sediment sampling

During Phase I the bottom sediment sampling programme from *Calypso II* was in the hands of Flood, Scoffin and Smith. Samples were collected both from inter-reef shelf areas and also from reef flats in the Howick Group. Gibbs also took dredge samples to extract bottom fauna. During Phase II Scoffin sampled in the area of Arlington Reef, Low Isles and the Hope Islands, again collecting shelf and reef flat material. Scoffin also took more samples during Phase III in the Turtle Islands. During the supplementary charter of *Calypso II* Orme sampled intensively in the area between Lizard Island and the mainland, and this area has the most intensive bottom-sampling coverage of the expedition. Most samples were split, one half being treated with peroxide and the other preserved in alcohol.

Geophysics

The geophysical programme was operated from *Odyssey* during Phase II, with two high resolution boomers, two streamers, an E. G. & G. recorder, and side-scan sonar. Positions were accurately fixed by using a Decca Hi-fix receiver supplied and installed in *Odyssey* by the Hydrographic Department, Royal Australian Navy, which also set up shore stations and seconded Petty Officer Clyde Hillsdon to the Expedition while the equipment was in use. The aim of this programme was to determine sub-bottom sedimentary structures and the contact between reef and non-reef rocks. In spite of difficult weather, several hundred kilometres of records were obtained, most comprehensively in the northern area between Lizard Island and the mainland; this is also the area of most intensive sediment sampling. The geophysical programme was operated by Webb, Kelland and Laundon during the first part of Phase II, and by Orme and Sargent during the second.

Dating programme

Particular attention was paid throughout the Expedition to the collection of critical material from cores, surface sediment samples, and surface rocks for radiocarbon dating, to determine more precisely the sequence of Holocene events and to give a time dimension to the geomorphological record. Polach of the Australian National University Radiocarbon Dating Laboratory joined the Expedition during Phase I to advise on sampling techniques, and also provided preliminary dates on samples while the Expedition was still in the field. Collection of samples was largely in the hands of McLean, Hopley and Bloom, with Thom responsible for the core samples.

Reef zonation and composition

Veron with Thom made initial collections and reef transects during Phase I in the Howick Group, but the main work on reef zonation took place from *James Kirby* during Phase III. Studies were made of ribbon reefs on the shelf edge at Great Detached Reef (11° 40′ S) and Tijou Reef (13° 08′ S): quantitative surveys using line transect methods were made on the reef front, reef flat and back-reef by Veron, Price and Hudson. Less intensive studies for comparative and faunistic purposes were made at other sites. Zell, Veron and Hudson made large collections of living corals during Phase III, to supplement other collections made during Phases I and II by Zell. Corals were collected from a total of 33 sites, the specimens from each site varying in number from 20 to 300.

Other benthic fauna

Gibbs studied the bottom fauna of the shelf from dredge collections, and also the infauna of unconsolidated sediments on reef flats. Collections were made of polychaetes, echinoderms, crustaceans, molluscs and other groups. Particular attention was paid to a resurvey of Low Isles and Three Isles to compare with faunistic surveys in 1928–29 and 1954. Price collected benthic algae during Phase III as part of the reef transect surveys.

Terrestrial flora

Stoddart collected vascular plants on 40 islands during the geomorphological surveys and mapping. The collections totalled over 1100 numbers, mostly in six sets. The first set has been retained by the Queensland Herbarium in Brisbane, where the material was sorted and dried as it arrived from the field, and the other sets are being distributed by the Department of Botany, National Museum of Natural History, Washington. Determination of the material has been carried out by staff of the Queensland Herbarium under its Director, Mr S. Everist, and by Dr F. R. Fosberg in Washington.

Prehistory

John Beaton from the Australian National University joined the Expedition for the first part of Phase III to carry out a reconnaissance survey for prehistoric habitation sites which could have geomorphological implications. Surveys were made on Lizard, Howick, Bewick, Noble and Pipon Islands, and were supplemented by other observations by McLean.

Clearly an expedition of such magnitude and complexity, spending such a high proportion of its total time in active fieldwork, would not have been possible without an immense amount of support from many individuals and organizations. The captains and crews of the vessels used made it possible to remain at work and carry out a crowded programme, often in very difficult conditions. The Department of Geology, University of Queensland, was the Expedition's main base in Australia, and not only lent much of the geophysical and other equipment but also seconded technical staff. Other geophysical equipment was loaned by the Natural Environment Research Council in England. The Royal Australian Navy generously cooperated with the Expedition during the geophysical surveys, and loaned Decca Hi-fix equipment and an operator for use in *Odyssey*. The Navy also allowed us the use of a store in Cairns for both equipment and an increasing quantity of specimens. Phase III of the Expedition was made possible by the provision of *James Kirby* by James Cook University of North Queensland. Drilling equipment during Phase I was loaned by the Department of Biogeography and Geomorphology, Research School of Pacific Studies, Australian National University.

The Expedition was only made possible by the generous financial support given by the Royal Society of London, the University of Queensland in Brisbane, and James Cook University in Townsville, as well as by many other institutions. I acknowledge in particular the support of the Southern Zone Research Committee of the Royal Society, which was initially responsible for planning the Expedition, the Expedition Committee established by the University of Queensland, which overlooked its affairs in Australia, and the officers of the Great Barrier Reef Committee, notably Dr Owen Jones and Dr Patricia Mather, for many kindnesses. All members of the Expedition thank Professor J. A. Steers and Sir Maurice Yonge, who, as members of the 1928–29 Expedition, supported this further study and whose work provided such a stimulus in the field.

Lastly, the smooth operation of the Expedition owed a very great deal to the tolerance and goodwill of all those involved in it, and their willingness to accept an often complex programme so that all the participants could complete their individual projects.

REFERENCES (Stoddart)

Beaglehole, J. C. 1955 (ed.) *The journals of Captain James Cook on his voyage of discovery*. vol. 1. *The voyage of the Endeavour, 1768–1771*. Cambridge: Hakluyt Society.

Bird, E. C. F. 1970 *J. trop. Geog.* **31**, 33–39.

Braithwaite, C. J. R., Taylor, J. D. & Kennedy, W. J. 1973 *Phil. Trans. R. Soc. Lond.* B **266**, 307–340.

Brandon, D. E. 1973 In *Biology and geology of coral reefs* (eds O. A. Jones & R. Endean), vol. 1, pp. 187–232. New York: Academic Press.

Coleman, F. 1971 *Frequencies, tracks and intensities of tropical cyclones in the Australian region November 1909 to June 1969*. (42 pages.) Canberra: Department of the Interior, Bureau of Meteorology.

Cook, P. J. & Polach, H. A. 1973 *Mar. Geol.* **14**, 1–16.

Easton, A. K. 1970 *Flinders University, Horace Lamb Research Centre, Research Papers*, **37**, 1–326.

Fairbridge, R. W. 1950 *J. Geol.* **58**, 330–401.

Fairbridge, R. W. 1961 *Phys. Chem. Earth* **4**, 99–185.

Gill, E. D. & Hopley, D. 1972 *Mar. Geol.* **12**, 223–233.

Hedley, C. & Taylor, T. G. 1908 *Rep. Aust. Ass. Adv. Sci. 1907*, 394–413.

Keyser, F. de & Lucas, K. G. 1968 *Bull. Bur. Min. Res. Geol. Geophys.* **84**, 1–254.

Lloyd, A. R. 1973 In *Biology and geology of coral reefs* (eds O. A. Jones & R. Endean), vol. 1, pp. 347–366. New York: Academic Press.

Maxwell, W. G. H. 1968 *Atlas of the Great Barrier Reef*. (258 pages.) Amsterdam: Elsevier.

Maxwell, W. G. H. 1973 In *Biology and geology of coral reefs* (eds O. A. Jones & R. Endean), vol. 1, pp. 233–272 New York: Academic Press.

Milliman, J. D. & Emery, K. O. 1968 *Science, N.Y.* **162**, 1121–1123.

Purdy, E. G. 1974*a* *Spec. Publs Soc. econ. Paleont. Miner. Tulsa*, **18**, 9–76.

Purdy, E. G. 1974*b* *Bull. Am. Ass. Petrol. Geol.* **58**, 825–855.

Queensland Department of Harbours and Marine 1973 *Tide tables and notes on boating for the coast of Queensland*. (72 pages.) Brisbane: Department of Harbours and Marine.

Richards, H. C. & Hill, D 1942 *Rep. Gt. Barrier Reef Comm.* **5**, 1–111.

Spender, M. A. 1930 *Geogrl J.* **76**, 194–214 and 273–297.

Steers, J. A. 1929 *Geogrl J.* **74**, 232–257 and 341–367.

Steers, J. A. 1937 *Geogrl J.* **89**, 1–28 and 119–139.

Steers, J. A. 1938 *Rep. Gt Barrier Reef Comm.* **4**, 51–96.

Stephenson, T. A., Stephenson, A., Tandy, G., & Spender, M. A. 1931 *Scient. Rep. Gt Barrier Reef Exped. 1928–29* **3**, 17–112.

Stoddart, D. R. 1969 *Biol. Rev.* **44**, 433–498.

Stoddart, D. R. 1973 *Geography* **58**, 313–323.

Thom, B. G. & Chappell, J. 1975 *Search* **6**, 90–93.

Thom, B. G., Hails, J. R. & Martin, A. R. H. 1969 *Mar. Geol.* **7**, 161–168.

Van Andel, Tj. H. & Veevers, J. J. 1967 *Bull. Bur. Min. Res. Geol. Geophys.* **83**, 1–173.

Veeh, H. H. & Veevers, J. J. 1970 *Nature, Lond.* **226**, 536–537.

Yonge, C. M. 1930 *A year on the Great Barrier Reef*. (245 pages.) London: Putnam.

Phil. Trans. R. Soc. Lond. A. **291**, 23–35 (1978)   [23]

Printed in Great Britain

Aspects of the geological history and structure of the northern Great Barrier Reef

By G. R. Orme†, J. P. Webb†, N. C. Kelland‡ and G. E. G. Sargent†

† *Department of Geology and Mineralogy, University of Queensland, Brisbane, 4067, Australia*

‡ *Sonar Marine Ltd, P.O. Box 7, Ashford, Middlesex, U.K.*

[Plates 1 and 2; pullouts 1 and 2]

Continuous high resolution seismic profiling of the continental shelf near Cairns between latitudes 16° 15′ S and 16° 55′ S to depths of 70–120 m revealed a sequence of varied lithologies with major and minor disconformities in both reef and non-reef accumulations. The most conspicuous sub-bottom reflector (I) is a complex surface formed through marine regression, shelf emergence, and subaerial erosion, which has been dissected by an ancient drainage system represented by sediment filled channels. It occurs at approximately − 67 m near Spur and Onyx reefs, has been deeply channelled below Trinity Opening, and generally rises towards the mainland. This surface marks a major disconformity representing a considerable hiatus in the development of the reef province, and has strongly influenced the distribution and thickness of both reefs and sedimentary accumulations formed during the succeeding marine transgression. While the age of reflector I is conjectural, the higher parts of this surface may have been emergent and non-depositional for longer periods than the lower parts, so that overlying sedimentary accumulations and reefs in contact with it may not be exactly contemporaneous.

A continental slope terrace at approximately − 116 m may reflect a late Würm low stand of sea level, and periods of still stand during the Holocene transgression, or minor regressions may be indicated by minor disconformities, and by the marine terraces and changes in slope which occur at several levels, the most conspicuous occurring at − 30 m on reefs of the marginal shelf.

The seismic profiles illustrate the form and internal structure of some sedimentary units, and indicate that the present viable outer reefs are merely remnants of more extensive precursors.

1. Introduction

Fairbridge (1950) and Maxwell (1968a) have provided comprehensive accounts of the Great Barrier Reef Province, and both authors discuss the probable influence of major tectonic features and postulated sea level changes on general reef distribution, in order to account for the major contrast in character of different regions of this reef province. There has been considerable debate regarding the significance of various bathymetric features in terms of the succession of sea level changes with reference to the Quaternary geological history of the Queensland continental shelf (see, for example, Thom & Chappell 1975).

In recent years, investigations of reef complexes in other parts of the world have resulted in a growing awareness of the influence of pre-Holocene surfaces on the form and distribution of Holocene coral reefs (see, for example, MacNeil 1954; Hoffmeister & Ladd 1944; Purdy 1974a), and it has been shown that Holocene sediment distribution patterns owe much to the influence of relict features (Purdy 1974b). On the Queensland shelf, relict sediments are considered to be widespread and important features of the present seabed (Maxwell 1968b; Maxwell & Swinchatt 1970).

Davies (1974), in estimating the position of the Holocene–Pleistocene junction beneath Heron Island Reef, has drawn attention to the possible significance of a solution unconformity at shallow depth, and there has been speculation regarding bathymetric evidence for ancient karst surfaces beneath Holocene reefs (see, for example, Davies 1975; Davies, Radke & Robinson 1976; Flood 1976a, b).

An opportunity to investigate these concepts, and other aspects of the geological history of the Great Barrier Reef in the Cairns area, with the use of high resolution reflexion seismic methods with accurate position fixing (Decca Hi-fix), was provided by the Royal Society (London) and Universities of Queensland Expedition to the Great Barrier Reef (1973).

Specific objectives of the seismic investigations were to determine lateral and vertical stratigraphic relations between reef masses and bedded sedimentary accumulations; define disconformities, and evaluate stratigraphic and geomorphological evidence for sea level changes; locate ancient drainage channels; determine reef thicknesses; ascertain the nature of reef foundations; determine the degree to which inherited features have influenced reef distribution and form; and to obtain a clearer picture of the geological history of this part of the Great Barrier Reef by assessing the sequence of palaeoenvironmental changes indicated by the stratigraphic record.

2. GEOLOGICAL SETTING

The present characteristics of the continental terrace, and adjacent Coral Sea Provinces, were initiated by Tertiary events. Extensive planation of the Australian continent in early Tertiary time was succeeded by widespread crustal warping and fracturing in the late Tertiary time; accompanied by arching of the Eastern Highlands and the subsidence along normal faults of adjacent areas to the east. This period of tectonic activity also promoted the development of the Queensland Trough and the separation of the Coral Sea Plateau from the Queensland shelf as a distinct reef province (Ewing, Hawkins & Ludwig 1970; Orme 1977) and subsequent subsidence of the Queensland shelf.

The major palaeoenvironmental changes which took place on the Queensland shelf are reflected in the six Great Barrier Reef Bores (Michaelmas Cay, Anchor Cay, Heron Island, Wreck Island, Aquarius No. 1, Capricorn 1A), which indicate in general a trend from non-marine conditions to the reefal environment of the present day. Reefal conditions '...similar to those existing along the inner parts of the Great Barrier Reefs first came into existence along the northeastern Australian shelf during the early Middle Miocene' (Lloyd 1973, p. 365). In the Capricorn Basin (Southern Region of the Great Barrier Reef) the change from non-marine to marine sedimentation occurred in the late Oligocene; subaerial erosion surfaces and submerged 'hardgrounds' on the upper continental slope suggest a period of uplift and non-sedimentation from late middle Miocene to Pliocene, followed, in the late Pliocene, by shallow-water sedimentation leading to reef-building conditions (Palmieri 1974). In the Cairns area the prevalence of a reefal environment on the marginal shelf since Pliocene times is implied by the Michaelmas Cay bore hole sequence (Richards & Hill 1942; Lloyd 1973).

Various bathymetric anomalies and shoreline features have often been correlated speculatively with previous stands of sea level. It is believed that with the onset of the Pleistocene, the emergence of the continental shelf and present coastal belt began, and that 'during the Riss and Würm glacials (120 000–105 000 years B.P. and 70 000–12 000 years B.P.) the greater part of the shelf was exposed' (Maxwell 1973, p. 266).

3. GENERAL PHYSICAL CHARACTERISTICS OF THE AREA

The area with which the present investigation is concerned occupies the continental shelf near Cairns, between latitudes 16° 15′ S and 16° 55′ S. It lies in the northern part of the Central Region of the Great Barrier Reef Province as defined by Maxwell (1968a), which is here 60 km wide, and displays a semi-meridional trend of physiographic zones and sediment facies. Unlike the Northern Region, there is here no continuous barrier of shelf edge (ribbon) reefs, the most extensive reefs (Batt and Tongue Reefs, and the Arlington Complex) occurring on the inner part of the marginal shelf.

Depths are generally less than 30 fathoms (55 m) except where the main channels (Grafton Passage and Trinity Opening) extend across the marginal shelf. Bathymetric zones recognized by Maxwell & Swinchatt (1970) are: the near-shore zone extending from the coast to five fathoms (9 m), the inner shelf extending to 20 fathoms (36.5 m) forming a reef-free channel, and the marginal shelf extending to the shelf break at 40–50 fathoms (73–91 m).

This is a region of high rainfall (2032–3048 mm per year), with a maximum tidal range of 3 m. The nature and distribution of the sediments have been studied by Maxwell & Swinchatt (1970). Fringing reefs are present north of Yule point, although there is evidence that they did not begin as fringing reefs, but have acquired this state through progradation of shoreline (Bird 1971).

The adjacent mainland rises rapidly to the Atherton Tableland over 950 m above sea level, and the main rivers discharging to the sea in this area are the Daintree, Mowbrey, and Barron Rivers.

4. METHODS

Continuous seismic reflexion profiling was carried out along traverses arranged as far as possible in two, more or less orthogonal, sets, namely roughly parallel to and at right angles to the edge of the continental shelf. This pattern was pursued in order to facilitate correlation of observed features. Decca Hi-fix control was employed as a means of accurate position fixing, fixes being taken at intervals of 5 min along the tracks.

For most traverses the equipment used comprised a 1kJ triggered capacitor bank driving either a Huntec ED.10 high resolution boomer, or a slightly smaller transducer of similar characteristics (Sargent 1969). The recording system comprised short 6- or 10-element hydrophone streamers coupled, via an input bandpass filter, to an E.G. & G. Model 254 graphic recorder. The sources normally operated at repetition rates of 2 pulses per second and the passband of the input filter was customarily set in the range 500–5000 Hz.

A magnetic tape recorder, run in parallel with the Model 254, recorded broadband seismic data, but the present interpretation is based entirely on the 28 cm wide Alfax monitor records produced by the graphic recorder.

The depth of sub-bottom penetration achieved by this system at operational speeds varied between 70 and 120 m, and the quality of the records produced was influenced strongly by sea state.

For deeper penetrations some traverses were undertaken with a 5 kJ sparker source. For these surveys the recording configuration remained unchanged, but the input filter low frequency cut off was adjusted to lower frequencies. The sparker records, while showing considerable loss in near-bottom resolutions, provided useful data on deeper reflectors.

5. RESULTS

In the discussion of the continuous seismic profiles it is (reasonably) assumed that the interpreted seismic reflexions are coincident either with significant geological interfaces referred to as 'surfaces' or with a layering (producing seismically reflective velocity contrasts) within consanguineous deposits. The sub-bottom velocity distribution in the area is not known, and sea water velocity has been used in reducing near-bottom interval times to thicknesses. Furthermore, no attempt has been made at this time to correct for the effects of recording system geometry and the apparent migration of dipping reflectors. As a result, the computed positions and depths of reflecting horizons are likely to be systematically displaced from their true spatial relations as may be revealed by confirmatory drilling. However, these effects will be generally small, and not of great significance in the framework of the present study.

Prominent reflectors may be produced by marked changes in lithology, varying degrees of cementation or solution, and mineralogical and textural changes. Hence, subaerially exposed,

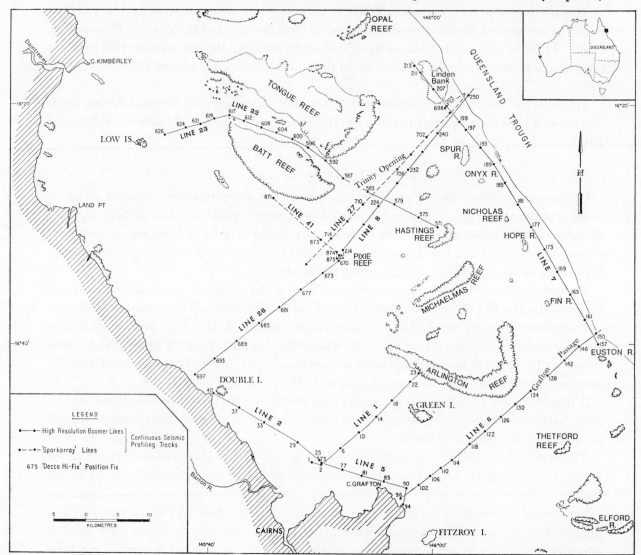

FIGURE 1. Map of the continental shelf near Cairns showing reef distribution and seismic profiling tracks.

weathered and cemented surfaces have marked reflectivity. Likewise, well sorted coarse-grained deposits may be differentiated, in some records, from those deposits which lack hydraulic classification and which contain quantities of carbonaceous matter, on the basis of their different acoustic properties. Similarly, well sorted quartz sands may be acoustically more transparent than carbonate shell sands. Owing to high attenuation of signals by reef masses it is often difficult to obtain definition of their internal structures, and consequently some of the major reflexions may lose clarity as they are traced into some of the thicker reef masses near the shelf edge.

Of the 47 profiles made in the Cairns Region during the 1973 Expedition, only 11 are considered in this paper (see figure 1), but these are sufficiently representative to demonstrate the broad geological characteristics of the area.

(a) Seismic reflectors

The seabed shows an interesting diversity of relief and acoustic properties, in consequence largely of the occurrence of both reef and non-reef facies. The seabed has very rugged topography where it rises abruptly over modern viable reefs, which are most numerous and extensive on the marginal shelf, in contrast to its flat or gently undulating nature over the sediment-covered inner shelf and inter-reef areas. Diversification of relief is also caused, locally, by ancient ridges and pinnacles which protrude above the seabed.

The greatest bathymetric relief (58 m) illustrated by the profiles is seen on profile 7 (plate 1) where Linden Bank rises to −15 m from the adjacent seabed, which at Trinity Opening lies at −73 m. On line 6, near the shelf edge the seabed is at −60 m; opposite the mouth of the Barron River on line 2 it lies at −12 m; and along line 22 (figure 2) it lies −37 to −40 m, rising to −15 m over the flanks of Batt and Tongue reefs, and falling to −53 m where the profile crosses Trinity Opening. Present seabed relief is influenced, therefore, not only by reef growth but also by ancient drainage channels.

Prominent sub-bottom reflectors are present, numbered I to III in increasing depth. All are shown on profile 22 (figure 2, pullout 1) and their form and relations with older and younger deposits indicate that they are erosion surfaces. Horizon 'S' is a subsidiary erosion surface present at shallow sub-bottom depth, particularly in channel filling deposits. There are also numerous minor reflexions within the reef masses, which may facilitate the reconstruction of the development of reef forms, though consideration of this aspect is beyond the scope of this paper.

(i) Reflector I

The most striking of all sub-bottom features is reflector I, which can be clearly seen in all seismic profiles, and which displays considerable relief. Beneath line 22 (see figure 2) it descends from a general level of −45 to −60 m beneath Trinity Opening, and rises to −30 m beneath Batt Reef.

Along line 7 the general level of reflector I is −67.5 m, rising to −47 m below some of the outer-shelf reefs, and falling to approximately −105 m at Trinity Opening (see figure 3, pullout 2 and plate 2). The irregular, channelled nature of surface I is also shown on profile 6 (plate 2). Along line 6, near the shelf edge, surface I occurs at a depth of 64 m below sea level and it is channelled to approximately −86 m.

It is apparent that reflector I, which in general rises towards the present shoreline, is an ancient erosion surface that transects both reef and non-reef bedded accumulations, and which has been dissected by an ancient drainage system into a series of ridges and small plateaux. In places on the marginal shelf it lies at shallow sub-bottom depth, especially where it occurs over eroded remnants of ancient reef masses, and, indeed, may locally form the present seabed.

(ii) *Reflectors II and III*

Reflector II, well defined in several of the profiles, lies 9–12 m below reflector I, but has been cut by the latter surface where ancient drainage channels occur, i.e. beneath Trinity Opening and below the northern end off Batt Reef (figures 2 and 3). Reflector III occurs at greater depth and has been detected only in profile 22 where it rises near the northern tip of Batt Reef. It is also probably an erosion surface.

(b) *Variations in thickness of sedimentary accumulations and reef masses*

The widely variable thickness of sediments overlying reflector I is well illustrated by profile 22, there being a cover of only 4 or 5 m over the higher, inter-reef areas of this surface, but a thickness of 34 m where line 22 crosses Trinity Opening. Locally there may be no overlying sediment where reflector I reaches the present seabed as a relict or exhumed ancient surface. Nearer to the shelf edge, along line 7, the thickness of sediment cover is in many places reduced to approximately 3 m, but beneath Trinity Opening deposits occupying an ancient drainage channel reach a thickness of 32 m (see figure 3 and plate 1, figure 2).

Along line 6 (plate 2), which completely crosses the continental shelf between the shelf edges at Grafton Passage and Cape Grafton, the thickness of sedimentary accumulations overlying reflector I varies from $2\frac{1}{2}$ m near the shelf edge to 21 m in the buried channel, which lies beneath the inner shelf zone at station number 115 (Decca Hi-fix position).

Wedging and lensing of sediment are commonplace and these phenomena can be seen particularly on lines 1 and 6 to be associated with reef masses, partly buried ancient ridges, and former drainage channels. In some areas wedges of sediment thin seawards from the present shoreline. Not only sediment source and dispersal factors, therefore, but also the configuration of surface I has influenced the variations in thickness of sedimentary accumulations above this reflector.

Surface I is clearly defined beneath many of the platform reefs, e.g. Tongue and Batt reefs, and it is noteworthy that the reefs developed on this surface pass upward from a lower massive

---

DESCRIPTION OF FIGURE 2

FIGURE 2. Continuous seismic reflexion profile (high resolution boomer) and interpretation along line 22, extending from the vicinity of Hastings Reef, along the marginal shelf to the northern tip of Batt Reef, viewed from the northeast. (Note that the interpretation sections begin at the 30 ms level, i.e. 22.5 m below sea level.)

A major disconformity occurs at reflector I which represents an extensive erosion surface formed during a period of shelf emergence. The channels at 583, 589 and 591 (Decca Hi-fix positions) are tributaries which unite to form the single channel shown on profile 7 (see figure 3 and plate 2) beneath Trinity Opening at the shelf edge. Facies changes between reef masses and inter-reef bedded sediments are evident, and rhythmically bedded accumulations are apparent above and below reflector I, particularly between 603 and 617 (Decca Hi-fix positions).

A figure of 1500 m/s, centrally bracketing the range of seawater velocity, has been used throughout this work. The reproduction of original seismic profiles have reference timing lines at 0.01 s intervals corresponding to an interpreted depth scale of 7.5 m per timing line interval. The velocity used is not confirmed by drilling or seismic measurements. However, on the basis of extensive (substantiated) experience this 1500 m/s figure is reasonable, and follows convention under the circumstances. An error of up to 10 % underestimation of sub-bottom thicknesses is the most that is envisaged in deeply buried sections of these records.

stage to a greatly diminished state at higher levels. This change occurs between the − 30 m and − 40 m levels, and is very noticeable at Pixie Reef, where a mere remnant of a much more massive and extensive reef structure rises above the − 30 m level. The thickness of Tongue and Batt reefs shown on profile 22 (figure 2) ranges between approximately 7.5 and 22.5 m, but adjacent to line 22 these reefs reach the present sea level, suggesting that their total thickness may be 30 or 38 m. Vestigial erosion surfaces have been detected within some platform reefs implying that post surface I reef development was not uninterrupted. However, more extensive scrutiny of this evidence will be necessary before the stratigraphic significance of such features can be properly assessed.

It is not possible to give accurate figures for the thicknesses of the reefs near the shelf edge (line 7), since in most cases reflector I loses definition and cannot be traced continuously through the reef masses with certainty. However, although the upward trend of reflector I within reef masses may be due partly to lateral seismic velocity changes, the indications are that towards the shelf edge, reef masses are thicker and represent a more continuous sequence than the platform reefs on the western part of the marginal shelf.

Profile 6 (figure 3 and plate 2) shows an ancient ridge (Decca Hi-fix station numbers 118–121) forming the foundation for a younger reef mass which has been partly buried by flanking sediments, and which rises above the present seabed as a number of pinnacles to reach 30 m below the present sea level. This feature does not appear on bathymetric charts of this area, for there is no reef at or near sea level. Its present 'drowned' state suggests its inability to maintain an upward growth rate commensurate with sea level rise.

(c) Buried drainage channels

Not all of the ancient drainage channels are reflected by the present bathymetry, for some have been partially concealed by subsequent reef growth. Some buried channels trend more or less NW–SE along the continental shelf until they meet others which cross directly to the shelf edge. Line 6 (plate 2) is channelled in several places on the marginal shelf, but a major buried channel occurs beneath the inner shelf. The three sub-bottom channels shown on profile 22 at 583, 589 and 591 (Decca Hi-fix positions), join to form the single incised channel beneath Trinity Opening at the shelf edge (line 7, plate 1).

DESCRIPTION OF FIGURE 3

FIGURE 3. Interpretations of parts of the seismic reflexion profiles along lines 1, 6 and 7, and the morphology of the upper continental slope as shown on line 27.

Line 7. The section shown here lies near the shelf edge extending from a point adjacent to Onyx Reef north-eastward to the northern flank of Linden Bank. Seismic reflectors are obscure beneath the thick reef masses, but I is clearly channelled beneath Trinity Opening to a depth of approximately 105 m. General facies changes between reefal and inter-reef accumulations are indicated (compare with plate 1, figure 2).

Line 6. This is part of the interpretation of the profile which is shown complete in plate 2. The erosion surface, reflector I, is quite clear. It is channelled at 115 (Decca Hi-fix position), and passes beneath a partially buried reef mass between 118 and 121. At approximately 122–124, buried terraces occur at approximately − 49 and − 53 m. General facies differences are expressed in the channel filling deposits by variations in their acoustic properties.

Line 1. This part of profile number 1 lies just to the north of Green Island Reef and terminates at Arlington Reef. It shows reef rock and flanking sediment wedges on the leeward side of Green Island Reef and a partially buried terrace at − 36 m. A narrow terrace is evident at − 30 m on the outer slope of Arlington Reef.

Line 27. Profile of the upper continental slope at Trinity Opening showing a break in slope at − 49 m and narrow terraces at − 97.5 and − 116.3 m.

On the marginal shelf, the channels, particularly those towards the shelf edge, are narrower, and simpler both in profile and in terms of channel filling deposits, than those detected beneath the inner shelf region. That shown on line 2 (plate 1) near the mouth of the Barron River for example, contains a sequence of rhythmically bedded sediments alternating with wedges and lenses of acoustically more uniform and more transparent sediment. There is bedding discordance, and differential compaction structures are evident where more recent sediment has compacted over an irregular erosion surface (see plate 1, figure 3).

(d) Marine terraces

Several marine terraces or marked changes of slope were encountered during the present investigation. On the continental slope, narrow terraces occur at approximately -98 m and -116 m (see figure 3). Terraces also occur on reefs near the shelf edge at -19.5, -45, -67, and -57 m, i.e. Onyx Reef (plate 1, figure 1); and adjacent to Euston Reef (line 6, plate 2) there is a leeward terrace of -48 m, and a seaward terrace at -60 m.

There is a marked planation at approximately -30 m on Pixie Reef, and at Arlington Reef there is a narrow terrace at similar depth. On Hastings Reef a marked break of slope occurs at -22.5 m. A partly buried terrace at -36 m occurs to leeward of Green Island Reef (figure 3, line 1), and completely buried terraces are evident on the continental shelf at -49 and -53 m (figure 3, line 6, and plate 2). The main terrace levels encountered during the present study are summarized in figure 4.

6. Interpretation

There is a major disconformity representing a considerable hiatus at reflector I, which is interpreted as an ancient surface produced by marine regression, shelf emergence, and subaerial erosion. Withdrawal of the sea resulted in the extension of rivers across the exposed shelf to the new shoreline near the present shelf break. A subsequent lowering of sea level and accompanying change of the base level of erosion, caused deeper channelling of the shelf edge to at least -116 m. At this time reef-rock of exposed shelf edge reefs may have presented a steep, probably cliffed, shoreline to the Coral Sea, breached in this area principally by rivers at Grafton Passage, Trinity Opening, and just to the north of Linden Bank. Consequently the exposed shelf surface developed considerable relief and a very diverse topography.

Where ancient drainage channels occur, the acoustic properties of the sequence immediately

---

Description of plate 1

Photographs of parts of the seismic reflexion profiles (high resolution boomer) along lines 2 and 7.

Figure 1. The shelf edge rising from the 90 ms (-67.5 m) level to 26 ms (-19.5 m) near Onyx Reef, viewed from the northeast (line 7). A conspicuous terrace level occurs at 58–60 ms (-43.5 to 45 m), with other terraces at 67 ms (ca. -50 m), 70 ms (-54.15 m) and 77 ms (-57.75 m). Reflector I is present, but indistinct beneath the thick reef mass.

Figure 2. The southern part of Linden Bank rising to 20 ms (-15 m) from Trinity Opening at 97 ms (-72.5 m). The buried channel (B), and the reflectors I and II are shown.

Figure 3. A buried channel beneath the seabed (SB) at about 16 ms (approximately -12 m) to the northeast of the mouth of the Barron River. Wedging and discordant bedding are evident in complex channel filling accumulations which comprise rhythmically bedded sediments and acoustically more transparent deposits. An erosion surface (E) occurs at a major disconformity, above which differential compaction structures are evident, particularly between 29 and 30 (Decca Hi-fix positions).

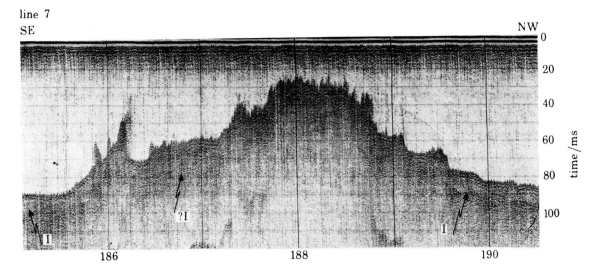

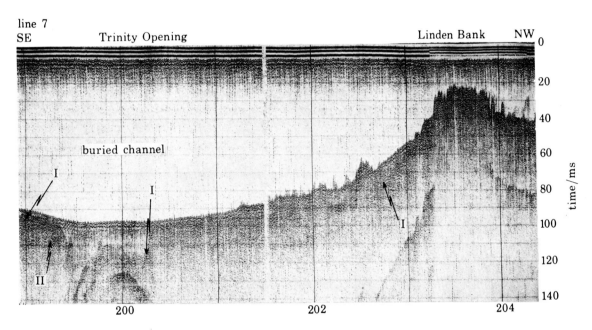

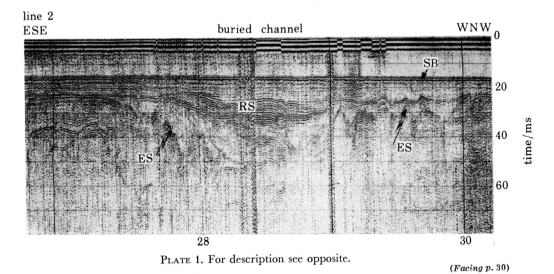

PLATE 1. For description see opposite.

Phil. Trans. R. Soc. Lond. A, volume 291

Orme et al., plate 2

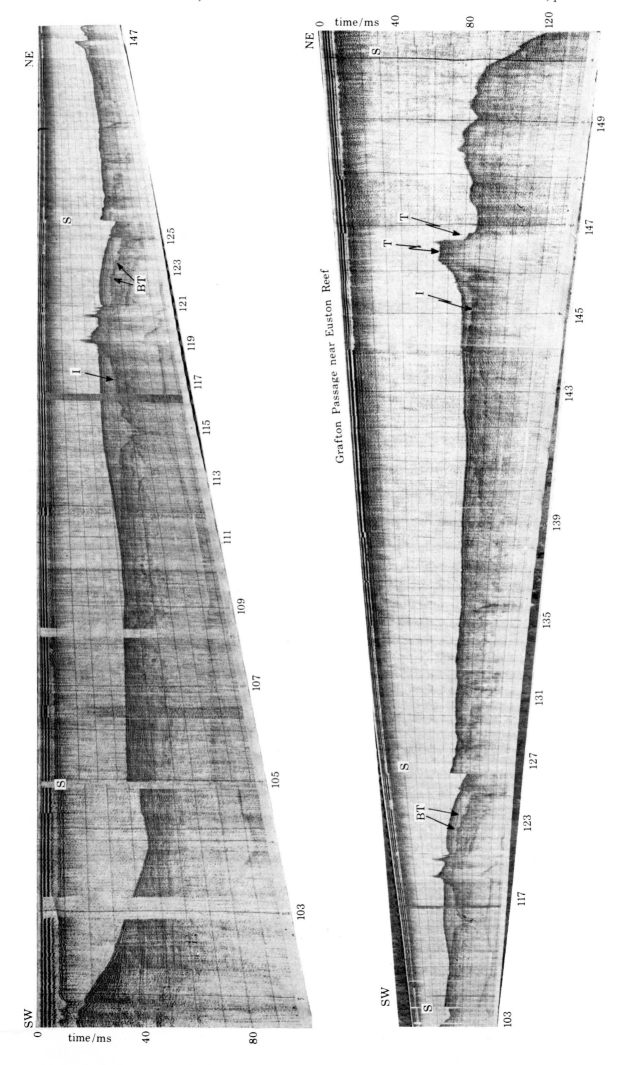

Grafton Passage near Euston Reef

overlying reflector I suggest that the sea level rise at the commencement of the ensuing transgression caused the backfilling of stream channels with poorly sorted, dominantly terrigenous sediment. As the transgression proceeded, the higher parts of the 'marginal shelf' probably formed karstified islands, separated from each other and from the mainland by progressively broadening channels. This environment gradually changed in character from a fluviatile régime to a transitional marine stage still strongly influenced by low salinity and high turbidity discharge from the major streams, which inhibited reef growth over most of the shelf. However, when the rise of sea level became sufficient to submerge completely the entire shelf the effects of the non-marine influences waned, and reefs developed by pronounced lateral accretion. That later stages of reef accretion were dominantly vertical is indicated by the seismic profiles, so that many reefs, especially near the shelf edge, are mere remnants of more extensive precursors. This evidence, together with 'drowned' reefs below approximately -30 m might suggest that a more rapid rate of sea rise had occurred during later stages of the transgression.

7. Discussion

The information provided by the seismic profiles permits the determination of a sequence of geological events; the deduction of a probable succession of palaeoenvironmental changes; and the assessment of some controls on reef distribution, thickness and form. Therefore, in an attempt to present this information objectively, no assumptions have been made regarding the ages of the various erosion surfaces referred to in the preceding sections of this paper. However, it is now appropriate to consider the results of the seismic investigation in relation to some previously published views concerning the geological history of this part of the continental shelf. The only positive data regarding the stratigraphy of the shelf in the Cairns area are provided by the Michaelmas Cay bore, which was drilled to 600 ft (182.9 m), passing through reef-rock (378 ft, 115.2 m) followed by quartz–foraminiferal sands (Richards & Hill 1942), and although it has not been possible to date precisely any part of the core, none of it is older than Pliocene (Lloyd 1973). However, by analogy with evidence from the Heron Island Reef, Davies (1974), considering the Michaelmas Cay bore hole log, postulated a 'solution unconformity' at a core depth of 27 m (98.8 ft), which he proposed as the Holocene–Pleistocene junction. Subsequently Davies (1975) related this interpretation to bathymetric evidence for a -17 m (-55.7 ft) platform beneath Michaelmas Cay and adjacent reefs. However, because in the Michaelmas Cay bore no core was recorded between 42 ft (12.8 m) and 90 ft (27.5 m) it seems unlikely that

---

Description of Plate 2

Photographs of the entire seismic reflexion profile (high resolution boomer) along line 6, extending from the shelf edge at Grafton Passage to Cape Grafton. The most conspicuous feature is a major disconformity marked by reflector I, which is a markedly channelled erosion surface. A partially buried reefal ridge occurs at 119–121 (Decca Hi-fix positions), and adjacent, buried, seaward terraces are indicated at approximately 65 and 70 ms (*ca.* -49 and -53 m). Terrace levels near the shelf edge adjacent to Euston Reef are indicated at 64 and 80 ms (-48 and -60 m).

Figure 1. An oblique view of profile 6 from the southwest.

Figure 2. Profile 6 viewed obliquely from the east.

The profile represents a continental shelf width of 48 km, and the vertical exaggeration is approximately ×20, disallowing the effect of perspective. T, terraces; BT, buried terraces; I, major disconformity; S, scale change.

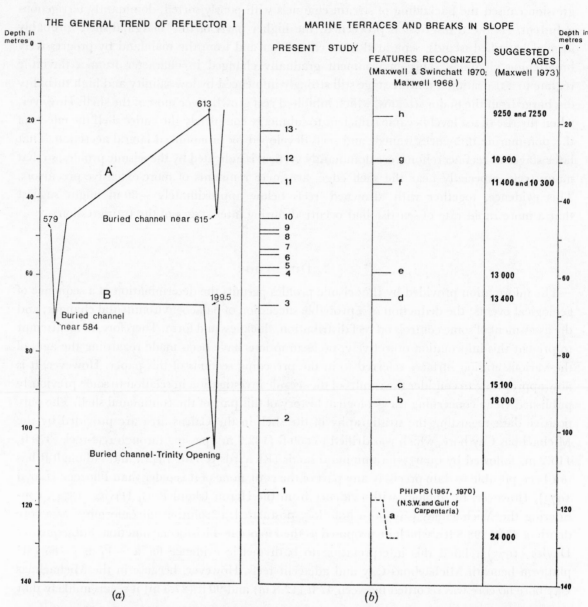

FIGURE 4. Diagrammatic summary of significant bathymetric levels. (a) The general trend of reflector I along line 22 (A) and along line 7 (B). (b) Marine terraces and breaks in slope. 1 (−116 m) and 2 (−97.5 m), narrow terraces on the upper continental slope; 3 (−67 m), 5 (−57 m), 6 (−45 m), 10 (−19.5 m), 4 (−60 m) and 9 (−48 m) terraces on reef masses near the shelf edge; 7 (ca. −53 m) and 8 (ca. −49 m), probable buried terraces on line 6; 11 (−36 m), partly buried terrace leeward of Green Island; 12 (−30 m), the well developed features on many reefs, for example at Pixie Reef and the Arlington Complex; 13 (−22.5 m), particularly evident break in slope at Hastings Reef. (a) (−119 m, −128 m) Recognized by Phipps (1967, 1970) in other areas (New South Wales and Gulf of Carpentaria); (b) (56 fathom†), (c) (48 fathom), (d) (36 fathom), (e) (32 fathom), (f) (20 fathom), (g) (16 fathom), (h) (10 fathom), features recognized on the Queensland Shelf (Maxwell 1968); e, f, and g are the conspicuous features recognized in the Arlington area by Maxwell & Swinchatt (1970).

† 1 fathom ≈ 1.83 m.

the position of the Thurber discontinuity (Thurber, Broecker, Blanchard & Potratz 1965) can be accurately determined.

Maxwell & Swinchatt (1970) pointed out that the most distinctive bathymetric features of the Arlington Reef Complex occur at the 16 fathom (*ca.* − 30 m) level, and suggest that uniform depths at this level of the interior of the complex together with breaks in slope on the outer reef walls indicate the presence of a pre-Holocene reef platform. The persistence of such bathymetric features at the − 30 m level in this general area is supported by the present investigation, but there is no universal correlation between this and the position of the major erosion surface (reflector I). The latter is the result primarily of erosion following a marine regression and shelf emergence, whereas the 30 m features are a consequence of events related to the succeeding transgression.

Of the seismic profiles considered in the present paper, that nearest to Michaelmas Reef is profile 22 (see figure 1), and at its closest approach the major disconformity, represented by reflector I, lies at a depth of about 40 m below sea level, where it occurs beneath approximately 5 m of sediment cover.

It is therefore likely that most of the Michaelmas bore sequence lies below the level of reflector I, and indeed, in the vicinity of the Arlington Reef Complex there may be a coincidence between the level of reflector I, the − 30 m platform, and the Holocene–Pleistocene junction. Furthermore, evidence of a coral reef habitat extending to a core depth of 476 ft (145 m), well below the limiting depth for hermatypic corals, indicates subsidence of this area (Richards & Hill 1942), particularly since changes in water depth due to changes of sea level may not have been adequate to account for this sequence.

Although there are obvious constraints in comparing Quaternary eustatic oscillation curves with bathymetric features on the Queensland continental shelf and slope, Fairbridge's (1961) curve indicates that sea level may have prevailed below − 40 m during the Riss, and both early and late Würm glacials. Consequently, under conditions of shelf stability during the later Pleistocene, part of the shelf may have been exposed for an extended period, and therefore the higher parts of reflector I may be the product of a considerable period of subaerial weathering and erosion. This erosion surface was submerged progressively during the succeeding, Flandrian transgression, and sediment accumulations and reef masses in contact with it may not be of uniform age, the hiatus represented by this surface varying considerably from place to place according its elevation in relation to sea level. In consideration of this interpretation, the presence of erosion surfaces in sediment accumulations and reef masses which overlie reflector I, and the absence of absolute dates for any part of I, there is little value in speculation regarding reef accretion rates or sedimentation rates for the sequence overlying reflector I.

The dates suggested by Maxwell (1973) for the most prominent bathymetric anomalies are 10900 a B.P. for the − 30 m features, and 13000 a B.P. for the 32 fathom (− 59 m) terrace. A late Würm date may be indicated by the 116 m continental slope terrace by comparison with the − 119 to − 128 m terrace detected in the New South Wales and Gulf of Carpentaria areas (dated 24000 a B.P.) by Phipps (1967, 1970), which probably represent an erosional episode at approximately 24000 a B.P. (Maxwell 1973).

Figure 4*b* summarizes the significant bathymetric levels encountered in the Cairns area, and compares these with features recognized by previous authors. Figure 4*a* illustrates the general trend of surface I in relation to the present sea level.

8. SUMMARY AND CONCLUSIONS

Among the prominent seismic reflectors revealed by this investigation, the most conspicuous (reflector I) is a major erosion surface representing a considerable hiatus in the development of the continental shelf as a reef province. This complex erosion surface, formed through shelf emergence and subaerial erosion has exerted a fundamental control on the attainment of the present physiographic characteristics of the Great Barrier Reef Province, strongly influencing both the distribution and thickness of reefs, and the distribution and thickness of sedimentary accumulations. Indeed, locally on the marginal shelf it may form part of the present seabed as a relict or exhumed surface. Evidence of an ancient fluviatile system which extended across this surface is provided by sediment filled channels.

The sequence overlying surface I provides evidence of the ensuing marine transgression during which reefs were regenerated near the shelf edge and re-established on inner parts of the marginal shelf. Periods of still-stand during the Holocene transgression or minor regressions are indicated by terraces or marked changes in slope, and by minor disconformities within both sedimentary accumulations and reef masses, the most marked bathymetric anomalies occurring at the − 30 m level.

Some fundamental aspects concerning reef development in this area have been revealed by the seismic data, namely:

(1) Large platform reefs of the marginal shelf have developed on foundations of lithified sediments and/or an ancient reef rock. In some cases, elevations formed by remnant reef masses have formed sites for more recent reef growth.

(2) The location of reefs on ridges or small plateaux on the marginal shelf suggests a delay in the establishment of marine conditions suitable for reef growth until later phases of the transgression.

(3) The thickness of the platform reefs appears to bear a relation to the relief of surface I from which they rise.

(4) Reefs near the shelf edge are thicker than those on the western part of the marginal shelf, which implies the prevalence of a more continuously reefal environment near the edge of the continental shelf. They are also more extensive at depth, which suggests that formerly there was a more continuous barrier of shelf edge reefs in this area.

(5) The − 30 m level is marked by changes of slope or by extensive terraces, and appears to make a significant period of still-stand during the Holocene Transgression, or alternatively it may be the result of a minor regression. Many platform reefs diminish upwards, especially above the − 30 m level, reflecting perhaps a late period of dominantly vertical reef accretion in response to an increased rate of sea level rise.

There is a major area for speculation regarding the age of the erosion surface represented by reflector I, parts of which have been emergent for longer periods than others, so that overlying sedimentary accumulations and reef masses in contact with it may not be contemporaneous. Furthermore, the presence of subsidiary erosion surfaces above surface I indicates that the stratigraphic record is incomplete.

Thus, while this seismically based approach has revealed some of the fundamental characteristics of this part of the Great Barrier Reef, it has also raised further questions about its geological history. The complexity of the stratigraphic record of reef accretion and sediment accumulation under the influence of oscillating sea levels poses interpretation problems which may be solved only through systematic drilling related to more intensive high resolution seismic programmes.

The authors gratefully acknowledge financial support from Royal Society of London, the University of Queensland, and Mount Isa Mines Ltd, for this phase of the work of the 1973 Expedition. They are indebted to the Royal Australian Navy for the provision and operation of Decca Hi-fix equipment, and for logistic support, and, in particular, express their gratitude to Captain J. H. S. Osborn, Lieutenant Commander R. K. Doolan, Commander P. V. Blackman, and Petty Officer Clyde Hillsdon.

The seismic equipment was provided by the Department of Geology and Mineralogy of the University of Queensland, in part through an Australian Research Grants Committee grant to Dr J. P. Webb. Supplementary equipment was lent by Natural Environment Research Council (U.K.) through the Institute of Oceanographic Sciences, Taunton, England.

Maintenance and operation of the seismic equipment under frequently very trying conditions was carried out chiefly by Mr E. Laundon.

REFERENCES (Orme *et al.*)

Bird, E. C. F. 1971 *Aust. geogr. Stud.* **9**, 107–115.
Davies, P. J. 1974 *Proc. 2nd Intern. Symp. Coral Reefs, 1973*, vol. 2, pp. 573–578.
Davies, P. J. 1975 *Habitat* **3**, 3–7.
Davies, P. J., Radke, B. M. & Robison, C. R. 1976 *B.M.R. Jl Aust. Geol. Geophys.* **1**, 231–246.
Ewing, M., Hawkins, L. V. & Ludwig, W. J. 1970 *J. geophys. Res.* **75**, 1953–1962.
Fairbridge, R. W. 1950 *J. Geol.* **58**, 330–401.
Fairbridge, R. W. 1961 *Phys. Chem. Earth* **4**, 99–185.
Flood, P. G. 1976a In *The Great Barrier Reef, 25th International Geological Congress*, Excursion Guide No. 6, A.C. (ed. J. S. Jell), pp. 14–20. Canberra: Progress Press.
Flood, P. G. 1976b *25th International Geological Society Abstracts*, p. 496. Geological Society of Australia.
Hoffmeister, J. E. & Ladd, H. S. 1944 *J. Geol.* **52**, 388–402.
Lloyd, A. R. 1973 In *Biology and geology of coral reefs* (eds O. A. Jones & R. Endean, vol. 1 (Geology 1), pp. 347–366. New York: Academic Press.
MacNeil, F. S. 1954 *Am. J. Sci.* **252**, 402–427.
Maxwell, W. G. H. 1968a *Atlas of the Great Barrier Reef*. Amsterdam: Elsevier.
Maxwell, W. G. H. 1968b *Aust. J. Sci.* **31**, 85–86.
Maxwell, W. G. H. 1973 In *Biology and geology of coral reefs* (eds O. A. Jones & R. Endean), vol. 1 (Geology 1), pp. 233–272. New York: Academic Press.
Maxwell, W. G. H. & Swinchatt, J. P. 1970 *Bull. geol. Soc. Am.* **81**, 681–724.
Orme, G. R. 1977 In *Biology and geology of coral reefs* (eds O. A. Jones & R. Endean), vol. 4 (Geology 2), pp. 267–306. New York: Academic Press.
Palmieri, V. 1974 *Rep. geol. Surv. Qld.* **86**, 1–14.
Phipps, C. V. G. 1967 *A.P.E.A. Jl* **7** (2), 44–49.
Phipps, C. V. G. 1970 *Aust. J. Sci.* **32**, 329–330.
Purdy, E. G. 1974a *Soc. econ. Palaeontol. Miner. spec. Publ.* **18**, 9–76.
Purdy, E. G. 1974b *Bull. Am. Ass. Petrol. Geol.* **58**, 825–855.
Richards, H. C. & Hill, D. 1942 *Rep. Gt Barrier Reef Comm.* **4**, 1–111.
Sargent, G. E. G. 1969 *9th Commonwealth Mining and Metallurgical Congress*, Paper 28. London: Institution of Mining and Metallurgy.
Thom, B. G. & Chappell, J. 1975 *Search* **6**, 90–93.
Thurber, D. L., Broecker, W. S., Blanchard, R. L. & Potratz, H. A. 1965 *Science, N.Y.* **149**, 55–58.

Phil. Trans. R. Soc. Lond. A. **291**, 37–54 (1978)   [37]

Printed in Great Britain

Drilling investigation of Bewick and Stapleton islands

By B. G. Thom†, G. R. Orme‡ and H. A. Polach§

†*Department of Geography, Faculty of Military Studies, University of New South Wales, Duntroon, Australia* 2600

‡*Department of Geology and Mineralogy, University of Queensland, Brisbane,* 4067, *Australia*

§*Radiocarbon Laboratory, Australian National University, Canberra,* 2600, *Australia*

[Plate 1]

The stratigraphy and sediments of Bewick and Stapleton cays were investigated by rotary drilling. Emphasis has been placed on core collected from the hole, 30 m deep, drilled on the leeward side of Bewick Island. Core recovery using a variety of techniques was variable. However, sufficient material existed for chemical, mineral and petrological analyses, as well as for general biological and fabric description. Samples were also selected for radiocarbon dating. Data suggest three and possibly four disconformities in the Bewick core. The youngest hiatus occurs at *ca.* 3–4 m below low water mark separating Holocene reef and reef flat sediments from pre-Holocene carbonates. An attempt has been made to interpret changes in environments of deposition in the subsurface, and diagenetic effects.

Introduction

Until the 1973 Royal Society–University of Queensland Expedition, drilling on the Great Barrier Reef has been limited to six holes. Four holes have been drilled in the southern Great Barrier Reef Province, only one of which recovered sufficient samples in the top 200 m to permit any detailed work. This was the Heron Island hole drilled by the Great Barrier Reef Committee (Richards & Hill 1942; Maxwell 1962; Davies 1974). In the northern Great Barrier Reef Province the Great Barrier Reef Committee was also responsible for a bore hole at Michaelmas Cay (Richards & Hill 1942). Initial interpretation of the upper sequence at Heron Island was of post-Pleistocene age, because none of the fauna showed the effects of Pleistocene ice ages (Richards & Hill 1942). Maxwell (1973, p. 266) states: 'The major disconformity between Pleistocene foraminiferal limestone and Holocene coralline limestone is marked in the Heron Island bore by the quartzose sands of the interval 289–308ft'.¶ However, Davies (1974) has reinterpreted the stratigraphy on the basis of a mineralogical change at −20 m, and a marked planar resistivity contrast at this depth. He suggests that at 20 m there exists a prominent solution surface, which could be correlated with the first subsurface 'unconformity' at −27 m on Michaelmas Cay. This interpretation is compatible with dated reef sequences reported from various places in the Pacific (see Stoddart 1969; Tracey & Ladd 1974).

As part of the 1973 Expedition to the northern Great Barrier Reef Province an investigation of the stratigraphy of several reef islands with the use of shallow coring techniques was envisaged. Logistical difficulties limited drilling to two sites, Bewick (14° 26′ S, 144° 49′ E) and Stapleton islands (14° 19′ S, 144° 51′ E). The primary objective at these sites was to test the hypothesis that islands in this part of the Great Barrier Reef have been formed by incremental addition of reefal sediment during successive Quaternary marine transgressions. For this

¶ 1 foot (ft) = 0.3048 m.

purpose, drill performance was recorded in detail and all pieces of core were megascopically described. Chemical, mineralogical, petrologic and geochronologic (^{14}C) analyses were conducted on samples from the core to ascertain if time as well as facies discontinuities existed in the sequence. A secondary objective was to determine if environmental and diagenetic changes were recorded by the petrology of the rocks. This paper represents a preliminary report of findings which resulted from the limited drilling programme.

METHODS

(a) Field

The technique of drilling involved diamond tungsten bit coring by means of a drill unit of low mass (Thom 1976). As employed on Bewick Island, the trailer-mounted Gemco drill was winched from a barge stranded on the low-tide reef flat onto the upper beach-rock surface. Two coring techniques involving pumped seawater circulation were used: the retractable barrel, and wireline equipment. Recovery with the retractable barrel, especially when coupled with tungsten carbide faced bits, was reasonable although variable. When the wireline barrel was used the core runs retrieved little or no material.

On Stapleton Island, the drill was sited on the upper beach face. Drilling commenced on loose sand requiring casing to 5.8 m. Solid auger rods were employed to insert the casing and recover the sample in wet unconsolidated sand. This hole was terminated at the first discontinuity at 14.6 m after obtaining relatively poor recoveries from five core runs. Wireline equipment was used exclusively at Stapleton.

Core logging procedures followed those reported by the U.S. Geological Survey in their drilling of Bikini Atoll (Emery, Tracey & Ladd 1954, p. 77). After extrusion from the core barrell, all core pieces, fragments or sandy detritus were described and packed in sequence in core boxes, being orientated correctly where possible. Each core piece was labelled.

In general, the cores are believed to provide a reliable indication of the stratigraphy, particularly the Bewick hole. As noted by many authors in their descriptions of drilling operations on coral reefs, there are considerable difficulties in achieving high core recovery rates. The patchy nature of the Bewick core can be attributed to four factors:

(1) occurrence of cavities in the section;

(2) interbedded consolidated and weakly consolidated materials with much of the latter being washed away in the circulating water;

(3) brittle coral rubble which was either pushed aside or easily fell to the bottom of the hole during core barrel recovery;

(4) operator inexperience involving experimentation and error.

The third factor became apparent during laboratory analyses when material from the upper part of several core runs were shown to be out of sequence. Fortunately, excessive caving did not occur and contaminated material could be easily isolated.

(b) Laboratory

A group of 126 samples from the Bewick core and 15 samples from the Stapleton core were analysed in the chemical laboratory of the Department of Biogeography and Geomorphology, A.N.U. (J. Caldwell, analyst). An infrared spectrophotometer was used to determine the proportion of calcite, aragonite and non-carbonate impurities in the Bewick samples.

The content of strontium and iron in each sample was determined by using an atomic absorption spectrophotometer. X-ray analysis was conducted on the Bewick samples to (i) cross-check the infrared method on calcite–aragonite determinations; (ii) determine if the calcite occurred as low or high-magnesium type; and (iii) look for the occurrence of dolomite.

TABLE 1. RADIOCARBON DATES FROM DRILL CORES

island	sample no.	laboratory number	^{14}C age a B.P.	approximate depth below h.w.s.t. m	material	percentage aragonite	percentage calcite	comment
Bewick	BE-D1	ANU-1386	$2\,030 \pm 70$	0	*Tridacna* in beach-rock	—	—	sample collected from surface of beach-rock 100 m ENE of drill site
	2–3–5	ANU-1284	$6\,920 \pm 130$	5.2	coral fragment	99	—	
	2–4–1	ANU-1395	$6\,380 \pm 120$	5.9	coral fragment	65	35	high-magnesium calcite sp.
	2–7–1	ANU-1283	$6\,610 \pm 130$	7.9	coral (*Porites*?) fragment	90	10	displaced by drilling below discon-formity
	2–9–1	ANU-1282	$>37\,300^{+2200}_{-1700}$	9.1	coral and coralline algae fragment	—	99	
	2–13–2	ANU-1281	$>34\,400^{+2500}_{-1900}$	11.9	coral fragment	2	98	
	2–15–4	ANU-1280	$>30\,350 \pm 1150$	13.3	fine-grained calcarenite	—	94	
Stapleton	1–1	ANU-1663	$3\,130 \pm 80$	7.9	carbonate sand	48	52	
	1–2–3	ANU-1664	$3\,160 \pm 90$	10.1	reef calcarenite	5	95	high-magnesium calcite sp.
	1–3–3	ANU-1721	$5\,260 \pm 130$	12.2	coral fragment	70	40	high-magnesium calcite sp.

Petrographical inspection of the core involved several stages. The first was a description of each piece as seen through a binocular microscope. Secondly, selected pieces of the Bewick core were impregnated with plastic in a vacuum chamber. Thirdly, the prepared blocks were etched and stained by using Alizarin Red S, potassium ferricyanide and Feigl's solution, following the technique described by Davies & Till (1968). Finally, from the blocks peels were taken and thin sections cut. The soft and unevenly lithified nature of the core resulted in peels and thin sections of variable quality. The relative abundance of constituents was visually estimated. It was considered that the thin sections and peels were too inadequate for point count analysis.

Radiocarbon assays were undertaken at the A.N.U. laboratory under the supervision of H. Polach. Six dates were obtained from the Bewick core and three from the Stapleton core (table 1). Bewick core samples of uncrystallized and recrystallized material were submitted for ^{14}C determination. It was apparent that dates on calcite-rich samples were meaningless due to post-depositional mineralogical change (see Chappell, Broecker, Polach & Thom 1974; Chappell & Thom 1978). Unfortunately, no aragonite-rich corals of Pleistocene age were recovered below the first discontinuity, thus eliminating the possibility of accurate uranium series dating.

DRILL SITE CHARACTERISTICS

The northern province of the Great Barrier Reef is characterized by a narrow shelf with prominent ribbon reefs on the shelf edge (Maxwell 1968). Between the shelf edge and the mainland are a number of patch reefs on some of which are 'low wooded islands' (Steers 1929). Bewick Island is such a feature, but Stapleton Island is essentially a sand cay on the leeward side of an extensive reef flat. The depth of water between the islands rarely exceeds 20 m.

Stapleton Island was the first site drilled. This hole was used to test the wireline drilling technique. Landing the drill trailer on this island was quite difficult because of loose sand. The only site available was a leeward-projecting sand beach with deep water nearshore. Drilling was conducted at the high spring tide (h.s.t.) level. Core recovery at Stapleton was poor.

On Bewick Island the drill hole was located on the leeward side on a beach-rock surface. The collar elevation of the hole was 3 m above low water spring tide (l.w.s.t.) datum. The site was inundated by 10–15 cm of water on two occasions during high spring tide. A narrow unvegetated reef flat surrounds Bewick Island. High winds and seas prevented drilling on the windward side of the island. The interior of the Bewick reef flat is heavily vegetated by mangroves, preventing access for drilling. Thus the site selected for drilling was the only one accessible at this time of year (July–August). Furthermore, the barge remained on the reef flat near the drill enabling water to be pumped even at low tide. Another advantage was that the surface on which drilling commenced was consolidated.

Plans to drill on Noble Island closer to the mainland and Waterwitch Island on the ribbon reef were abandoned for a variety of reasons, including rough seas, lack of finance, and other commitments for the drill and barge.

CHEMICAL AND MINERALOGICAL PROPERTIES

It is not possible in this brief paper to report all data obtained by chemical analyses of the Bewick core (figure 1). However, the salient points can be summarized.

1. All the samples are calcium carbonates relatively low in magnesium carbonates. Only a trace of dolomite was detected in a few samples (e.g. 2–6–9).

2. Taking the core as a whole, calcite is the dominant mineral. Aragonite is only contained in the upper 6–7 m in any significant quantities (figure 1). Below that depth, high aragonite values are associated with the uppermost sample of any core run. These are probably cave-in materials.

3. The aragonite content of the upper 7 m ranges from 30 to 80 %. Below this depth aragonite rarely exceeds 10 % and mostly is less than 3 % (figure 1).

4. Figure 1 also shows the variation in calcite content down the profile at Bewick. Below 7 m

the proportion of calcite dramatically increases to greater than 90%. Many samples are composed of 100% calcite. High calcite values appear to be independent of constituent type. Corals with well preserved structure are composed of calcite (e.g. 2–9–1; see table 1).

5. Above 7 m the calcite mineral could be classed as a high magnesium type, but below that depth only low magnesium calcite occurs.

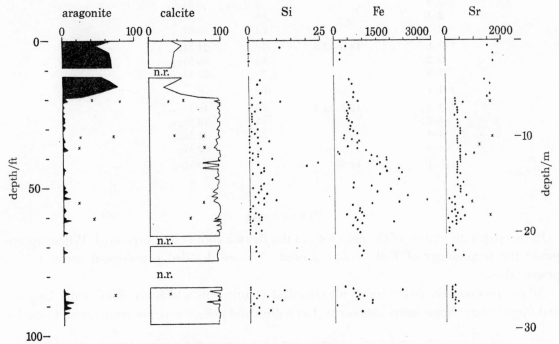

FIGURE 1. Chemical and mineralogical data for the Bewick core; aragonite, calcite and Si are given as percentages, Fe and Sr as parts/10^6.

6. Impurities including silica are mostly below 10% of total sample. Above 12.2 m the impurity content, with few exceptions, does not exceed 4% (figure 1). However, below that depth the impure silica values are quite variable ranging from 1 to 24%.

7. Variation in the strontium content of Bewick samples is closely related to the aragonite pattern (figure 1). The upper 7 m of the core contains 1000–2000 parts strontium/10^6, whereas below that depth strontium does not exceed 1000 parts/10^6, except in samples at the top of some core runs.

8. The amount of iron in the samples shows considerable 'down the profile' variation. Above 12.2 m the iron content only once in 49 samples attains a value of 1000 parts/10^6. Below that depth the amount varies from 340 to 3300 parts/10^6. It is possible to recognize layers in which the carbonates are relatively enriched in iron (e.g. 12.2–14.3 m).

Only limited mineralogical analyses have been conducted on Stapleton material (table 2). Above 14.6 m the sediment is predominantly mixed aragonite–calcite in composition; one sample (1–4–1) is 100% aragonite, the only 'pure' aragonite value obtained from the Great Barrier Reef drill materials. Two samples at the base of the core contained 1–2% aragonite and 92–99% calcite. The amount of magnesium in the samples showed that with the exception of the two lowest samples, which are low magnesium calcites, the calcite species could be described as a high magnesium type.

TABLE 2. X-RAY DIFFRACTION ANALYSIS OF STAPLETON IS CORE SEDIMENTS

sample no.	depth below h.w.s.t./m	percentage aragonite	percentage calcite sp.	quartz
1–1 auger	7.9	48	52 Mg_7	—
1–2–2	8.5–11.6	81	19 Mg_{10}	—
2–3		5	95 Mg_{13}	—
2–4		91	9 Mg_7	—
2–5		62	38 Mg_7	—
1–3–1	11.6–12.5	76	24 Mg_{13}	—
3–2		62	38 Mg_7	—
3–3		70	30 Mg_{13}	—
1–4–1	—	100	0	
1–4–2	12.5–14.6	52	48 Mg_{17}	—
4–3		85	16 Mg_{10}	—
4–4		56	44 Mg_3	—
4–5		2	92 Mg_3	6
1–5–1	14.8	1	99 Mg_0	—

PETROLOGY

In this paper an outline of the petrology of the Bewick core only is presented. Where appropriate the terminology of Folk (1965) is used. A more detailed petrological account is in preparation.

All samples contain recognizable allochemical constituents which are dominantly biogenic and fragmentary. Some intraclasts occur, but oolites and pellets were not recognized. Coralline

---

DESCRIPTION OF PLATE 1

Photomicrographs illustrating some aspects of the petrology of the Bewick core. (All photomicrographs are in plane-polarized light.)

FIGURE 1. Biosparite (lithologic unit 1). Abraded, rounded allochems are cemented by a well developed fringe of aragonite needles. Radially arranged acicular aragonite completely fills some pore spaces (X) and only partly fills others (Y). (Magn. × 56.)

FIGURE 2. Well developed acicular aragonite crystals, the common cement fabric above the 7 m level. (Magn. × 225.) An enlargement of area (X), figure 1.

FIGURE 3. Unconsolidated biogenic sand, part of lithologic unit 11. Many of the uncemented allochems are angular to subangular, and the largest allochem in the photomicrograph clearly shows the original aragonite fabrics of the mollusc shell, no inversion having occurred. (Magn. × 56.)

FIGURE 4. Part of a gastropod shell which has suffered inversion to calcite. The original shell layers have been 'replaced' by a coarser calcite mosaic. Micrite veneers the internal surfaces of the whorls. (Magn. × 56.)

FIGURE 5. Calcite mosaics formed by aggrading neomorphism in a biomicrite. Such fabrics are common in samples from below the 7 m level, especially from the lower half of the core. (Magn. × 90.)

FIGURE 6. Haematite included in calcite mosaics filling foraminifer chambers. The haematite may have replaced pyrite which commonly occurs at such sites in parts of lagoons where circulation is restricted. (Magn. × 225.)

FIGURE 7. An inverted coral fragment from the lower part of the core. Original aragonite fabrics have completely disappeared but the general architecture of the coral is preserved in calcite. (Magn. × 90.)

FIGURE 8. Part of a fossiliferous micrite containing numerous quartz grains of very fine sand-and-silt grade, which are common below the 12.2 m (40 ft) level. The blotchy texture is due to the random concentrations of iron oxides and to some neomorphism of the matrix. (Magn. × 90.)

FIGURE 9. Coralline algal fragment showing recrystallization in places. Through recrystallization, original calcite fabrics have been replaced in some areas, for example by coarser calcite mosaics. (Magn. × 56.)

Phil. Trans. R. Soc. Lond. A, volume 291

Thom et al., plate 1

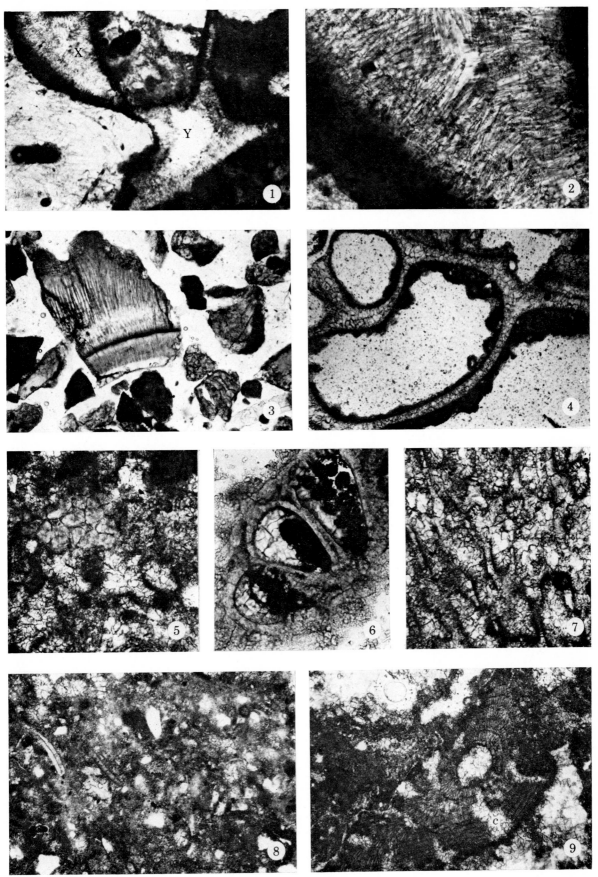

PLATE 1. For description see opposite.

(Facing p. 42)

algae are present throughout much of the core, as are molluscs and foraminiferans. Corals, *Halimeda*, and bryozoans were identified in many samples, but were restricted to certain levels. Minor constituents are echinoid plates and spicules. Biosparite (0–3 m) and unconsolidated biogenic sands and gravels are the common types above 7 m, whereas below this level micro-crystalline calcite matrix is a major component of many core samples. Owing to variations in size and abundance of allochems, there is a natural gradation in some parts of the core between fossiliferous micrite, fine biomicrite, biomicrite, and biomicrudite. Some samples have reached a more advanced state of recrystallization, *sensu lato*, than others. The appearance of pseudo-breccia at certain levels is due to the irregular distribution of neomorphic mosaics and/or the uneven concentration of impurities such as iron oxide particles. Below 12.2 m most samples contain significant amounts of silt grade to very fine sand grade quartz. On the basis of composition and texture it is possible to recognize a number of petrographic types.

(a) Petrographic types

Biosparite

This rock is light brown to creamy grey in colour, and is composed of well sorted, rounded allochems which have been poorly to well cemented by acicular aragonite (plate 1, figures 1 and 2). The allochems are rounded fragments of corals, molluscs, coralline algae (crustose and articulated types), benthonic foraminiferans, bryozoans, abundant *Halimeda*, and a few intra-clasts. Carbonate mud fills chambers in some allochems, e.g. foraminiferans, but is quantitatively unimportant. There is no evidence of inversion. This rock type represents a carbonate beach sand which became cemented to form 'beach-rock'.

Unconsolidated biogenic sand

Fragments of molluscs (pelecypods and gastropods), coralline algae (articulated and crustose types), corals, a few benthonic foraminiferans, and some *Halimeda* plates are the constituent allochems of this coarse to fine carbonate sand (plate 1, figure 3). Many of these biogenic fragments are angular to subangular, but some of the larger pelecypod fragments show a higher degree of rounding; none show any signs of inversion. Acicular aragonite occupies cavities in coral fragments.

Unconsolidated coral/algal gravel (reef rubble)

This sediment consists entirely of bored, subangular pebbles of corals and coralline algae (crustose type). Gastropod shells which are partly filled with acicular aragonite, the needle-like crystals projecting radially into the open spaces of whorls, are incorporated in some coralline algal pebbles. Acicular aragonite also occurs between coral septa, but other cavities are occupied by limonite-stained carbonate mud. Evidence of inversion of the coral debris is local and slight. This sediment, like the unconsolidated biogenic sand, is a reef flat deposit.

Biolithite

Certain sections of the core consist entirely of corals, with here and there an association of cavity-filling biomicrite. Although the gross skeletal forms of the corals have been preserved, they have suffered inversion to calcite and show no evidence of their original aragonitic fabrics. It is possible that this represents part of a reef frame in a growth position.

Coarse algal biomicrite

This type is light grey with poorly sorted allochems dominated by conspicuous, large, white, angular to subangular fragments of crustose coralline algae, and large mollusc fragments. Fragments of corals, some bryozoans and small foraminiferans are also present. Inversion of the mollusc and coral fragments has occurred (plate 1, figures 4 and 7), and the fine calcitic matrix shows evidence also of aggrading neomorphism. The nature and distribution of some calcite mosaics indicate that in parts of this rock type, allochems may have been enclosed originally by cement rather than by carbonate mud. A sheltered, low to moderate energy depositional environment, in which fine sediment was occasionally washed from the deposit, is therefore suggested.

Biomicrite

Here the sediments are creamy grey to light brownish grey in colour. The allochems are numerous, large, angular fragments of coralline algae, foraminiferans, and inverted coral and mollusc fragments, in a very fine calcitic matrix. Molluscs, particularly gastropods, dominate in some areas. In some samples echinoid plates, and in others bryozoans, are minor constituents. Both the microcrystalline matrix and some of the calcitic allochems have been affected by aggrading neomorphism (plate 1, figures 5 and 9), especially in samples from lower parts of the core. This rock is similar to the coarse algal biomicrite, but it has a more homogeneous texture due to better sorted allochems. Very fine sand and silt grade angular quartz grains are common. A sheltered, low energy depositional environment is indicated.

Fine biomicrite

A creamy grey to light brown, fine-grained rock containing a wide variety of allochems, many of which are represented merely as 'ghosts'. The most common allochems are foraminiferans and small mollusc fragments, but fragments of coralline algae and corals are present. Calcite mosaics, which in some cases contain small haematite particles, occupy foraminiferal chambers (plate 1, figure 6). The calcitic matrix is very fine-grained, especially where heavily limonite stained, and includes clay particles, haematite particles and small, angular quartz grains. Mosaics resulting from aggrading neomorphism of the micrite are common. Although in some areas the rock looks like a fossiliferous micrite, 'ghosts' of former allochems indicate that this apparent lack of allochems is due to their almost complete obliteration by recrystallization *sensu lato*. The original sediment probably accumulated in a low energy lagoonal situation with patch reefs.

Recrystallized biomicrite (pseudo-breccia)

Creamy grey to light brown, this type is iron-stained rock with a very heterogeneous texture. The apparent coarse texture, approaching a calcirudite, is superficial, being largely due to neomorphic mosaics and iron staining. The recognizable allochems are angular fragments of pelecypods, foraminiferans and some algae, and the calcitic matrix is a complex of mosaics resulting from aggrading neomorphism. A striking feature is the abundant quartz (silt and fine sand grade) scattered throughout the ground-mass (plate 1, figure 8). It is likely that the original deposit accumulated in a relatively low energy lagoonal situation.

Fossiliferous micrite

Only a few large allochems, namely gastropod fragments and benthonic foraminiferans, with small calcite spicules, some silt and very fine sand grade quartz grains are present in this type. Echinoid plates are a minor component in some samples. The rock is mainly microcrystalline calcite with some clay and randomly distributed haematite particles. Some of the quartz grains show evidence of authigenic growth. This deposit accumulated in a very low energy lagoonal environment.

Biomicrudite

Poorly sorted, large subangular fragments of corals, crustose coralline algae, pelecypods and gastropods associated with smaller allochems including bryozoan fragments are enveloped by a very fine-grained calcitic matrix, throughout which silt grade quartz is distributed. In some samples patchy pigmentation imparts a superficial appearance of a breccia, but the rock is really a biomicrudite. Short transportation, poor sorting, and deposition in a protected area near a reef are suggested.

Calcitic gravels and 'breccias'

In the lower half of the core there are thin layers of creamy grey allochems which are usually coated with reddish brown iron oxide particles, or, in the consolidated form, are enclosed in a reddish brown microcrystalline calcitic matrix. At megascopic level these may be described generally as unconsolidated calcitic gravel or calcitic breccia respectively. However, there are several varieties. In some the allochems are dominantly angular fragments of corals and coralline algae, but unlike those of the unconsolidated coral/algal gravel, they are inverted and usually finer grained. The consolidated variety could therefore qualify for the term biomicrudite. In other types, fragments of micrite and fossiliferous micrite are enclosed in a ferruginous, calcitic matrix and could aptly be named intramicrudite, especially since the assortment of small biogenic grains in the intraclasts and enclosing matrix are similar. In such deposits the angularity of the allochems indicates brief transportation to a shallow protected area of accumulation. One conspicuous layer has some aspects of a 'dismicrite' suggesting mild disturbance of a fossiliferous micrite in a low energy environment. The cavities and cracks created by the disturbance subsequently became filled with iron-stained carbonate sediment. Limonite is the chief iron stain, but irregular concentrations of haematite particles occur in the matrix.

(b) Diagenesis

Most core samples are light in colour, being creamy grey to light brownish grey. Darker pigmentation is due to iron oxide particles incorporated in matrix and cement fabrics, occurring as a coating of allochems, or as a superficial limonite stain introduced after lithification. Haematite particles have also been noted.

The effects of diagenesis are apparent throughout the core, ranging from the simple acicular aragonite cement of the upper levels to the aggrading neomorphic mosaics which are common below 7 m. The effects of diagenesis also vary with depth and to some extent may reflect disconformities, for instance the change from aragonite to calcite at 7 m, which was shown by the mineralogical analysis, can be observed also in stained peels and in thin sections. This inversion from aragonite to calcite in molluscs and corals is accompanied by the 'replacement' of the characteristic aragonite fabrics by coarser calcite mosaics, but 'ghosts' of original lamellar shell structure survive in some allochems. Micrite envelopes (Bathurst 1966) enclosing drusy calcite mosaics are present in parts of the core.

Some samples have undergone so much aggrading neomorphism that many allochems have been almost completely obliterated and are represented merely as 'ghosts' in coarse neomorphic mosaics.

The authigenic growth of quartz is suggested by its delicately spired boundaries and calcite inclusions, whereas evident mechanical wear and rounding indicates the terrigenous nature of other quartz grains.

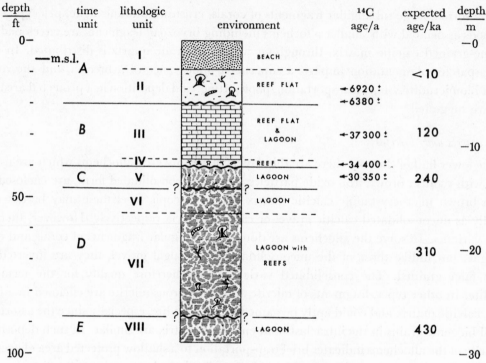

FIGURE 2. Stratigraphic units and possible absolute ages of Bewick core.

STRATIGRAPHY OF THE CORES

(a) Bewick

The Bewick core can be divided into eight lithologic units, labelled I–VIII in figure 2, and a maximum of five time units. Lithologic and time units are recognized on the basis of the following criteria: (i) radiocarbon date hiatus; (ii) sharp breaks in mineralogy and/or chemistry; (iii) composition of allochems; (iv) textural and fabric characteristics of core samples; (v) existence of features in core samples, including colour, which could reflect weathering or solution effects; and (vi) change in drilling performance. A lithologic unit can be bounded by either a disconformity or a facies change.

Table 3 summarizes the general characteristics of each lithologic unit. The contact between unit I, the biosparite, and unit II, the reef flat deposits at 3 m below ground surface, is sharp, and is considered to represent a facies change. The reef flat of 6000–7000 years ago has been overwhelmed by cay sands, which at Bewick are at least 4000 years old (R. F. McLean, personal communication). Lithologic units I and II constitute time unit A, of Holocene age (tables 1 and 3; figure 2).

The discontinuity at − 7 m below ground surface must have the status of a major disconformity. Although visible signs of weathering on the top of unit III are slight, the recrystallized nature of the limestone and the vastly older ^{14}C ages point to this contact representing the 'Thurber Discontinuity' (Stoddart 1969). Unit III consists of coarse algal biomicrite, biomicrite and some biolithite. Unit IV is a thin bed and is clearly separated from the biomicrite above on the basis of its high coral content, all of which are recrystallized. The change from unit III to unit IV is interpreted as a facies change.

The hiatus at 12.2 m was observed within core run 13. In this run creamy white coral (biolithite) of unit IV overlies an oxidized light brown limestone, which petrologically is a biomicrite interbedded with a fine biomicrite (unit V). The sharp nature of the break between units IV and V suggests a disconformity. The fabric of many samples below the contact shows somewhat more intense recrystallization (e.g. neomorphic mosaics and 'ghosts'). Furthermore, core samples show, under the binocular microscope, oxidized crevices, cavities and grain faces. Unit V also contains noticeable terrigenous material. Time unit C is therefore regarded as beginning at the top of lithologic unit V at 12.2 m below ground surface, and besides biomicrite and fine-biomicrite, consists also of fossiliferous micrite and recrystallized biomicrite.

The break between units V and VI at 14.3 m is also interpreted as a disconformity separating time unit C from time unit D. Samples from the upper part of VI under the binocular microscope have a cavernous appearance. Many of the cavities are filled with a brown micrite containing terrigenous quartz silt. Limonitic staining is apparent and haematite particles occur in the matrix. It could be argued that given the thinness of V (2 m), the weathering features in the upper part of VI are not part of a solution disconformity, but an expression of deeper weathering in V. Until this question is resolved a query should be placed on the disconformity purporting to separate time units C and D.

A facies change separates lithologic units VI and VII at − 17 m. The former is mainly fossiliferous micrite containing terrigenous silt and clay, whereas VII consists of several interbedded types (table 3). Within VII, core samples show a wide range of composition and texture. Ferruginous and argillaceous components are common. Fossiliferous micrite, biomicrite, with 'gravels' and 'breccias' of inverted reef rubble, and some recognizable intraclasts suggest reworking of reef and lagoon sediment which accumulated in a protected lagoon situation. The contact between VI and VII is transitional.

At the base of the core, biomicrite and biomicrudite were identified in thin section (unit VIII). The environment of deposition of materials recovered from depths below 26.8 m is not very different from the materials immediately above the contact. However, the core samples in unit VIII are much harder, perhaps more recrystallized (especially the corals), and certainly more iron stained. Iron values jump from 800 to 1300 parts/10^6 at the contact. Again, samples in the fine-grained limestone below the discontinuity possess solution cavities partially filled with brownish argillaceous carbonate mud. It is suggested, but not confirmed, that a time break exists at 26.8 m separating time unit D from time unit E.

(b) Stapleton

Three lithologic units were recognized at Stapleton, designated upper, middle and lower (figure 3). To 8.5 m below ground surface there exists a carbonate sand, unconsolidated, well

TABLE 3. BEWICK CORE, SUMMARY OF STRATIGRAPHY

time unit	depth below h.w.s.t. m	lithologic unit	dominant petrographic types	dominant components and texture	diagenetic features	postulated depositional environments	representative samples
A	0–3	I	biosparite	rounded, well sorted allochems (corals, molluscs, coralline algae, benthonic forams, bryozoans, *Halimeda*, intraclasts)	acicular aragonite cement	beach	2-1-1 2-1-3 1-2-4
	3–6	II	unconsolidated biogenic sand	coarse to fine sand; angular to subangular allochems (molluscs, coralline algae, corals, some *Halimeda*, (benthonic forams))		reef flat	2-3-1
			unconsolidated coral/algal gravel (reef rubble)	bored, subangular pebbles of coral and coralline algal fragments; some gastropod shells present	acicular aragonite in cavities in corals and gastropod shells		2-3-3 2-3-6
disconformity B	6–11.6	III	coarse algal biomicrite	poorly sorted allochems (coralline algae, molluscs, corals, bryozoan and foram fragments)	inversion of mollusc and coral fragments to calcitic mosaics; evidence for cement in some samples	low-moderate energy reef flat with lagoonal areas	2-4-3 2-4-6 2-5-5 2-5-6
			biomicrite	molluscs, coralline algae, corals, forams, with some detrital quartz in the matrix	mosaics resulting from aggrading neomorphism present in the matrix		2-10-3 2-11-1
	11.6–12.2	IV	biolithite	corals	inversion of coral fabrics to calcitic mosaics	corals in growth position	2-13-2 2-13-5

	Depth	Zone	Lithology	Composition	Diagenesis	Environment	Samples
disconformity C	12.2–14.3	V	varieties of biomicrite (fine biomicrite, recrystallized biomicrite) and fossiliferous micrite	molluscs, forams, some coralline algae and corals; silt and very fine sand grade quartz in a fine matrix, with clay and haematite particles	iron staining and neomorphic mosaics extreme in places, producing pseudobreccia; allochems present in some samples as 'ghosts'	low energy lagoon with patch reefs receiving terrigenous sediment	2–13–7 2–13–11 2–14–4 2–15–2 2–16–1
disconformity D	14.3–16.8	VI	fossiliferous micrite	few large allochems algae and corals, with foram and mollusc fragments. Abundant silt to fine sand grade quartz and also ferruginous particles	cavernous appearance; limonite staining of allochems; authigenic growth on some quartz grains	shallow protected lagoonal area receiving terrigenous sediment	2–18–1
	16.8–26.8	VII	interbedded fossiliferous micrite, biomicrite, and calcite gravels and breccias	coralline algae, coral, forams, and molluscs; silt-sized quartz; ferruginous and argillaceous components	iron stained allochems and matrix; inverted reef rubble	shallow lagoon adjacent to patch reefs (limited transportation)	2–22–3 2–23–8 2–25–3 2–28–3
disconformity or transitional zone E	26.8–29.3	VIII	biomicrite, biomicrudite and breccia	poorly sorted, large subangular fragments of corals, molluscs, etc., in fine-grained matrix	solution cavities filled with brown calcilutite; more severe iron staining; allochems 'recrystallized'	shallow lagoon quite protected but with reef patches	2–35–10 2–36–2 2–37–3

sorted and containing a variety of constituents. The middle unit appears to contain inter-bedded unconsolidated sand mixed with cemented layers (containing molluscs, coralline algae and foraminifers) and coral fragments. On the basis of aragonite/calcite ratios and ^{14}C dates, this unit is Holocene in age. At the base of the hole at 14.6 m occurs a biomicrite with a calcite mineralogy. This was also a hard layer difficult to penetrate. It is considered to represent the Holocene–Pleistocene ('Thurber') discontinuity.

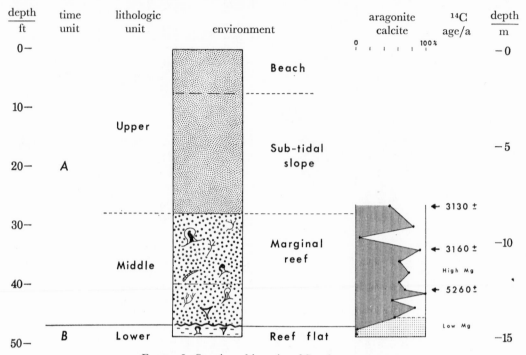

FIGURE 3. Stratigraphic units of Stapleton core.

STRATIGRAPHIC MODEL

From Bewick and Stapleton shallow drill results, it is possible to construct a conceptual model of island genesis in the Late Quaternary. This model is diagrammatically depicted in figure 4, which shows a number of stages in the development of islands such as Bewick.

The model is based on three assumptions. (i) Sea level since the Brunhes-Matuyama magnetic reversal of *ca.* 700 000 years ago has oscillated primarily in response to glaciations in the Northern Hemisphere. The amplitude of major oscillations is 100–120 m with a period of 100 000–120 000 years (Shackleton & Opdyke 1973). Essentially, sea level has spent 90 % of the upper Quaternary below its present position. (ii) The northern Queensland continental shelf has slowly subsided throughout the Quaternary, probably since the mid-Tertiary. The thick pile of Cainozoic limestones documented in deeper Great Barrier Reef drill holes supports this assumption. Furthermore, no Pleistocene sediments are exposed above sea level on the continental shelf, and only last interglacial (?) materials are recorded on the mainland (D. Hopley, personal communication). (iii) Biogenic activity continued at or slightly below sea level during each major transgression.

The model requires the return of sea level on a slowly subsiding shelf for periods of 5000 to 20 000 years at least six or seven times in the last 700 000 years. During each transgression, reef and associated lagoon sediments developed on remnants of older reef–lagoon complexes. During marine regressions the newly deposited limestone was exposed to subaerial processes. It is possible that a karst morphology was developed on much of the exposed terrain during the relatively long intervals when sea level would have oscillated below the level of the former islands. The extent to which these islands would have protruded in the form of tower karst hills is unknown. However, it is expected that the exposed surface would have been subjected to cavernous weathering under a combination of freshwater vadose and phreatic conditions.

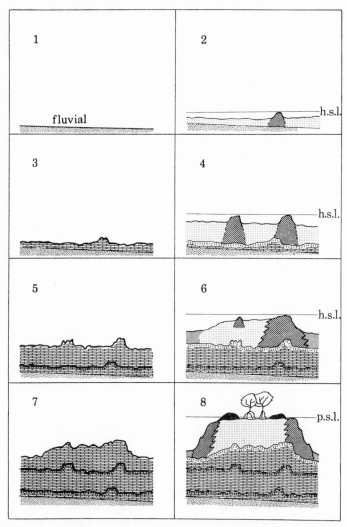

FIGURE 4. Idealized model of the development of a platform reef with windward and leeward cays during the late Quaternary (based on Bewick core data).

The model also suggests a change in the environment of deposition from Pleistocene time units B–E compared with the Holocene unit A. High energy reefal conditions appear to have dominated the Holocene, but below the upper disconformity more sheltered environments have been interpreted. The biomicrite and fossiliferous biomicrite of the Bewick core reflect shallow lagoonal conditions, and the thin interbedded coarse deposits (gravels, breccias, etc.) indicate occasional disturbances and delivery of coarse reef debris. The shelf itself could have had lower relief or there could have been more protected areas, including lagoons, within a larger reef complex. Without further drilling either view is tenable. However, the increase in terrigenous material below 12.2 m suggests greater opportunity of transport from the mainland, probably across a shallower shelf than now exists.

One implication of the Bewick–Stapleton drilling is illustrated in figure 5. Variation in the depth of the Pleistocene–Holocene disconformity may strongly influence the degree of development of patch reefs and associated cay and mangrove cover. Well developed 'low wooded islands' (e.g. Bewick) could possess a shallow Holocene reef cap compared with reefs which contain little or no cay development. The mechanism responsible for such a pattern involves a more or less constant rate of vertical reef growth which lagged behind the rate of sea level rise before the 'still-stand' of *ca.* 6000 a B.P. Variation in the depth of the Holocene–Pleistocene surface, drowned at different times during the transgression, probably interacts with other factors, such as configuration of the surface (tilted, flat, centrally depressed, etc.), its areal extent, and hydrodynamic and biologic factors, to produce the variety of reef morphologies characteristic of the Great Barrier Reef.

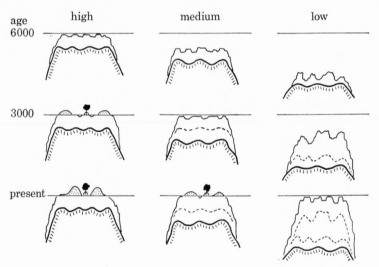

FIGURE 5. Expected influence of variations in the depth of Pleistocene–Holocene disconformity below present sea level on the age and degree of development of reef flats and cays in the northern region of the Great Barrier Reef.

CONCLUSION

Two drill holes undertaken as part of the 1973 Expedition yielded information on the development of reef island complexes in the northern Great Barrier Reef Province in the late Quaternary. The hole on Bewick Island, in particular, suggests the presence of several discontinuities in the subsurface. Here eight lithological units were identified, separated in three instances by facies transitions, and in four other cases by disconformities. The uppermost disconformity is confirmed by several lines of evidence, including radiocarbon dating, and clearly separates Holocene from late Pleistocene sediments. Other time breaks are not as strongly supported, especially the lowest two. At Stapleton, drilling stopped at the upper discontinuity.

The data are interpreted as showing episodic growth of a limestone 'pile' during Quaternary marine transgressions associated with periodic continental deglaciations, i.e. interglacial intervals. The thin nature of each limestone unit is consistent with expected rates of reef growth and associated limestone deposition during the limited time period when sea level reached its high levels relative to the land (P. J. Davies, personal communication). However, this view remains untested and requires detailed geochronologic study of uninverted material of Pleistocene age. It does appear likely that reefal limestones were exposed to subaerial weathering and were karstified during Quaternary marine regressions. The cavernous, recrystallized (*sensu lato*) low-magnesium calcite rocks below 7 m are consistent with this inference.

The Bewick core also shows marked changes in environmental conditions with time. The upper part of the core contains sediments deposited under relatively high energy conditions. Below the upper disconformity the dominant rock types are biomicrite and fossiliferous micrite which suggest somewhat sheltered, relatively shallow lagoonal conditions within which patch reefs grow. The occurrence of terrigenous clastic grains below 12.2 m perhaps requires a shallower shelf, and/or a nearer source of terrigenous material, than is near the Bewick drilling site at present.

The discontinuous stratigraphic sequence at Bewick and Stapleton is similar in some respects to that interpreted from Heron Island and Michaelmas Cay (Davies 1974). The shallow nature of the Holocene as observed at Bewick, where it extends to only 4 m below low water springs, is seen in other reef complexes in the Pacific (e.g. Mururoa; Labeyrie, Lalou & Delibrias 1969). The data from Bewick and Stapleton cores are not inconsistent with the general model of barrier reef development proposed by Purdy (1974).

Finally, it should be apparent that interpretations of reef structure from a limited number of shallow drill holes must be treated with considerable caution. The need to drill at a number of places across the continental shelf in order to test various models of shelf behaviour in relation to sea level change, as well as to examine subsurface characteristics in different environmental conditions, was an objective not attainable in the time available to the 1973 Expedition. Even more frustrating, but equally urgent, was the desire to drill at more than one site on the same island. Furthermore, it was not possible to tie in our drilling with continuous seismic profiling. Given these limitations, it is possible that the information made available by the two holes has only partially answered some questions, but also it has raised a number of new questions, the answers to which will require further drilling efforts.

References (Thom *et al.*)

Bathurst, R. G. C. 1966 *Geol. J.* **5**, 15–32.

Chappell, J., Broecker, W. S., Polach, H. A. & Thom, B. G. 1974 In *Proc. 2nd International Symposium on Coral Reefs, 1973*, vol. 2, pp. 563–572.

Chappell, J. & Thom, B. G. 1978 In *Coral reefs: research methods* (eds D. R. Stoddart & R. E. Johannes). Monographs on Oceanographic Methodology, Unesco.

Davies, P. J. 1974 In *Proc. 2nd International Symposium on Coral Reefs, 1973*, vol. 2, pp. 573–578.

Davies, P. J. & Till, R. 1968 *J. sedim. Petrol.* **38**, 234–237.

Emery, K. O., Tracey, J. I. & Ladd, H. S. 1954 *U.S.G.S. Professional Paper* 260-A.

Folk, R. L. 1965 In *Dolomitization and limestone diagenesis – a symposium* (eds L. C. Pray & R. C. Murray), pp. 14–48. Society of Economic Paleontologists and Mineralogists, Special Publication No. 13.

Labeyrie, J., Lalou, C. & Delibrias, G. 1969 *Cah. Pacif.* **13**, 59–73.

Maxwell, W. G. H. 1962 *J. geol. Soc. Aust.* **8**, 217–238.

Maxwell, W. G. H. 1968 *Atlas of the Great Barrier Reef.* Amsterdam: Elsevier.

Maxwell, W. G. H. 1973 In *Biology and geology of coral reefs* (eds O. A. Jones & R. Endean), vol. 1, pp. 233–272. New York: Academic Press.

Purdy, E. G. 1974 In *Reefs in time and space* (ed. L. F. Laporte), pp. 9–76. Society of Economic Paleontologists and Mineralogists, Special Publication No. 18.

Richards, H. C. & Hill, D. 1942 *Rep. Gt Barrier Reef Comm.* **4**, 1–111.
Shackleton, N. J. & Opdyke, N. D. 1973 *Quat. Res.* **3**, 39–55.
Steers, J. A. 1929 *Geogrl J.* **74**, 232–257 and 341–367.
Stoddart, D. R. 1969 *Biol. Rev.* **44**, 433–498.
Thom, B. G. 1976 *Coral reefs: research methods* (eds D. R. Stoddart & R. E. Johannes). Monographs on Oceanographic Methodology, Unesco.
Tracey, J. I. & Ladd, H. S. 1974 In *Proc. 2nd International Symposium on Coral Reefs, 1973*, vol. 2, pp. 537–550.

Phil. Trans. R. Soc. Lond. A. **291**, 55–71 (1978)   [55]

Printed in Great Britain

Reefal sediments of the northern Great Barrier Reef

By P. G. Flood† and T. P. Scoffin‡

†*Department of Geology and Mineralogy, University of Queensland, St Lucia, Queensland, 4067, Australia*
‡*Grant Institute of Geology, University of Edinburgh, West Mains Road,*
Edinburgh EH9 3JW, U.K.

with an appendix by A. B. Cribb

[Plates 1 and 2]

Skeletal carbonate sediments collected from nine reef flats have been analysed by using multivariate statistical techiques to determine the inter- and intra-reefal variations of sediment texture and composition. Q-mode cluster analysis of granulometric analysis data allows the combined collection of approximately 200 samples to be grouped into four sediment types. The same data were re-analysed by using Q-mode factor analysis techniques which showed that three factors will explain more than 90 % of the variations exhibited by the sediments. The factor scores illustrate the relative influence that individual particle sizes have on each factor. R-mode cluster analysis shows three distinct grouping of sizes which are interpreted as the individual population of sizes that are subjected to differing modes of transportation (traction, saltation and suspension load) which form in response to the prevailing hydrodynamic régime. The distribution of particle size within these three populations is modified by the presence of six skeletal modes within the sand size range. The organic group contributing to each skeletal mode has been identified with the aid of a scanning electron microscope.

1. Introduction

This paper describes the unconsolidated sedimentary deposits associated with reefs in the northern region of the Great Barrier Reef (figure 1). A subjective account of the sediment bodies and a statistical account of their component and grain size compositional variability is provided. Eight of the reefs display the range of variation characteristic of the inner shelf reefs. They are:

low wooded island type:	Low Isles, Three Isles, Pipon, Watson,
(cf. Maxwell's (1968) high reef of inner shelf)	and Ingram–Beanley Reefs
platform type:	Mid and Megaera Reefs
lagoonal platform type:	Stapleton Reef

One shelf edge platform reef immediately to the south of Waterwitch Passage was visited. More than 200 sediment samples were examined.

Steers (1929, 1930), Stephenson, Stephenson, Tandy & Spender (1931) and Stoddart, McLean & Hopley (1978, part B of this Discussion) provide detailed accounts of the morphological features and ecological zonation characteristic of reefs in this region (figure 2, plate 1).

2. Prevailing physical conditions

The reader is referred to Maxwell (1968) for details of the physical environment and hydrographic setting of the Northern Province of the Great Barrier Reef.

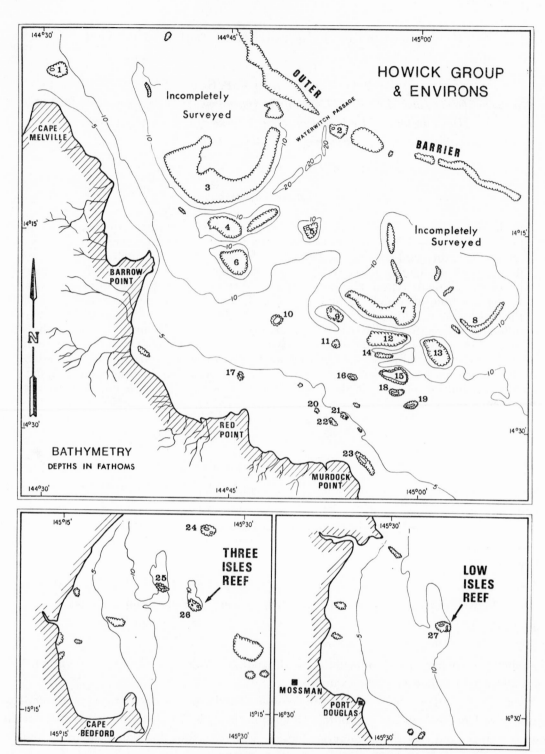

FIGURE 1. Location of the nine reefs studied: 1, Pipon; 2, 'Waterwitch'; 5, Stapleton; 9, Ingram–Beanley; 11, Watson; 12, Mid; 14, Megaera; 26, Three Isles; 27, Low Isles.

During the expedition (July–November 1973) the trade winds blew steadily from the south-east. Only four calm days were recorded. The wind, normally 15–40 km/h, produced swells of 1–3 m amplitude in the inner shelf seas and plunging breakers on the windward sides of patch reefs up to 2 m in height. All the reefs under study had emergent reef flats at low water and even at high water, waves did not break on the central parts of reef tops.

Three main types of currents influence the reef-top sediments. Tidal currents seldom exceed 2 km/h even though there is a 3 m tidal range. The drainage on a falling tide is crudely radial though the impervious ramparts at the windward and flanking rims of the reef tend to channel the water approximately parallel to the reef edge for some distance until it can escade over a sill or through a gap. The bulk of the reef flat water drains to leeward during low tide simply because the leeward side of the reef is lower than the windward.

The wind usually blows the surface water to lee, and during low tide the water level on the reef flat is sufficiently shallow for the effect of this surface current to reach the bottom and agitate sand grains.

The waves impinging on the windward front of the reef refract in various patterns according to reef shape, proximity of neighbouring reefs, orientation of the reef front with respect to the wind, and the surface morphology (particularly of emergent features such as ramparts, platforms or mangroves). Generally there is a consistent refraction around the margins of the reef flats which results in a long-shore transport of grains from windward to leeward. Where the two lateral sets of waves converge at the leeward edge the current energy is lost and sediment is deposited.

3. Reefal sediment bodies

There are several distinct unconsolidated sediment bodies occurring on the tops of most of the inner shelf reefs. They are: (*a*) rim deposits of shingle ramparts and boulder tracts; (*b*) blanket sands of reef flat or lagoon; and (*c*) leeward sand cay.

(a) Rim deposits

At the perimeter there is normally a rim of broken coral. On the windward side the dominant components of this rim are *Acropora* branches which are piled up by waves to form large asymmetric ridges (often also called ramparts) 50–100 m in wavelength and 0.5–3 m in amplitude. The seaward slope is usually less than 5° and slopes towards the reef centre about 60°. These ramparts have a crescent-shaped outer margin paralleling the reef edge and a cuspate inner margin from which tongues of steeply banked shingle project on the reef flat. Commonly several sets of ramparts occur on one reef and seawater is ponded between adjacent sets forming moats at low tide. Thickets of branching *Acropora* corals living in shallow water on the windward side of the reef are the source of the gravel to boulder sized coral fragments of the ramparts.

Towards the leeward flanks of the reef the shingle ramparts give way to a linear pile of massive boulders. Each boulder is normally one massive coral colony (predominantly *Porites*) which formerly grew on the leeward flanks of the reef in shallow water as coral heads or 'bommies'. A few isolated boulders are irregularly scattered along the leeward rim of the reef or they may be so numerous that they form a continuous coarse conglomerate deposit: the boulder tract.

These rim sediments are produced by wave action which removes skeletal debris and deposits them at the margins of the shallow reefs when the wave loses its energy. The grains are too coarse to be moved by wind or tidal currents and these sediments remain as intertidal lag deposits.

The very large boulders on the reef are not deposited by normal wave action but during cyclones, as are the ramparts. The occurrence of distinct sets of ramparts and boulder tracts on any one reef flat and the analysis of their spatial distributions over the years (Stoddart *et al.* 1978, part B of this Discussion) suggest that the main movement of reef rim deposits is during storms.

(b) *Blanket sands*

The reef flat sediment cover is normally only a few centimetres thick. The thickest deposits occur in hollows on the underlying solid reef top. Those parts of the reef flat that do not dry completely at each low tide commonly have a sparse cover of soft vegetation. *Thalassia* grass and some algae species occur only where the sediment thickness is greater than 5 cm. The typical concentric variation in sediment thickness across the reef flats plays a part in producing the marked concentric zones of vegetation on the margin of reefs (algal specimens collected along a radial transect across Ingram–Beanley Reef were kindly identified by Dr Alan Cribb of the Botany Department, University of Queensland, Brisbane; see the appendix).

The sediments on the reef flats and in shallow lagoons consist essentially of sand sized grains of corals, benthonic Foraminifera, and green and red calcareous algae. The coral sand is supplied by the mechanical and biological breakdown of corals growing mainly on the reef front (though a few corals grow on the reef flat). The benthonic Foraminifera are attached to the soft plants on the reef flat. *Marginopora* is particularly abundant on the *Thalassia* grass and *Baculogypsina* and *Calcarina* occur in vast quantities attached to the short fronds of *Laurencia* algae on the windward margin of the reef top just seaward of the ramparts. In this region the intertidal zone is the area of maximum biological corrosion of carbonate. Here, boring activities of *Lithophaga* bivalves, sponges and filamentous algae are important in the breakdown of coarse coral and mollusc skeletons.

The distribution of silt and clay sizes is governed by the distribution of stabilizing networks such as mangrove roots, grass blades, algal fronds, porous shingle ramparts, all of which act as baffles, locally lowering the current velocities and allowing fine sediment to settle.

The sediments on the reef flats are moderately to poorly sorted. The grains can be whole or broken, subangular to subrounded. Organisms play a part in influencing texture; for example, grass thickets produce poor sorting with finer sediment whereas burrowing crustaceans and worms mix and sometimes separate grain sizes. On the bare areas of reef flats, the sediments are mobile and shallow asymmetric ripples with well sorted, well rounded grains occur. Experiments using dyed sand on Ingram–Beanley Reef flat showed that the net movement of sand over several tidal cycles was to leeward.

(c) *Leeward sand cays*

Wave refraction around the reefs allows the deposition of sand towards the leeward edge. These deposits include the cays, bars and spits. Wind removes the sand and finer sizes from a beach at low tide to develop a cay that is emergent above the high water mark. The cay is commonly stabilized by freshwater vegetation, whereas the intertidal rim deposits, particularly the inner portions of ramparts, are colonized by mangroves. Both types of intertidal deposit are fairly readily cemented (see Scoffin & McLean 1978, this volume). The components of the cay sands are essentially similar to those of the mobile parts of the reef flat consisting of coral,

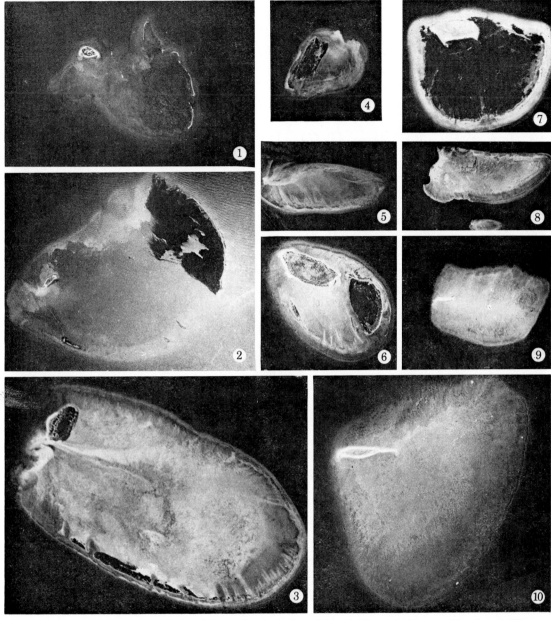

FIGURE 2. Vertical aerial photographs of the reef tops. 1, Low Isles; 2, Pipon; 3, Ingram–Beanley; 4, Watson;
5, Megaera; 6, Three Isles; 7, Bewick (site of diamond drill hole: Thom, Orme & Polach 1978, this volume);
8, Mid; 9, 'Waterwitch'; 10, Stapleton (site of diamond drill hole). For a comparison of sizes see figure 3.
Published with permission of the Director of National Mapping, Australia.

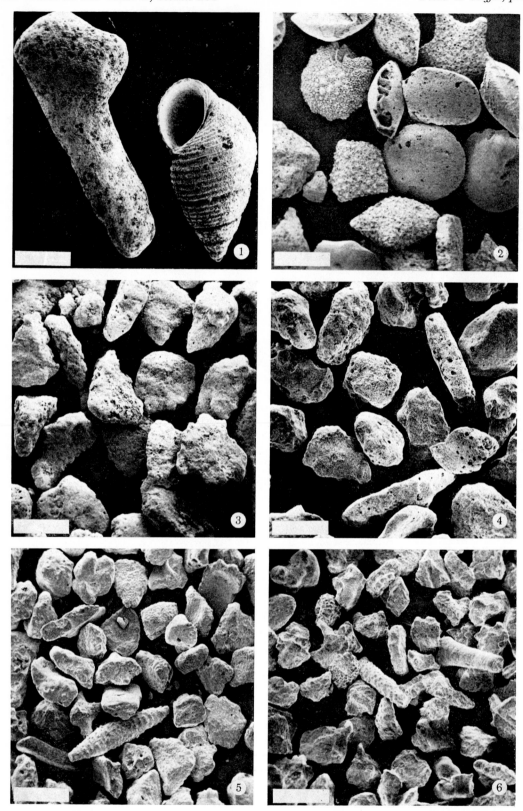

FIGURE 9. Scanning electron microphotographs of selected ¼ φ fractions: (1) coral and shell, size −0.25 φ, bar scale 1 mm; (2) benthonic Foraminifera, size 0.25 φ, bar scale 0.8 mm; (3) coral, *Halimeda* and others, size 1 φ, bar scale 0.8 mm; (4) size 2 φ, bar scale 0.5 mm; (5) size 2.75 φ, bar scale 0.5 mm; (6) size 3.25 φ, bar scale 0.2 mm.

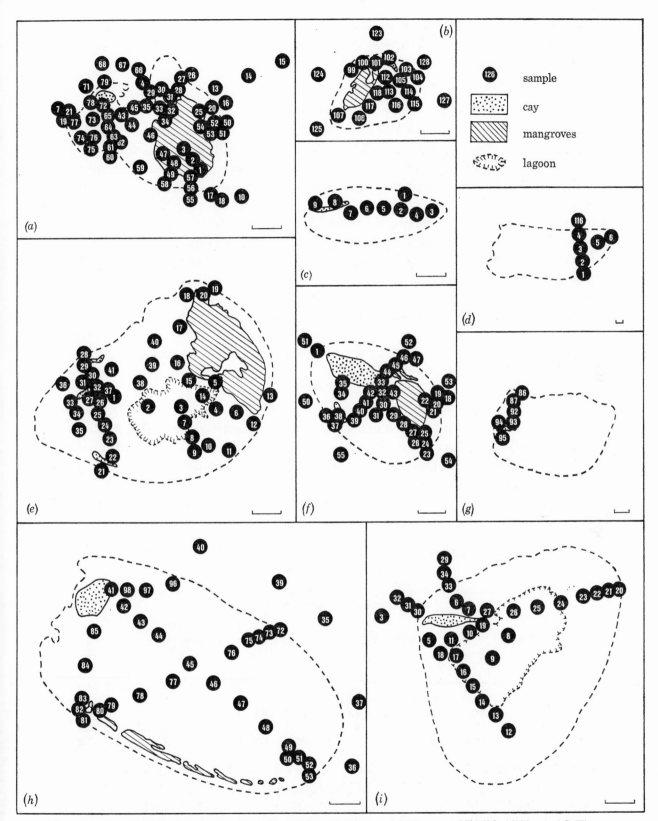

FIGURE 3. Location of sediment samples. (a) Low Isles; (b) Watson; (c) Megaera; (d) Mid; (e) Pipon; (f) Three; (g) 'Waterwitch'; (h) Ingram–Beanley; (i) Stapleton. The bar scales represent 0.5 km.

Foraminifera and coralline algae. These grains have been under the influence of wave action on the beaches of the leeward sand cay and therefore normally show improved rounding, sorting and polishing (see McLean & Stoddart 1978, this volume).

4. THE SEDIMENTS

Sampling of the reef flats (figure 3) was accomplished by traversing on foot during periods of low tide. Tidal conditions and the Expedition programme generally dictated that any one reef flat could be examined for only a few hours and on one occasion only. The sediments were collected by hand and placed into plastic bags. Later each sample was split into two equal parts: one was treated with 10 % hydrogen peroxide to remove organic material, the other with alcohol to preserve it. A representative fraction of the treated sample was sieved to determine the particle size distribution. Another representative fraction was impregnated with epoxy resin for making a thin section for point counting to determine the component composition. Multivariate statistical programs have been used to analyse the sediment data and to provide objective groupings of the samples, components and grain size variables. Component and grain size data were analysed using Davis's (1973) cluster and Klovan & Imbrie's (1971) factor programs.

Representative samples are housed at the University of Queensland, Australia and at the University of Edinburgh, Scotland.

(a) Components

Coral, Foraminifera, *Halimeda*, molluscs and coralline algae are quantitatively the most important skeletal components present in the sediments. The component analysis data obtained from the Low Isles and Three Isles sediments was processed in a manner similar to that used by Imbrie & Purdy (1962). The Q-mode cluster analysis dendrogram shows five distinct groupings (cf. reaction groups) which represent discrete sediment types (see figure 4).

The mean and standard deviation of each component present within the types 1–5 are shown in table 1.

TABLE 1. REEF FLAT SEDIMENTS: COMPOSITIONAL STATISTICS FOR LOW ISLES
AND THREE ISLES REEFS ONLY

(Mean ($\bar{x}$) and standard deviation (s); value in parentheses indicates number of samples;
* denotes dominant compositional group.)

	coral		molluscs		coralline algae		*Halimeda*		benthonic Foraminifera		*Margino-pora*		others	
	$\bar{x}$	s	$\bar{x}$	s	$\bar{x}$	s	$\bar{x}$	s	$\bar{x}$	s	$\bar{x}$	s	$\bar{x}$	s
type 1 (10)	27.7*	4.1	9.7	2.7	7.6	3.4	11.0	5.0	29.6*	6.7	1.3	1.3	12.7	4.3
type 2 (15)	40.0*	14.2	11.4	3.9	6.3	4.1	12.3	4.3	11.7	4.7	2.7	2.0	12.8	4.9
type 3 (15)	29.1*	4.0	10.5	2.7	7.0	2.9	21.2*	4.4	15.0	4.1	3.8	2.1	13.4	4.9
type 4 (5)	14.8	5.5	5.8	3.2	4.8	2.9	11.6	5.8	52.0*	7.9	2.2	1.7	9.2	5.7
type 5 (10)	18.5	5.2	11.3	2.6	4.0	2.2	34.2*	6.0	21.5	6.1	2.2	1.3	12.4	3.2

Q-mode factor analysis, which conveniently bypasses the problem of closure of data, showed that 97 % of the component variation could be explained by three factors. The relative component contribution with respect to each factor (i.e. the factor score) is shown in figure 5. The normalized varimax triangular plot shows the relation of the samples with respect to the factors I, II and III (39, 26 and 31 % variance respectively) and illustrates the approximate fields occupied by the five cluster groupings. The results indicate that the variation exhibited by the majority of sediments can be expressed in terms of three components: coral (and to a lesser extent coralline algae), benthonic Foraminifera and *Halimeda*. Combinations of these components produce the five sediment types.

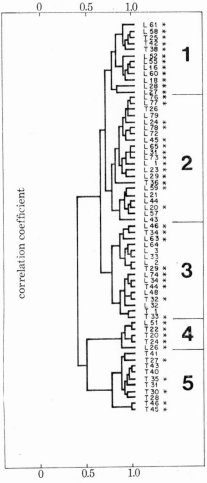

FIGURE 4. Q-mode cluster dendrogram of component analysis of reef flat sediments from Low and Three Isles Reefs. Five groups (1–5) are recognizable. Asterisk denotes that the sample is common to the component and textural analyses.

(b) Texture

Grain size analyses were made using a Ro-Tap shaker, U.S. standard sieves and a sieving time of 15 min. The results, expressed as percentages by mass in each $\frac{1}{4}$ ϕ ($-\log_2$ millimetre transformation) interval of the sand size range, were processed by using the same programs as used for the component analyses. Because the number of samples exceeded the maximum that could be processed by the cluster program, analysis was performed on two subgroups: the Low Isles and Three Isles samples, and samples from the reefs of the Howick Group respectively.

The Q-mode cluster dendrograms (figure 6) show three distinct groupings (1–3). Statistics of the grain size relations within the three groups are given in table 2. Typical textural parameter values expressed in verbal terms are:

Type 1: usually coarse or very coarse sands and gravel; moderately to poorly sorted; strongly fine skewed or fine skewed; kurtosis variable.

Type 2: usually very coarse sand or coarse sand, with minor gravel; moderately to moderately well sorted, occasionally well sorted; strongly fine skewed, fine skewed or near symmetrical; kurtosis variable.

Type 3: usually coarse and medium sands; moderately to poorly sorted; fine skewed, kurtosis variable.

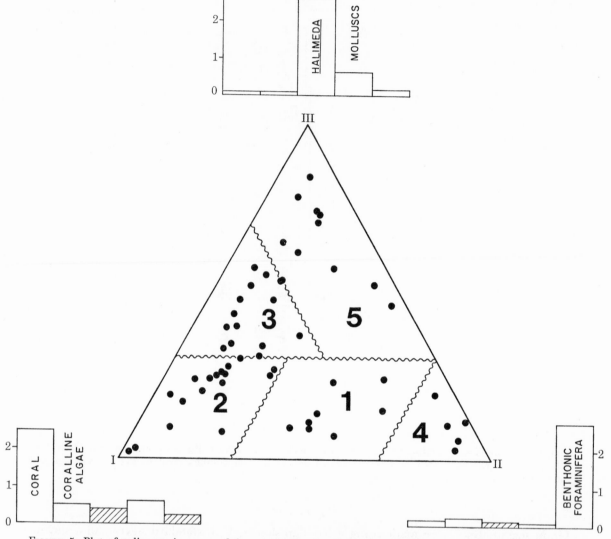

FIGURE 5. Plot of sediments in terms of three normalized varimax factors. The areas occupied by component groupings (1–5) are indicated. The relative influence of components to each factor is shown as the plot of factor scores. Hatched areas are those showing negative influence.

Q-mode factor analysis (figure 7) indicates that 95% of the grain-size variation can be explained by three factors. The normalized varimax triangular plot shows the relation of the samples with respect to factors 1, 2 and 3 (40, 40, and 15% variance respectively) and illustrates the fields occupied by the three cluster groupings. The relative grain size contribution with respect to each factor is shown by the factor score.

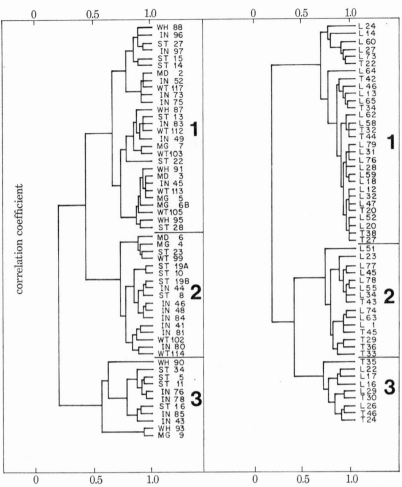

FIGURE 6. Q-mode cluster dendrograms of grain size analysis of the reef flat sediments. Three clusters (1–3) are recognizable. The other cluster relates to sediments from the mangrove area or from the inter-reef area.

TABLE 2. REEF FLAT SEDIMENTS: GRAIN SIZE STATISTICS

Percentage by mass retained at $\frac{1}{4}$ φ intervals (22 values)

type 1 (55)

$\bar{x}$	26.2*	6.0	7.3	7.9	7.0	9.6	6.4	5.0	4.7	4.0	2.9	(a)
s	11.1	3.1	3.3	2.7	2.6	3.9	2.1	1.8	2.1	2.1	1.9	
size (φ)	gravel	−0.75	−0.50	−0.25	0.00	0.25	0.50	0.75	1.00	1.25	1.50	
$\bar{x}$	2.5	2.2	1.3	1.7	1.3	0.8	0.6	0.4	0.2	0.3	0.9	(b)
s	1.8	1.7	1.1	1.5	1.3	0.8	0.6	0.4	0.2	0.4	1.5	
size (φ)	1.75	2.00	2.25	2.50	2.75	3.00	3.25	3.50	3.75	4.00	mud	

type 2 (24)

$\bar{x}$	7.5	3.6	5.9	9.2	10.7*	17.2*	11.9*	8.7	7.3	5.2	3.2	(a)
s	3.4	1.6	2.7	3.1	4.1	6.9	3.5	2.4	2.9	3.0	2.1	
$\bar{x}$	2.0	1.3	1.0	1.4	1.0	0.7	0.4	0.3	0.1	0.1	0.6	(b)
s	1.7	1.5	1.2	1.7	1.3	1.0	0.6	0.4	0.1	0.2	0.6	

type 3 (17)

$\bar{x}$	4.5	2.1	3.2	4.7	4.7	7.8	7.3	7.5	9.2*	9.9*	8.3*	(a)
s	4.2	1.5	2.1	2.1	1.8	2.7	2.1	1.8	3.7	4.9	4.0	
size/mm	gravel	1.68	1.41	1.19	1.00	0.84	0.71	0.59	0.50	0.42	0.35	
$\bar{x}$	6.2*	5.7*	4.7*	4.6*	3.1*	1.9	1.3	0.7	0.3	0.3	0.9	(b)
s	2.5	3.0	2.1	1.9	1.8	1.5	1.1	0.7	0.3	0.3	0.6	
size/mm	0.30	0.25	0.21	0.17	0.15	0.125	0.105	0.088	0.074	0.062	mud	

$\bar{x}$, mean; s, standard deviation; * denotes dominant grain size. (n), number of samples.

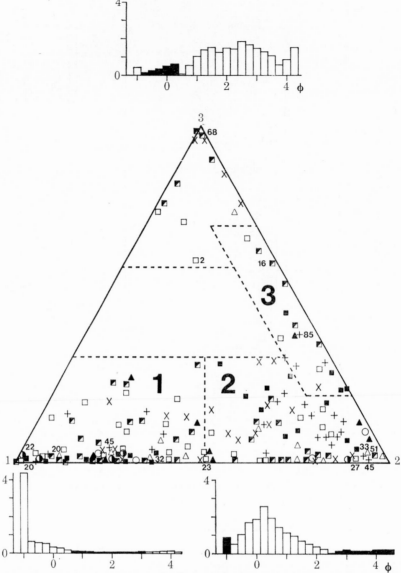

FIGURE 7. Plot of sediments in terms of three normalized varimax factors. The areas occupied by each grain size group (1–3) are indicated. The relative influence (black for negative score) of each $\frac{1}{4}$ ϕ interval is shown on the plot of the factor scores (-1 ϕ = 2 mm; 0 ϕ = 1 mm; 1 ϕ = 0.5 mm; 2 ϕ = 0.25 mm; 3 ϕ = 0.125 mm; 4 ϕ = 0.062 mm). ■, Three Isles; ◪, Low Isles; +, Ingram; ×, Stapleton; □, Pipon; △, Watson; ○, Mid; ◑, Megaera; ▲, 'Waterwitch'.

(c) Relations between component and grain size

The complex relation between organic components and grain size, produced by the irregular nature of the disintegration of individual skeletons (Sorby Principle: Folk & Robles 1964), has been investigated in two ways. The component and grain size data of each sediment sample were combined, and R-mode cluster analysis performed; representative fractions of each of the distinct modes indicated on the grain size frequency histograms were examined using the scanning electron microscope. The following associations are indicated within the sand size range:

coral: sizes -0.75 to -0.25 ϕ (1.68–1.19 mm) and 3.25–4.00 ϕ (0.105–0.062 mm);

benthonic Foraminifera: sizes 0.00–0.75 φ (1–0.59 mm) (larger species) and 2.75–3.00 φ (0.149–0.125 mm) (smaller species);

coralline algae: sizes − 1.00 to − 0.25 φ (2–1.19 mm);

Halimeda: sizes 0.00–3.00 φ (1.0–0.125 mm);

molluscs: sizes − 0.75 to − 0.25 φ (1.68–1.19 mm) and 2.25–2.50 φ (0.210–0.177 mm).

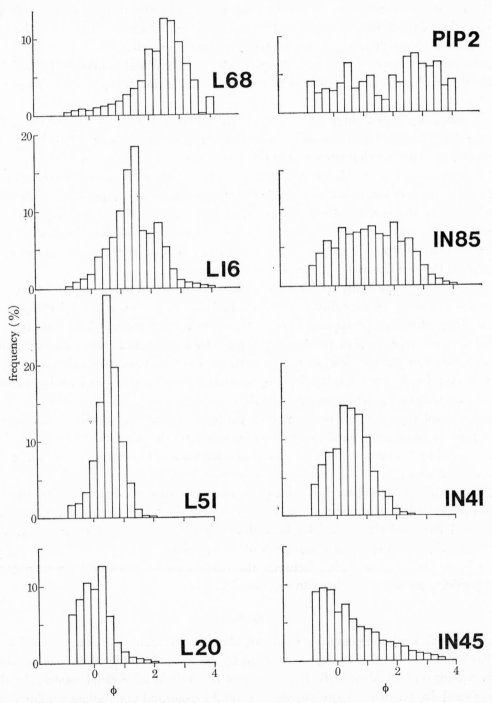

FIGURE 8. Grain size histograms for selected sediments. Several skeletal modes are obvious. Samples L16 and L68 are from the reef slope area and PIP2 is from a lagoonal area (conversions from φ to millimetres are given with figure 7).

(The range of sizes mentioned for each component group does not preclude the possible occurrence of the component in any other size.)

(d) Distribution of the sediment types

The sediments represent individual skeletons or the finer sediment derived from their breakdown. The distribution patterns of the component and textural types are controlled both by the location of the living organisms and by the movement of their detritus from the growth areas to areas of deposition. The following patterns were observed:

Reefal component type 1 (high percentage content of coral and benthonic Foraminifera), occurs at the windward outer margins of the reef flat and reef rim. It represents mixing of types 2 and 4.

Reefal component type 2 (high percentage content of coral) occurs toward the leeward part of the reef flat or forming the coral shingle ramparts. It can occur in pools adjacent to areas of prolific coral growth and characteristically represents any shallow lagoonal area.

Reefal component type 3 (high percentage of coral and *Halimeda*) occurs in the more protected areas of the reef flat, usually on a rocky substrate. The coral component is in the process of moving across the reef flat from windward to leeward and the *Halimeda* component is produced by the breakdown *in situ* of that organism. It represents mixing of types 2 and 5.

Reefal component type 4 (high percentage of benthonic Foraminifera) occurs at the seaward extremities of the reef flat and reef rim on the windward side.

Reefal component type 5 (high percentage of *Halimeda*) occurs in the more protected environments such as the lee of mangroves where contamination by other skeletal components is minimal. This sediment type appears to be restricted to reefs of low wooded island type.

Reefal textural type 1 (high percentage of particles coarser than sand size and varying percentages of other particle sizes) occurs towards the outer windward margin of the reef flat and on the reef rim. It represents lag deposits remaining after fine particles have been removed to the off-reef area or leeward across the reef flat.

Reefal textural type 2 (high percentage of particles of coarse to medium sand size and varying percentage of other particle sizes) occurs either on the windward side of the shingle ramparts and ridges or towards the central and leeward parts of the reef flat. It is characteristic of the sand flat areas.

Reefal textural type 3 (no gravel, varying percentages of coarse, medium, or fine sand) occurs towards the leeward parts of the reef flat in places where the tidal currents are reworking the medium and fine sand sizes. Particles finer than 3ϕ (i.e. very fine sand and smaller) are characteristically absent on the exposed parts of the reef flats.

There is no obvious correlation between the component and textural types, individual skeletal particles not being restricted to any one size.

(e) Synthesis

The organic skeletal components provide an almost continuous spectrum of particle sizes which are modified into distinct size populations by the hydrodynamic conditions (waves and tides) prevailing on individual reefs. Irrespective of the individual skeletal modes, the distribution of sand size particles (figure 10) shows several log-normal populations similar to those recognized in clastic sediments (Moss 1962, 1963; Visher 1969) suggesting that the size populations present in the sand-sized carbonate sediments may be explained in a similar manner regardless of inherent differences in shape, relative density, etc. (cf. Folk & Robles 1964; Force 1969).

Reefal sediments would therefore be considered to consist of any of the following:

A coarse population (approximately 1 φ (0.5 mm) and coarser), representing either lag deposits of particles too large to be removed from the outer part of the reef, where it was deposited by wave action, or particles that can be moved as a traction load by translatory wave action or tidal currents.

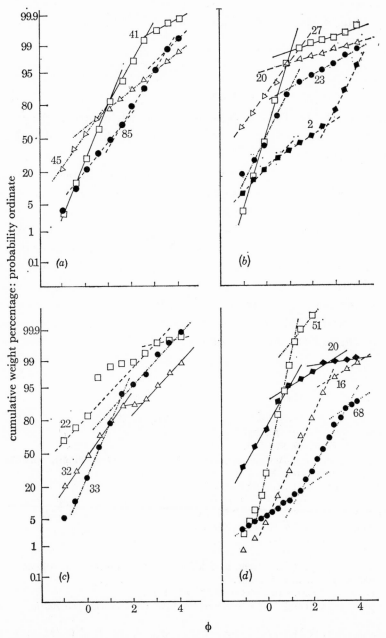

FIGURE 10. Cumulative curves of selected sediments illustrating the presence of distinct log-normal population (cf. Visher 1969). (a) Ingram–Beanley; (b) Pipon; (c) Three Isles; (d) Low Isles (conversions from φ to millimetres are given with figure 7).

A medium to fine (1–3 φ; 0.5–0.125 mm) population representing material eroded from the lag deposit and which is being deposited as the energy of the eroding agent (currents or waves) decreases.

A very fine population (3 φ (0.125 mm) and smaller) capable of being removed as suspension load from the reef to be deposited in the lagoonal or inter-reef areas.

Several populations representing material being moved by a combination of one or more of the modes of transportation (traction, saltation, or suspension). The particular size at which the mode of transportation changes depends upon the prevailing energy conditions of the waves and/or tidal currents.

These results are in harmony with Marshall & Orr's (1931) finding relating to the movement of sediment on the reef flat at Low Isles.

6. Conclusion

The occurrence and stability of sediment bodies is primarily related to the prevailing energy conditions (wave and tidal currents) and only secondarily are they related to biological factors (availability of skeletal particles).

The statistical analyses have shown that 95% of the variation displayed by the reefal sediments can be explained either in terms of three skeletal component groups: coral plus coralline algae, *Halimeda*, and benthonic Foraminifera; or alternatively in terms of three grain size populations: 1 φ (0.5 mm) and coarser, 1–3 φ, and 3 φ (0.125 mm) and finer. A textural gradient is displayed as follows: gravel and coarser sand on the windward part of the reef, medium sand on the central part of the reef flat and fine sand on the leeward part of the reef. Such gradients from windward to leeward are a direct response to diminishing amounts of the wave energy available to transport skeletal material from the areas of maximum productivity (the windward sector) to leeward. Tidal current can modify these textural differences.

The numerical results confirm the visual observations.

References (Flood & Scoffin)

Davis, J. C. 1973 *Statistics and data analysis in geology.* New York: John Wiley.

Folk, R. L. & Robles, R. 1964 *J. Geol.* **72**, 255–292.

Force, L. M. 1969 *J. sedim. Petrol.* **39**, 902–904.

Imbrie, J. & Purdy, E. G. 1962 In *Classification of carbonate rocks.* (ed. W. E. Ham), pp. 253–272. (American Association of Petroleum Geologists, Mem. 1.)

Klovan, J. E. & Imbrie, J. 1971 *Math. Geol.* **3**, 61–77.

McLean, R. F. & Stoddart, D. R. 1978 *Phil. Trans. R. Soc. Lond.* A **291**, 101–117 (this volume).

Marshall, S. M. & Orr, A. P. 1931 *Scient. Rep. Gt Barrier Reef Exped. 1928–29* **1**, 93–133.

Maxwell, W. G. H. 1968 *Atlas of the Great Barrier Reef.* Amsterdam: Elsevier.

Moss, A. J. 1962 *Am. J. Sci.* **260**, 337–373.

Moss, A. J. 1963 *Am. J. Sci.* **261**, 297–343.

Scoffin, T. P. & McLean, R. F. 1978 *Phil. Trans. R. Soc. Lond.* A **291**, 119–138 (this volume).

Steers, J. A. 1929 *Geogrl J.* **74**, 232–257, 341–367.

Steers, J. A. 1930 *Scient. Rep. Gt Barrier Reef Exped. 1928–29* **3**, 1–15.

Stephenson, T. A., Stephenson, A., Tandy, G. & Spender, M. A. 1931 *Scient. Rep. Gt Barrier Reef Exped. 1928–29* **3**, 17–112.

Stoddart, D. R., McLean, R. F. & Hopley, D. 1978 *Phil. Trans. R. Soc. Lond.* B **284**, 39–61 (part B of this Discussion).

Thom, B. G., Orme, G. R. & Polach, H. A. 1978 *Phil. Trans. R. Soc. Lond.* A **291**, 37–54 (this volume).

Visher, G. S. 1969 *J. sedim. Petrol.* **39**, 1074–1106.

Appendix. Algae collected on Ingram–Beanley Reef

By A. B. Cribb

Botany Department, University of Queensland, St Lucia, Queensland, Australia 4067

Location near sample 42, thin sand cover

Cyanophyta

Microcoleus lyngbyaceus (Kuetz.) Crouan

Chlorophyta

Cladophora sp.

Dictyosphoeria versluysii Weber–van Bosse

Enteromorpha clathrata (Roth) Grev.

Halimeda tuna (Ell. & Sol.) Lamour.

Valonia aegagropila C. Ag.

Valoniopsis pachynema (Mart.) Boerg.

Phaeophyta

Dictyota divaricata Lamour.

Padina gymnospora (Kuetz.) Vickers

Sargassum crassifolium J. Ag.

Rhodophyta

Centroceras clavulatum (C. Ag.) Mont.

Chondria sp.

Hypnea sp.

Jania adhaerens Lamour.

Laurencia flexilis Setch.

Laurencia obtusa (Huds.) Lamour.

Leveillea jungermannioides (Mart. & Her.) Harv.

Lophosiphonia scopulorum (Harv.) Wom.

Location near sample 44 on intertidal sand spit

Chlorophyta

Boodlea composita (Harv.) Brand

Caulerpa racemosa (Forssk.) J. Ag. var. *occidentalis* (J. Ag.) Boerg.

Dictyosphaeria versluysii Weber–van Bosse

Phaeophyta

Hydroclathrus clathratus (Bory) Howe

Sargassum crassifolium J. Ag.

Rhodophyta

Griffithsia sp.

Herposiphonia pacifica Hollenberg

Laurencia obtusa (Huds.) Lamour.

Laurencia papillosa (Forssk.) Grev.

Laurencia sp.

A. B. CRIBB

Location between samples 44 and 96, thin sand cover

Cyanophyta

 Microcoleus lyngbyaceus (Kuetz.) Crouan

Chlorophyta

 Acetabularia clavata Yam.

 Boodlea composita (Harv.) Brand

 Caulerpa racemosa (Forssk.) J. Ag. var. *occidentalis* (J. Ag.) Boerg.

 Caulerpa racemosa (Forssk.) J. Ag. var. *peltata* (Lamour.) Eubank

 Cladophora fuliginosa Kuetz.

 Dictyosphaeria versluysii Weber–van Bosse

 Halimeda opuntia (L.) Lamour.

 Valonia aegagropila C. Ag.

Phaeophyta

 Hydroclathrus clathratus (Bory) Howe

 Padina gymnospora (Kuetz.) Vickers

 Turbinaria ornata (Turn.) J. Ag.

Rhodophyta

 Amphiroa fragilissima (L.) Lamour.

 Centroceras clavulatum (C. Ag.) Mont.

 Champia parvula (C. Ag.) Harv.

 Gelidiopsis sp.

 Hypnea sp.

 Jania adhaerens Lamour.

 Laurencia flexilis Setch.

 Laurencia obtusa (Huds.) Lamour.

 Laurencia papillosa (Forssk.) Grev.

 Laurencia sp.

 Leveillea jungermannioides (Mart. & Her.) Harv.

 Spyridia filamentosa (Wulf.) Harv. in Hook.

 Tolypiocladia glomerulata (C. Ag.) Schmitz in Schmitz & Hauptfleisch

Location near sample 96, towards reef edge

Cyanophyta

 Hormothamnion enteromorphoides Grunow ex Born. & Flah

 Microcoleus lyngbyaceus (Kuetz.) Crouan

Chlorophyta

 Boodlea composita (Harv.) Brand

 Caulerpa lentillifera J. Ag.

 Caulerpa racemosa (Forssk.) J. Ag. var. *racemosa*

 Dictyosphaeria versluysii Weber–van Bosse

 Halimeda opuntia (L.) Lamour.

 Halimeda simulans Howe

Phaeophyta

 Cystoseira trinodis (Forsskal) C. Ag.

Hydroclathrus clathratus (Bory) Howe
Padina gymnospora (Kuetz.) Vickers
Rhodophyta
  Acanthophora spicifera (Vahl) Boerg.
  Amphiroa fragilissima (L.) Lamour.
  Chondria sp.
  Hypnea sp.
  Laurencia papillosa (Forssk.) Grev.
  Spyridia filamentosa (Wulf.) Harv. in Hook.
  Tolypiocladia glomerulata (C. Ag.) Schmitz in Schmitz & Hauptfleisch

Location near sample 97, rocky substrate

Cyanophyta
  Microcoleus lyngbyaceus (Kuetz.) Crouan
  Schizothrix calcicola (C. Ag.) Gom.
Chlorophyta
  Halimeda opuntia (L.) Lamour.
Phaeophyta
  Hydroclathrus clathratus (Bory) Howe
Rhodophyta
  Centroceras clavulatum (C. Ag.) Mont.
  Ceramium sp.
  Griffithsia subcylindrica Okam. ?
  Hypnea cornuta (Lamour.) J. Ag.
  Laurencia papillosa (Forssk.) Grev.
  Tolypiocladia glomerulata (C. Ag.) Schmitz in Schmitz & Hauptfleisch

Phil. Trans. R. Soc. Lond. A. **291**, 73–83 (1978)   [73]

Printed in Great Britain

An analysis of the textural variability displayed by inter-reef sediments of the Impure Carbonate Facies in the vicinity of the Howick Group

By P. G. Flood†, G. R. Orme† and T. P. Scoffin‡

†*Department of Geology and Mineralogy, University of Queensland, St Lucia, Queensland, 4067, Australia*
‡*Grant Institute of Geology, University of Edinburgh, West Mains Road, Edinburgh EH9 3JW, U.K.*

[Plates 1 and 2]

Recent surface sediments from the continental shelf in the vicinity of the Howick Group have been analysed to determine their textural, chemical and skeletal composition. There is a decrease in terrigenous particle size from quartz sand in the nearshore area to mud offshore. Carbonate particle size decreases from gravel to mud moving away from the reef. Multivariate statistical analysis of grain size data has determined the influence of various sources. Q-mode cluster and factor analysis distinguished four sediment types and indicated that three factors are responsible for 94 % of the textural variation. R-mode cluster analysis showed three distinct size populations. They result from intermixing of material from different sources and do not necessarily reflect distinct modes of transportation. The distribution of the textural sediment types displays concentric elliptical patterns around each platform reef in a zone approximately 2 km wide, within which there is a size gradient from coarse sand and gravel near the windward reef edge to fine sand near the leeward margin.

1. Introduction

The apparent homogeneity of the inter-reef sediments (most of which belong to the Impure Carbonate Facies) which were collected in the vicinity of the Howick Group highlights the lack of sensitivity on a local scale of the criteria used by previous researchers (Maxwell & Maiklam 1964; Maxwell 1968, 1973; Maxwell & Swinchatt 1970) for regional facies differentiation within the Great Barrier Reef Province. An account of the textural variations exhibited by these sediments is presented. This investigation supplements Maiklem's (1970) study of sediments of the High Carbonate Facies from the Capricorn Group, and Frankel's (1974) investigation of sediments of the Transitional and Terrigenous Facies within Princess Charlotte Bay. Together, they provide a more comprehensive understanding of the complex interplay of the terrigenous, *in situ*, and reefal sources which determine the regional facies pattern throughout the province.

2. The regional facies pattern

Maxwell (1968, 1973) has outlined a dual classificatory scheme for shelf sediments from the Great Barrier Reef Province based upon the carbonate:non-carbonate ratio or the mud:sand ratio. This scheme is useful on a regional scale for delineating areas where the sediments reflect the dominance of either the mainland terrigenous source or the reefal carbonate sources. Some, however, display variable influence from multiple sources and the Howick Group is one such area. In the present study the carbonate content is used for distinguishing facies. The mud content (terrigenous plus carbonate) used to differentiate subfacies. The resultant pattern

TABLE 1. INTER-REEF SEDIMENTS OF THE HOWICK GROUP AND NEIGHBOURING AREA: DATA

sample no.	position S lat.	E long.	depth	gravel	mud	sand	acid insoluble gravel +sand (%)	mud (%)	total (%)	median size (φ)	textural classification
1	14° 29′	144° 55′	16	5	29	66	21	85	39	1.9	slightly gravelly muddy sand
2	14° 29′	144° 58′	24	10	6	84	25	69	28	1.3	gravelly sand
3	14° 19′	144° 51′	10	12	0	88	0	0	0	1.3	gravelly sand
4	14° 20′	144° 50′	20	7	48	45	10	60	34	4.3	slightly gravelly sandy mud
29	14° 18′	144° 51′	6	0	3	97	2	36	3	2.3	sand (moderately sorted)
35	14° 25′	144° 53′	28	15	17	68	14	75	24	0.5	gravelly muddy sand
36	14° 26′	144° 54′	28	12	20	68	10	62	20	0.8	gravelly muddy sand
37	14° 25′	144° 54′	26	11	20	69	11	59	21	1.1	gravelly muddy sand
38	14° 24′	144° 54′	25	12	5	83	5	44	7	0.7	gravelly sand
39	14° 24′	144° 53′	17	11	24	65	10	56	21	2.3	gravelly muddy sand
40	14° 24′	144° 53′	18	12	36	62	2	64	25	2.8	gravelly muddy sand
56	14° 25′	144° 52′	8	6	1	93	1	1	1	0.2	gravelly sand
57	14° 25′	144° 51′	26	10	18	72	11	67	22	1.7	gravelly muddy sand
59	14° 24′	144° 54′	10	9	7	84	3	36	5	1.3	gravelly sand
60	14° 26′	144° 48′	11	7	11	82	1	60	2	2.1	gravelly muddy sand
61	14° 27′	144° 48′	14	6	0	94	1	0	1	0.6	gravelly sand
62	14° 27′	144° 49′	16	8	24	68	20	72	33	2.3	gravelly muddy sand
63	14° 28′	144° 50′	14	10	25	65	18	75	33	2.3	gravelly muddy sand
64	14° 28′	144° 51′	13	6	31	63	22	77	39	2.6	gravelly muddy sand
65	14° 29′	144° 52′	13	9	24	67	21	65	31	2.2	gravelly muddy sand
66	14° 30′	144° 53′	13	11	28	61	23	96	43	2.3	gravelly muddy sand
67	14° 30′	144° 54′	18	7	31	62	24	77	40	2.6	gravelly muddy sand
68	14° 31′	144° 54′	10	8	14	78	25	77	33	2.1	gravelly muddy sand
69	14° 31′	144° 55′	20	8	18	74	25	77	34	1.5	gravelly muddy sand
70	14° 31′	144° 56′	15	10	17	73	23	73	32	1.1	gravelly muddy sand
71	14° 30′	144° 57′	14	7	34	59	4	73	28	2.7	gravelly muddy sand
112	14° 26′	144° 55′	20	7	36	57	11	52	26	3.0	gravelly muddy sand
113	14° 26′	144° 57′	25	11	9	80	27	59	30	1.6	gravelly muddy sand
114	14° 26′	144° 59′	30	7	29	64	22	82	39	2.3	gravelly muddy sand
115	14° 26′	145° 00′	22	6	32	62	14	74	33	2.6	gravelly muddy sand
116	14° 27′	144° 59′	12	0	2	98	2	40	3	1.4	sand (moderately sorted)
117	14° 31′	144° 58′	24	10	24	66	57	74	60	1.8	gravelly muddy sand
120	14° 30′	145° 02′	25	15	13	72	11	80	20	0.3	gravelly muddy sand
121	14° 29′	144° 58′	15	8	16	76	43	59	46	1.8	gravelly muddy sand
122	14° 29′	144° 57′	10	1	6	93	1	50	4	2.4	sand (moderately sorted)
123	14° 27′	144° 53′	10	6	5	89	1	61	4	0.2	gravelly sand
124	14° 28′	144° 53′	14	12	23	65	8	80	25	1.2	gravelly muddy sand
125	14° 29′	144° 53′	20	6	14	80	21	67	27	1.8	gravelly muddy sand
126	14° 28′	144° 54′	13	6	29	65	19	89	39	2.8	gravelly muddy sand
127	14° 28′	144° 54′	22	12	19	69	13	84	26	1.8	gravelly muddy sand
128	14° 27′	144° 53′	20	3	17	80	1	75	13	0.6	slightly gravelly muddy sand
130	14° 25′	144° 47′	14	7	41	52	10	89	42	3.0	gravelly muddy sand
131	14° 25′	144° 46′	15	4	47	49	12	75	42	4.0	slightly gravelly muddy sand
132	14° 25′	144° 45′	15	2	60	38	27	85	61	—	slightly gravelly sandy mud
133	14° 25′	144° 43′	14	0	75	25	25	80	66	7.0	sandy mud
135	14° 25′	144° 41′	13	1	64	35	19	79	57	6.0	slightly gravelly sandy mud
136	14° 25′	144° 39′	10	0	86	14	—	80	—	9.0	sandy mud
137	14° 25′	145° 01′	26	4	18	78	53	64	54	2.2	slightly gravelly muddy sand
138	14° 27′	145° 02′	22	8	7	85	28	69	31	1.8	gravelly sand
139	14° 29′	144° 57′	10	12	14	74	13	62	20	0.9	gravelly muddy sand
140	14° 25′	144° 47′	13	15	36	49	5	64	26	2.0	gravelly muddy sand
141	14° 25′	144° 48′	14	13	31	56	12	67	29	1.1	gravelly muddy sand
142	14° 25′	144° 50′	18	9	44	47	18	65	39	3.2	gravelly muddy sand
143	14° 27′	144° 54′	19	10	29	61	19	72	34	1.7	gravelly muddy sand
144	14° 29′	144° 55′	16	7	23	70	6	50	16	3.3	gravelly muddy sand

TABLE 1 (*cont.*)

sample no.	position S lat.	E long.	depth	gravel	mud	sand	acid insoluble gravel + sand (%)	mud (%)	total (%)	median size (φ)	textural classification
145	14° 29′	144° 57′	25	9	6	85	32	67	34	1.6	gravelly sand
146	14° 23′	144° 50′	22	42	2	56	6	4	7	−0.7	sandy gravel
147	14° 22′	144° 54′	17	6	53	41	5	59	34	6.0	gravelly mud
148	14° 18′	144° 53′	25	4	38	58	7	82	35	5.0	slightly gravelly muddy sand
149	14° 15′	144° 53′	27	25	8	67	0	32	3	−0.3	gravelly muddy sand
150	14° 16′	144° 51′	30	1	49	50	3	43	23	4.5	slightly gravelly muddy sand
151	14° 23′	144° 50′	17	10	50	40	5	74	40	4.5	gravelly mud
152	14° 32′	144° 53′	8	6	10	84	44	85	48	5.0	gravelly muddy sand
153	14° 34′	144° 51′	5	7	1	92	66	37	67	1.7	gravelly sand
154	14° 34′	144° 50′	4	12	8	80	46	2	43	0.3	gravelly sand
155	14° 32′	144° 50′	6	11	3	86	28	63	29	0.9	gravelly sand

Notes:
1. Positioning of all samples given to nearest minute.
2. All percentages quoted to nearest whole number.
3. Median size is measured in φ units and based on sieve analyses at $\frac{1}{4}$ φ intervals.
4. Textural classification after Folk (1954, *J. Geol.* **62**, 345–351).
5. Depths are stated in metres below i.s.l.w. (chart datum, Marine Chart AUS 823).
6. Sample nos 1–128, grab samples collected July–August 1973.
7. Sample nos 130–139, grab samples collected October 1973.
8. Sample nos 140–156, dredge samples collected July–August 1973.
9. Samples were split, one treated with hydrogen peroxide, the other preserved in alcohol.
10. Representative samples are housed at the University of Queensland and the University of Edinburgh.

obtained from analyses of inter-reef sediments collected from the Howick Group (see figure 1) is shown in figure 2. It consists of:

1. High carbonate (greater than 80% carbonate) facies
2. Impure carbonate (60–80% carbonate) facies
   (*a*) Very high mud (40–60% mud) subfacies
   (*b*) High to moderate mud (10–40% mud) subfacies
3. Transitional (40–60% carbonate) facies
   (*a*) Low mud (less than 10% mud) subfacies
   (*b*) High to moderate mud (10–40% mud) subfacies
4. Terrigenous (less than 40% carbonate) facies.

3. INTER-REEF SEDIMENTS OF THE HOWICK GROUP

The bathymetric character of the area of the continental shelf around the Howick Group is illustrated in figure 4.

The Impure Carbonate Facies consist of material derived from at least three sources:

(1) Terrigenous material (predominately mud) derived from a variety of igneous and sedimentary rock types which occur on the mainland and from the igneous rocks on Howick Island.

(2) Carbonate material representing the calcareous skeletons of *in situ* organisms such as: *Halimeda*, molluscs, bryozoans, echinoderms, ostracods, corals, benthonic foraminiferans, etc. The Foraminifera *Marginopora vertebralis* and *Alveolinella quoyi* are ubiquitous (and frequently blackened) in sediments of this facies and together with the planktonic foraminiferal species they are unquestionable indicators of the inter-reef shelf environment.

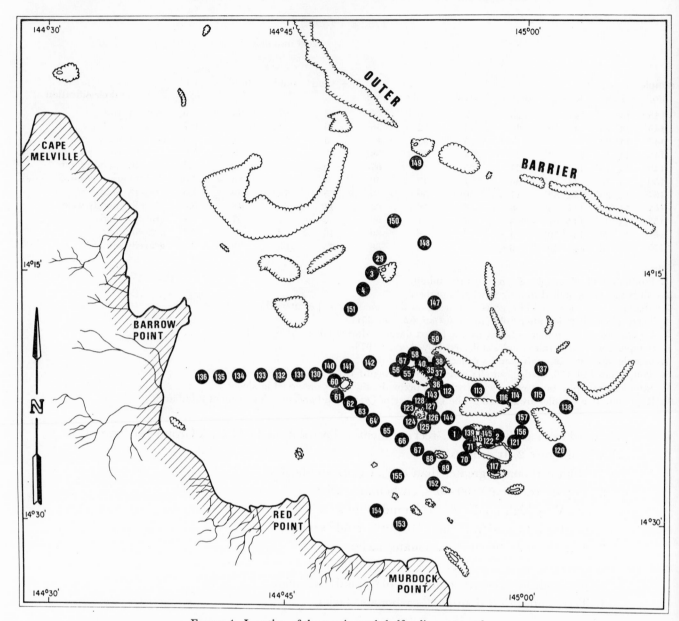

FIGURE 1. Location of the continental shelf sediment samples.

(3) The reef derived material includes significant contributions from corals, molluscs, *Halimeda* spp., and the benthonic foraminiferans *Calcarina hispida* and *Baculogypsina sphaerulata*. The reefal material is often distinguishable from *in situ* skeletal material by its abraded appearance.

(a) Texture

Q-mode (i.e. samples) cluster analysis (Davis 1973) of the grain size data (sieve analyses) shows four textural groupings (see figure 5) and Q-mode factor analysis (Klovan & Imbrie 1971) shows the 95 % of the variation is explained by three factors only. The triangular plot of the normalized varimax factors (factors 1, 2, and 3 explains 22, 24, and 49 % variance respectively) and the plot of the relative contribution of each $\frac{1}{4}\phi$ interval to the factors (i.e. factor score) is shown in figure 6. (Conversions from ϕ to millimetres are included in the legend to figure 6.)

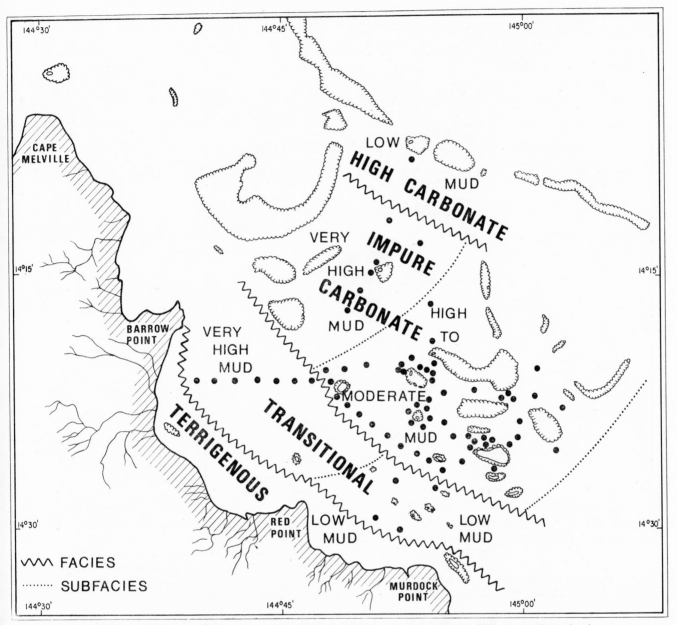

FIGURE 2. Regional facies pattern on the continental shelf, based on the results of the present investigation.

Textural type 1: greater than 50 % by mass coarser than 1 φ; less than 10 % finer than 4 φ; less than 20 % finer than 3 φ; consists of coarse sand with a varying percentage of gravel; moderately to poorly sorted; strongly fine skewed; variable kurtosis.

Textural type 2: greater than 50 % by mass in the range 1–4 φ; less than 10 % finer than 4 φ; less than 40 % finer than 3 φ; consists of medium sand; poorly sorted; near symmetrical to coarse skewed; commonly leptokurtic.

Textural type 3: 10–30 % finer than 4 φ, 20–40 % finer than 3 φ; less than 40 % of 1–3 φ range; consists of medium sand with a varying percentage of gravel or finer sand and mud; poorly to very poorly sorted; fine skewed; very leptokurtic.

Textural type 4: 30–60 % finer than 4 φ; 40–60 % finer than 3 φ; variable percentage gravel

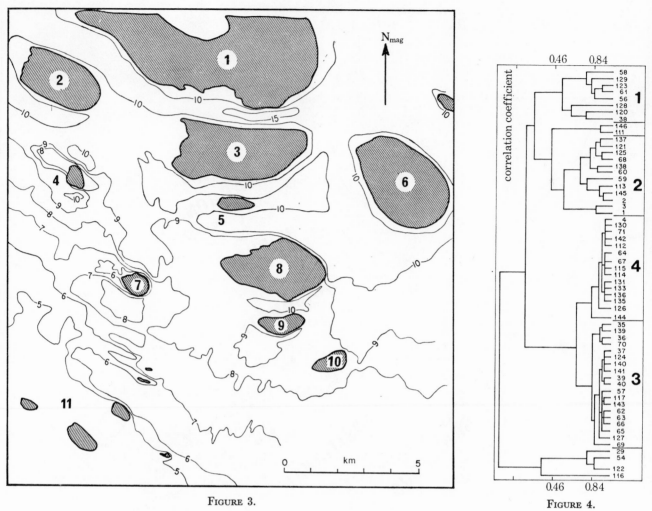

FIGURE 3.   FIGURE 4.

FIGURE 3. Bathymetric character of the continental shelf in the vicinity of the Howick Group. Based on survey by
H.M.A. Survey Ship *Paluma* in 1965; Marine Chart AUS 832, 1971; and echo soundings made during the
Expedition. Reef names: 1, Combe; 2, Ingram–Beanley; 3, Mid; 4, Watson; 5, Megaera; 6, Snake; 7,
Newton; 8, Howick; 9, Houghton; 10, Coquet; 11, Cole Islands. Contours in fathoms (1 fathom ≈ 1.83 m).

FIGURE 4. Q-mode cluster dendrogram of grain size analysis of the inter-reef sediments. Four groups are
recognizable. The four negatively correlated samples are reef slope sediments.

or sand; consists of fine to very fine sand with a high percentage of mud sizes; poorly to very
poorly sorted; strongly coarse skewed; very leptokurtic.

The R-mode (i.e. variables) cluster analysis dendrogram (figure 6) displays the behavioural
relation between each $\frac{1}{4}\phi$ size interval, the gravel (G) and the mud (M). Three grain size
populations are present: I, gravel and coarse sand (1ϕ and coarser); II, medium to find sand;
and III, very fine sand and mud (3ϕ and smaller). Individual sediments may consist of any
combination of the three populations (see figure 7). For example, sample 29 consists predominantly of the medium to fine sand population; sample 61 the coarse sand and gravel population; sample 2 the coarser populations; sample 37 a combination of the three populations;
sample 135 the very fine sand and mud population.

The relative proportions of each population with respect to the textural types are as follows:
type 1: I ≫ II > III, type 2: II ≫ III > I, type 3: III > II > I, type 4: III ≫ II > I.

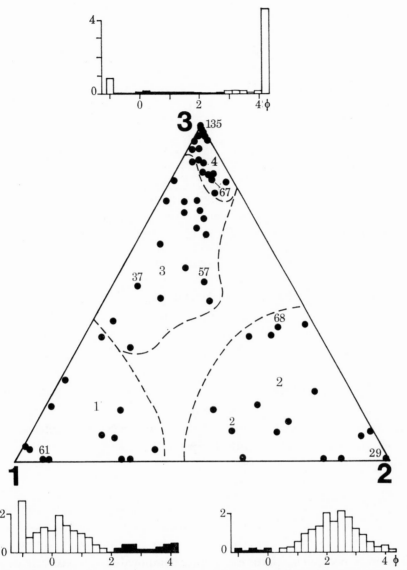

FIGURE 5. Plot of sediments in terms of three normalized varimax factors. The areas occupied by the grain size groupings (1–4) are indicated. The relative influence (black for negative) of each $\frac{1}{4}$ φ interval is shown on the plot of the factor scores (-1 φ = 2 mm, 0 φ = 1.0 mm, 1 φ = 0.5 mm, 2 φ = 0.25 mm, 3 φ = 0.125 mm, 4 φ = 0.062 mm).

These populations do not necessarily reflect distinct modes of transportation (traction, salta-tion or suspension; cf. Visher 1969) nor do they represent depositional populations (designated A ≡ framework, B ≡ interstitial, and C ≡ contact; cf. Moss 1972). They result from inter-mixing of material from the different sources. The gravel fraction represents either the *in situ* fauna which is usually coarser than 2 mm (-1 φ) and which compares with the size of the *in situ* fauna in other reefal provinces (see Swinchatt 1965; Milliman 1973, 1974), or the coarse carbonate detritus (usually coral) derived from the reef proper. The coarse sand (coarser than 1 φ) represents either particles formed by the breakdown of the *in situ* organisms (see figure 8*b*, plate 1), or the inter-reef and reefal contribution made by benthonic foraminiferans (see figures 9*a*, plate 2 and 8*a*). The medium to fine sand (see figure 8*c* and *d*) represents a compositionally

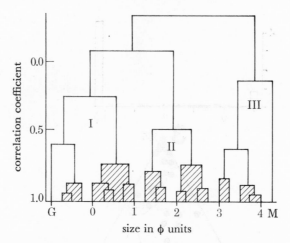

FIGURE 6. R-mode cluster dendrogram showing the presence of three distinct size populations. The gravel (G) fraction is correlated with the coarse sand sizes, and the mud (M) content is correlated with the fine sand fraction (conversions from φ to millimetres are given with figure 5).

complex population consisting of reef-derived skeletal debris, terrigenous quartz particles, *in situ* skeletal breakdown material, and *in situ* organisms such as ostracods, benthonic foraminiferans, molluscs, etc. The very fine sand (see figure 8*e* and *f*) represents an exceptionally heterogeneous admixture, consisting of the finer particles which are removed in suspension from the reef flat (Flood & Scoffin 1978, this volume), the *in situ* contribution, and the calcareous and siliceous tests of a variety of planktonic organisms. The mud fraction is predominantly terrigenous in origin; the carbonate contribution rarely exceeds 30%.

Figure 9 represents stereoscopic pictures of randomly selected sediments (mud fraction removed) representative of the four textural types of sediments present within the Impure Carbonate Facies.

(b) Distribution of the textural types

The relatively uniform terrigenous and *in situ* carbonate contribution throughout the area is overshadowed in the vicinity of individual reefs by a series of concentric zones characterized by higher carbonate content, increasingly coarser grain size and improved sorting (reflecting the predominance of one source). This pattern is further complicated in areas of dense reef development where overlapping of the concentric arrangement occurs. A somewhat idealized presentation of the distribution of the textural types is shown in figure 10.

Type 1 surrounds each reef and resembles either of the reefal textural types 1 or 2 (Flood & Scoffin 1978, this volume). It consists of a predominance of coarse coral detritus to the windward and benthonic foraminiferans to leeward and represents a talus slope type of accumulation of reef-derived skeletal debris at the base of the reef slope.

Type 2 is restricted to the sediment cone which is developed to the leeward of the platform reefs. It resembles reefal textural type 3 and consists of the finer sizes (1–4 φ) of skeletal detritus that may be moved across the reef flat to be deposited in the deeper water.

Type 3 surrounds each reef and represents an admixing of material derived from either the reef mass (finer than 3 φ), the *in situ* organisms (coarser than −1 φ) and the terrigenous mud component (finer than 4 φ). In general the reefal contribution is restricted to particle sizes smaller than 3 φ capable of being removed in suspension by wave or tidal currents from the reef flat to the inter-reef area.

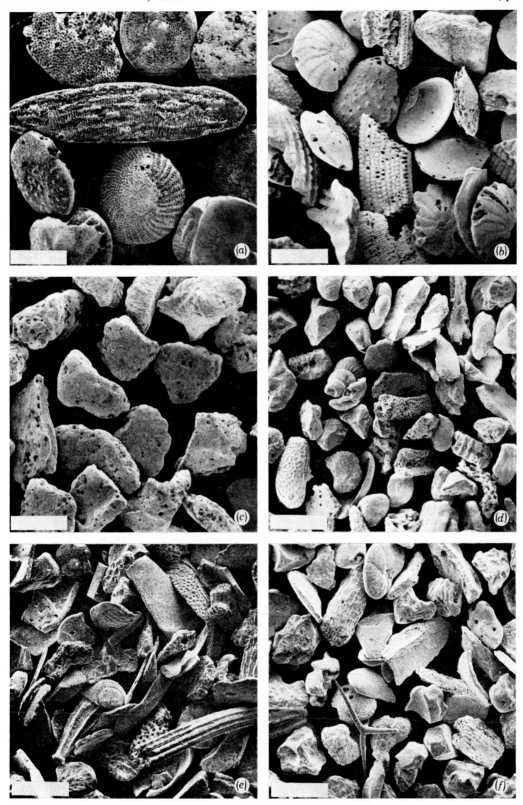

FIGURE 8. Scanning electron micrographs of various grain sizes showing the complex contribution of skeletal components: (*a*) size −0.25 φ, scale 1.3 mm, various benthonic Foraminifera; (*b*) size 1.0 φ, scale 0.6 mm, benthonic Foraminifera and molluscans; (*c*) size 1.5 φ, scale 0.5 mm, *Halimeda* and coral fragments; (*d*) size 2.75 φ, scale 0.5 mm, quartz particles, ostracods, molluscans, bryozoans, Foraminifera, etc; (*e*) size 3.0 φ, scale 0.5 mm, benthonic and planktonic Foraminifera, molluscan fragments, bryozoans, etc.; (*f*) size 3.5 φ, scale 0.5 mm, quartz particles, Foraminifera, siliceous sponge spicules, and a variety of other skeletal particles.

Phil. Trans. R. Soc. Lond. A, volume 291

Flood et al., plate 2

FIGURE 9. Stereoscopic views of scanning electron micrographs of randomly selected sediments (mud removed) representing the textural types: (*a*) 4 (sample 67); (*b*) 3 (57); (*c*) 2 (2); (*d*) 1 (61). Bar scale in (*d*) is 1 mm and applies to all micrographs.

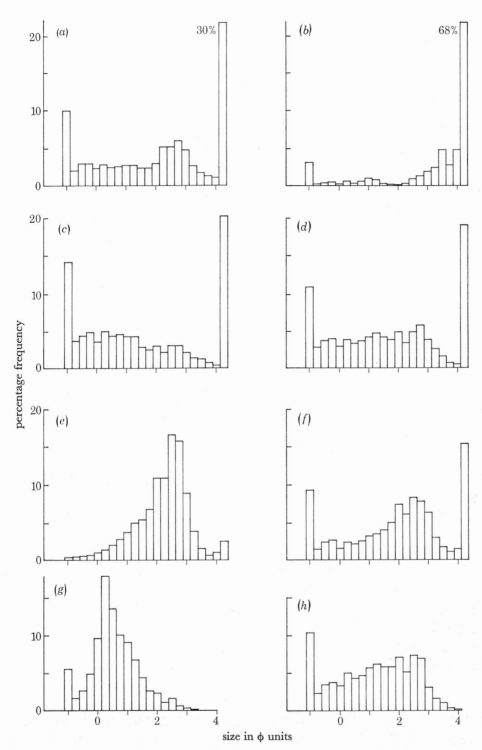

FIGURE 7. Frequency histograms of selected sediments illustrating the presence of distinct sand size modes (conversions from ϕ to millimetres are given with figure 5). (a) Sample 67; (b) 135; (c) 37; (d) 57; (e) 29; (f) 68; (g) 61; (h) 2.

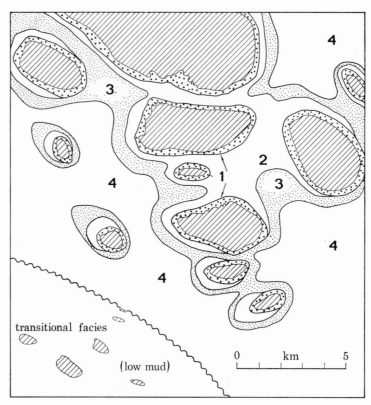

FIGURE 10. Idealized distribution of the textural types in the area of the Howick Group. The influence of the reefs is restricted to within 2 km of individual reef masses. Type 1 circumscribes the entire reef; 2 is restricted to the lee; 3 occurs in the inter-reef areas; 4 characterizes the remaining part of the shelf in the vicinity of the Howick Group.

Type 4 represents the textural type present on the shelf in areas which received a relatively significant contribution of terrigenous mud and material of coarse gravel sizes representing the skeletons of *in situ* organisms. Subordinate amounts of medium to very fine sand and mud sizes are derived from the reef masses. The distance of reefal influence rarely exceeds 2 km from any reef.

To summarize: the textural relation within the Impure Carbonate Facies can be qualitatively expressed in terms of the varying influence of the sediment sources.

Type 1: reefal (R) $\gg$ *in situ* (IS) > terrigenous (T) (R size is coarser than 1 ϕ).

Type 2: R > IS > T (R size ranges from 1 to 3 ϕ).

Type 3: IS > T > R (R size is smaller than 3 ϕ).

Type 4: T > IS $\gg$ R.

4. CONCLUSION

This study has shown that the Impure Carbonate Facies is divisible into four textural types that reflect the varying influence of reefal, *in situ*, and terrigenous sources. The distribution of the textural types displays concentric elliptical patterns around each platform reef in a zone approximately 2 km wide, within which there is a size gradient from coarse sand and gravel near the windward edge of the reef to fine sand near its leeward margin. More complex textural patterns are produced in areas of the shelf which have a denser pattern of reef development.

REFERENCES (Flood *et al.*)

Davis, J. C. 1973 *Statistics and data analysis in geology*. New York: John Wiley.

Flood, P. G. & Scoffin, T. P. 1978 *Phil. Trans. R. Soc. Lond.* A **291**, 55–71 (this volume).

Frankel, E. 1974 *Proc. 2nd Int. Coral Reef Symp.*, vol. 2, pp. 355–369.

Klovan, J. E. & Imbrie, J. 1971 *Math. Geol.* **3**, 61–77.

Maiklem, W. R. 1970 *J. sedim. Petrol.* **40**, 55–80.

Maxwell, W. G. H. 1968 *Atlas of the Great Barrier Reef*. Amsterdam: Elsevier.

Maxwell, W. G. H. 1973 In *Biology and geology of coral reefs* (eds O. A. Jones & R. Endean), vol. 1 (Geology 1), pp. 299–345. New York: Academic Press.

Maxwell, W. G. H. & Maiklem, W. R. 1964 *Univ. Queensland Geol. Dept Pap.* **5** (11), 1–21.

Maxwell, W. G. H. & Swinchatt, J. P. 1970 *Bull. geol. Soc. Am.* **81**, 691–724.

Milliman, J. D. 1973 In *Biology and geology of coral reefs* (eds O. A. Jones & R. Endean), vol. 1 (Geology 1), pp. 1–50. New York: Academic Press.

Milliman, J. D. 1974 *Marine carbonates*. New York: Springer-Verlag.

Moss, A. J. 1972 *Sedimentology* **18**, 159–219.

Swinchatt, J. 1965 *J. sedim. Petrol.* **35**, 71–90.

Visher, G. S. 1969 *J. sedim. Petrol.* **39**, 1074–1106.

Phil. Trans. R. Soc. Lond. A. **291**, 85–99 (1978)   [85]

Printed in Great Britain

Sedimentation trends in the lee of outer (ribbon) reefs, Northern Region of the Great Barrier Reef Province

By G. R. ORME, P. G. FLOOD AND G. E. G. SARGENT

Department of Geology and Mineralogy, University of Queensland, St Lucia, Queensland, 4067, Australia

[Plates 1 and 2; pullout 1]

A programme of surface sediment sampling, continuous high resolution seismic reflexion profiling, and side-scan sonar surveys, carried out in the Northern Region of the Great Barrier Reef Province between latitudes 14° 31′ S and 14° 45′ S, extending between the continental shelf edge and a line of mid-shelf continental islands, has demonstrated the diversity of sedimentation factors controlling sediment distribution patterns, and revealed complex petrological variations and unsuspected stratigraphic relations.

The granulometric characteristics of many inter-reef sediments are imparted chiefly by mixing, whereas sorting under high energy conditions, and the effectiveness of the 'Sorby Principle' are primarily responsible for the granulometric characteristics of the reef-top sediments. Terrigenous end-members of mainland and/or continental island provenance are important on the southwestern margin of the area, but *Halimeda* debris dominates sediments coinciding with luxuriant *Halimeda* growth on back-reef 'banks' and ridges.

A major disconformity indicating marine regression and shelf emergence is the most prominent seismic reflector, which outcrops near Cook's Passage to coincide with the rocky, current swept, seabed at −45 m, and which is incised near the shelf edge to −69 m. The succeeding transgression resulted in the filling of the incised channels with sediment, followed by rapid sedimentation in the back-reef area and the establishment of platform reefs on this surface. Periods of still-stand or minor regressions are indicated by marine terraces at −22.5 m and −30 m, and by erosion surfaces within back-reef sedimentary deposits and reef masses. The seismic profiles suggest that the Holocene sedimentation pattern is a repetition of an ancient trend.

1. INTRODUCTION

A general picture of the nature and distribution of sediments throughout the entire Great Barrier Reef Province has been provided by Maxwell (1968a). Factors effecting and affecting sedimentation, reef distribution and form, have been discussed in general terms by a number of workers (see, for example, Maxwell 1968a, 1973; Maiklem 1970; Fairbridge 1950, 1967; Maxwell & Swinchatt 1970; Frankel 1974; Orme, Flood & Ewart 1974). These include the apparently restricted dispersal of reef-derived sediment, the importance of the mixing of sediments from different sources in determining sediment facies characteristics and distribution, and the influence of relict sediments (Maxwell 1968b) related to eustatic changes of sea level. The implications of the granulometric characteristics of sediments from both the reef-top (Maxwell, Jell & McKellar 1964; Orme *et al.* 1974; Flood & Scoffin 1978, this volume) and inter-reef environments (Flood, Orme & Scoffin 1978, this volume) have received some consideration. Elsewhere, namely British Honduras, the influence of inherited features on reef configurations and present sediment facies patterns has been demonstrated (Purdy 1974a, b).

In the area here investigated, the narrow continental shelf is characterized by a well developed almost continuous barrier of shelf edge reefs, a mid-shelf zone of continental islands, platform reefs, a bathymetrically complex back-reef area, and a general meridional arrangement of sediment facies, which reflects the relative importance of terrigenous sediment from the mainland and carbonate production on the outer shelf. Consequently a complex interrelation between bathymetric, hydrological, ecological and provenance factors could be expected.

The primary objectives of this study were, therefore, to investigate the petrology of the sediments, to determine the nature and importance of provenance and dispersal factors, to obtain a more precise picture of the bathymetric diversity of the area, to examine any evidence of eustatic sea level changes and assess the importance of inherited features upon present facies distribution,

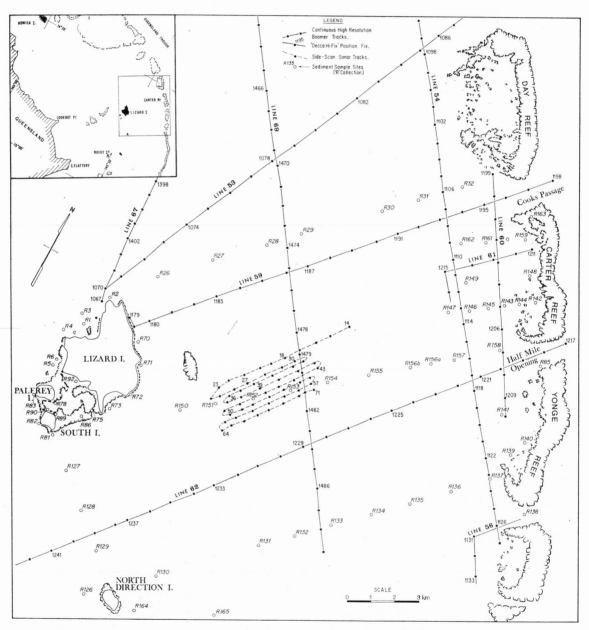

FIGURE 1. Map of the outer shelf near Lizard Island showing sediment sample sites, the locations of continuous seismic profiling traverses, and side-scan sonar tracks.

and finally, to determine whether present sedimentation follows trends established during pre-Holocene times.

This paper presents preliminary results from part of the area investigated. It is concerned with the outer shelf area which lies between 14° 31′ S and 14° 45′ S, extending between the continental shelf edge and a line of continental islands at approximately 145° 25′ E (see figure 1).

2. GENERAL CHARACTERISTICS OF THE AREA

A line of shelf edge reefs, comprising Day, Carter, and Yonge reefs, together with an unnamed reef to the south, forms an almost continuous barrier between the Queensland Trough and the continental shelf, and restricts the exchange of Coral Sea and shelf waters. These ribbon reefs are separated by deep, narrow passages, namely, One-mile Opening, Cook's Passage, Half-mile Opening and Cormorant Pass.

Carter Reef is approximately 5.5 km in length and 1–1.5 km in width; its dimensions are similar to those of the adjacent ribbon reefs. All descend steeply to the Queensland Trough and show a zonation which includes outer moat, reef crest, inner moat, sanded zone, and zone of coral heads, running parallel with their shelf edge orientation (Fairbridge 1950).

Fringing reefs are developed around the continental islands and extend between Lizard, Palfrey and South Islands to enclose a shallow lagoon.

Shelf relief between Lizard Island and the shelf edge is diversified due to reefal shoals (e.g. Petricola, Stewart) and submarine banks and ridges. These banks occur at a fairly uniform depth of approximately 25 m with intervening channels at −37 m. The deepest part of the shelf lies at −69 m, where the rocky seabed is channelled at the lagoonal approaches to Cook's Passage. To the northeast the continental slope descends very steeply to the Queensland Trough.

3. METHODS AND TECHNIQUES

Sediment samples were systematically collected from the seabed, spaced at approximately 1.5 km intervals along traverses, which coincided with some of the seismic profiling tracks. Closer sampling was undertaken in the vicinity of reefs and islands in the expectation that in such bathymetrically and environmentally diverse areas more subtle variations in sediment composition might be disclosed. In shallower areas the samples were collected directly by diving, in the deeper areas mechanical grabs were used. Alcohol was added as a preservative to some samples. Others were kept under refrigeration until treated with hydrogen peroxide to remove carbonaceous matter in preparation for sieve analysis.

Each sample was sieved using a 0.25 ϕ set of U.S. Standard sieves and each fraction was scrutinized microscopically in order to ascertain whether preferential size grades were adopted by certain grain types, which might in turn relate to ecological and/or dispersal factors.

Continuous seismic profiling with the use of a high resolution boomer was carried out along tracks running over and along the shelf forming a grid pattern in order to examine sub-bottom facies relations and shallow internal structures. The leeward side of Carter Reef and the approaches to Cook's Passage were given particular attention.

The equipment used comprised a 3.5 kV, 500 J triggered capacitor power supply unit driving either a Huntec ED10 high resolution boomer type device (loaned by N.E.R.C.) or a slightly smaller similar device (Sargent 1969). The reflexion seismic records were produced in real time on wet Alphax 'A' paper 9 inches wide (*ca.* 23 cm). A 10 element E.G. & G. hydrophone streamer, or similar device (also loaned by N.E.R.C., London), was coupled to a Khron-hite Model 3100 filter, usually set to a band pass of about 0.5–5 kHz, in turn coupled to an E.G. & G. 254 graphic recorder. Trigger rates were usually 1–2 p.p.s. Both intense swell and surface chop generated by continuous SE winds prevailed throughout the survey, to the extent that a substantial proportion of the available sea time was not usable for seismic reflexion work.

A side-scan sonar survey was carried out to the southeast of Lizard Island, individual traverses being run at a spacing of 250 m intervals to produce a map of the seabed in an area known to be occupied by *Halimeda*-covered submarine banks and ridges. Decca Hi-fix control was employed as a means of accurate position fixing, fixes being taken at 5 min intervals along all seismic and some sediment sampling traverses.

4. SEDIMENTS AND SEDIMENTATION FACTORS

The boundaries of the area with which this paper is concerned lie mainly within the limits of the 'High Carbonate Facies' and 'Impure Carbonate Facies' (Maxwell 1968*a*), the 'Transitional Facies' and the 'Terrigenous Facies' being of limited distribution. This preliminary account incorporates data from 70 systematically collected sediment samples.

Cursory examination indicated the presence of a number of subfacies, those occurring in the 'High Carbonate Facies' containing usually less than 5% insolubles, but others, particularly those strongly influenced by mainland and/or 'Continental Island' provenance, containing up to 78% insolubles (e.g. R4 and R128).

The sediments display great diversity in terms of both grain types and grain size characteristics. Their end-members may be considered broadly in terms of the insoluble (terrigenous) and the soluble (biogenic-carbonate) components. The insoluble end-members comprise bluish-grey mud, which occurs in most sediment samples. This may constitute only a very small proportion of the 'High Carbonate Facies', but is dominant in samples from the southwestern margin of the area where it reaches 58%. It is believed to reflect mainland provenance. Terrigenous quartz, feldspar, mica, heavy minerals and rock fragments are quantitatively important near the granitic continental islands (e.g. R5).

The carbonate grains are biogenic and predominantly skeletal, consisting of *Halimeda*, gastropods, pelecypods, foraminiferans, coralline algae, corals, encrusting and branching bryozoans, echinoid plates and spines, worm tubes, and spicules. Although representatives of each of these allochem types occur in most sediment samples there are considerable variations within these broad categories in terms of numbers of individuals and species, especially in the case of foraminiferans and molluscs. Acquired grain morphology is also very variable in relation to grain size and distribution. Furthermore, some worn skeletal grains are strongly iron stained and show signs of solution.

An investigation of the complex variations in composition of the sediment in terms of grain types and grain size characteristics is in progress and the following general account illustrates the salient characteristics of the end-members, their distributions, and implications regarding sedimentation in this part of the Great Barrier Reef Province.

(a) Distribution of end-members

The fine-grained insoluble end-member, the bluish-grey mud, increases in quantity towards the southwestern margin of the area (see figure 2). Other detrital end-members, in which quartz is dominant, are conspicuous in sediments fringing the continental islands, especially where the masking effect of coral reef debris is not marked. They form dominant sediment components in areas to the northwest (leeward side of Lizard Island giving rise to a distinct quartz-rich terrigenous subfacies). Quartz is present also in the 'High Carbonate Facies' to the east of Lizard Island; it occurs even in samples taken from close to the shelf edge, but here it is present within

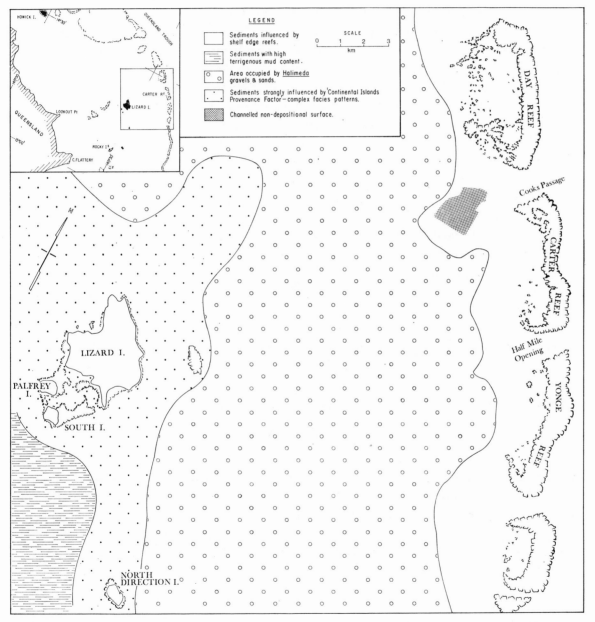

FIGURE 2. Map of the outer shelf near Lizard Island showing the general form of the sediment distribution pattern, and the salient source factors.

a restricted size range (2.5–3 ϕ), whereas a much wider size range of quartz (0–3.25 ϕ) occurs in sediment samples from the vicinity of Lizard Island.

The decrease in terrigenous quartz eastward is abrupt, so that over much of the 'back-reef' area it is sparse. Such a distribution pattern of igneous quartz shows the influence of continental island provenance, and of Lizard Island in particular. It also indicates the selective dispersal of the more readily transported quartz grades and the dilution effect of biogenic carbonate in the 'back-reef' zone.

Halimeda debris is ubiquitous, and is the principal constituent forming the sediments over most of the back-reef area. Unbroken plates are abundant in back-reef sediments and also in some reef top samples of the shelf edge. *Halimeda* debris is also present in the coarse fractions of the sediments fringing Lizard Island, and worn fragments occur in the terrigenous facies along the southwestern margin of the area. The acme of production of this component is associated with the extensive *Halimeda* 'meadows' which cover the submarine 'banks' of the back-reef area. Some *Halimeda* debris is probably transported away from this site by tidal currents.

Small gastropods are present in all samples and are well preserved in most. Molluscs are the dominant allochems in some areas, e.g. the windward side of Lizard Island (R71 and R75); they are also abundant in reef-top sediments and are present in gravels and sands of the *Halimeda* meadows. In most samples examined there is a general lack of abrasion of the small gastropods, which suggests that their occurrence more or less reflects an ecological situation.

The contribution made by foraminiferan tests is appreciable and represents a fauna which includes *Baculogypsina, Marginopora, Calcarina, Textularia, Alveolinella, Heterostegina, Amphistegina, Operculina*, and *Spiroloculina*. The foraminiferans from shelf edge reefs and from the fringing reefs of the continental islands are particularly varied and abundant. They are also conspicuous in sediment from the back-reef *Halimeda* 'banks', and some, particularly abraded *Marginopora* tests, are common in sediments from the terrigenous facies.

Coral-stick gravel is associated with reef-top deposits of the shelf edge and with the fringing reefs. However, coral debris is not common in the *Halimeda* sediments of the back-reef banks. Furthermore, many sediment grades in deposits adjacent to coral reefs are devoid of recognizable coral debris, which may be due to the discontinuous range of particle sizes produced by the breakdown of coral skeletons, the 'Sorby Principle' of Folk & Robles (1964).

The contribution made by coralline algae to sediment is small and becomes noticeable only on shelf edge reefs and the fringing reefs of continental islands.

(b) *Differentiation of sedimentary facies*

On the basis of the relative proportions of carbonate and terrigenous end-members present, a primary distinction may be made between High Carbonate Facies, Impure Carbonate Facies, Transitional Facies, and Terrigenous Facies, in the manner of Maxwell (1968a). Subdivision of these facies is possible according to dominant types of terrigenous particles and/or dominant carbonate allochems present. The frequency curves (figure 3) and cumulative curves (figure 4) illustrate the considerable granulometric differences which exist between representative samples of each subfacies.

(i) *High Carbonate Facies*

Halimeda *gravels and sands.* These represent the central part of the 'back-reef' environment, characterized by submarine banks and ridges, e.g. samples R154 and R133. This subfacies occupies most of the back-reef area and consists of whole and fragmented *Halimeda* plates, some gastropods and pelecypods, numerous and varied foraminiferans, together with minor components which include worm tubes, bryozoans, echinoid spines, spicules, and a minor component of blue-grey mud. A few quartz grains are also present. Variations occur between different samples of this subfacies according to the degree of wear displayed by allochems, the

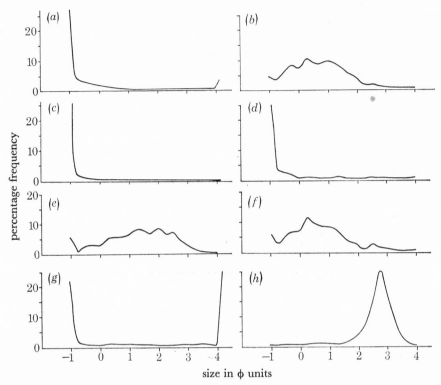

FIGURE 3. Frequency curves of representative samples of the main sediment types, illustrating the considerable variations in their granulometric characteristics. (*a*) R163: this sample is from a reef-top pool in the coral zone of Carter Reef. The bimodal curve represents a poorly sorted ($\sigma = 1.50\ \phi$) coral stick gravel to which *Halimeda* debris, foraminiferans and molluscs have been added; 71 % of the sample was coarser than $-1\ \phi$ ($M_z = -1.3\ \phi$, $Sk_I = +0.03$, $K_G = 2.66$). (*c*) R133 and (*d*) R154: these represent the common sediment type of the back-reef lagoon which is dominated by luxuriant *Halimeda* growth. They are coarse grained, unimodal, very well sorted to moderately sorted, negatively skewed *Halimeda* gravels. For R133, 93 % was coarser than $-1\ \phi$ ($M_z = -2.30\ \phi$, $\sigma_I = 0.03\ \phi$, $Sk_I = -0.89$, $K_G = 0.96$) and for R154, 80 % was coarser than $-1\ \phi$ ($M_z = -1.67\ \phi$, $\sigma_I = 0.65\ \phi$, $Sk_I = -0.22$, $K_G = 0.80$). They reflect the major *in situ* contribution of *Halimeda* debris in this region mid-way between the shelf edge and the mid-shelf continental islands (see figures 1, 2 and 5, and plate 1, figure 5). (*h*) R5: quartz-rich terrigenous subfacies. Unimodal, well sorted ($\sigma_I = 0.49\ \phi$), coarse-skewed ($Sk_I = -0.15$), leptokurtic ($K_G = 1.16$) sand ($M_z = 2.77\ \phi$). The mode in the fine sand and very fine sand grades reflects the importance of terrigenous grains, chiefly igneous quartz derived from Lizard Island (see plate 1, figure 6). (*g*) R128: terrigenous facies, high mud subfacies. The bimodality of this curve is a reflexion of the mixing of terrigenous components from the mainland with coarse grained carbonate debris from both *in situ* post-mortem contributions, and transported allochems; 21 % was coarser than $-1\ \phi$ and 57 % finer than $4\ \phi$. Sorting is poor ($\sigma_I = 6.8\ \phi$). R128 is representative of many samples from the southwestern margin of the area ($M_z = 3.5\ \phi$, $Sk_I = 0.00$, $K_G = 0.61$). (*e*) R71 and (*f*) R75. High Carbonate Facies, molluscan–foraminiferan subfacies (see plate 1, figures 1 and 2). R71 is from a site adjacent to the fringing reef on the windward side of Lizard Island, and R75 is from the entrance to Lizard Island Lagoon. Their polymodal curves relate mainly to grade sizes favoured by post-mortem contributions. R71 has an admixture of terrigenous quartz ($M_z = 1.23\ \phi$, $\sigma_I = 1.29\ \phi$, $Sk_I = +0.06$, $K_G = 1.09$). R75 ($M_z = 0.50\ \phi$, $\sigma_I = 1.11\ \phi$, $Sk_I = +0.38$, $K_G = 1.64$). (*b*) R148 is from the deeper water of the 'sanded zone' in the lee of Carter Reef. It has a polymodal curve in consequence of its diverse components consisting of coarse grained coral fragments, a variety of molluscs and foraminiferans, together with some *Halimeda* debris. ($M_z = 0.50\ \phi$, $\sigma = 0.95\ \phi$, $Sk_I = -0.11$, $K_G = 1.43$).

abundance and variety of foraminiferans, and in the relative proportions of minor constituents. All samples are very well sorted and unimodal, owing to the dominance of *Halimeda* debris, which points to the importance of ecological factors favouring the luxuriant growth of *Halimeda* in the back-reef area. The characteristics of this subfacies are promoted largely by *in situ* contributions from *Halimeda* (see figures 3 and 4, and plate 1, figure 5).

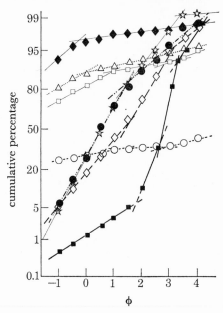

FIGURE 4. Cumulative curves for samples of the main sediment subfacies. The curves illustrate the considerable range of mean grain size and sorting, their wide separation indicating the great disparity between samples in terms of the relative importance of grain size populations. The high proportion of mud in R128 (○) ('Terrigenous Facies', 'high mud subfacies') distinguishes this from other curves. R5 (■) ('Terrigenous Facies', 'quartz-rich subfacies') shows marked inflexions at 2 φ and 3.1 φ, which may, to some extent, correspond with discrete grain size populations separated in response to different transportation modes, coarser than 2 φ representing the traction load, 2 φ to 3.1 φ representing saltation transport, and material finer than 3.1 φ approximating to the suspension load. The high gravel content of R133 (◆) and R154 (△) (the *Halimeda* dominated subfacies), and of R163 (□) (coral/algal subfacies) may be lag deposits, reflecting in part the influence of currents of removal. R75 (●) and R71 (◇), from the entrance to Lizard Island Lagoon, and a site near the windward fringing reef of Lizard Island respectively, are dominated by small mollusc shells and foraminiferan tests. ☆, R148.

DESCRIPTION OF PLATE 1

FIGURE 1. Sample R75, from the entrance to Lizard Island Lagoon showing characteristic allochems: small pelecypods, gastropods, *Alveolinella* and *Marginopora*. An echinoid spine and a few small *Halimeda* fragments are also present.

FIGURE 2. Sample R71, similar in composition to R75. Mollusc shells and foraminiferans are conspicuous. *Halimeda* and bryozoan fragments are also present, together with a few small quartz grains.

FIGURE 3. Sample R148, from the sanded zone: Carter Reef. Entirely composed of carbonate grains, the foraminiferans and small gastropods being conspicuous. *Halimeda* fragments are also present.

FIGURE 4. Sample R4 (Terrigenous Facies, quartz-rich subfacies). The coarse fraction consists of worn, fragmentary mollusc shells and foraminiferan tests. The majority of the grains are quartz.

FIGURE 5. Sample R133, from the *Halimeda* covered banks. Gastropod shells and bryozoan fragments are also present.

FIGURE 6. Sample R5 (Terrigenous Facies, quartz-rich subfacies). Well sorted sediment dominated by quartz. Some grains are iron stained.

The bar scale in figures 1 to 5 represents a length of 2 mm, and that in figure 6 represents 1 mm.

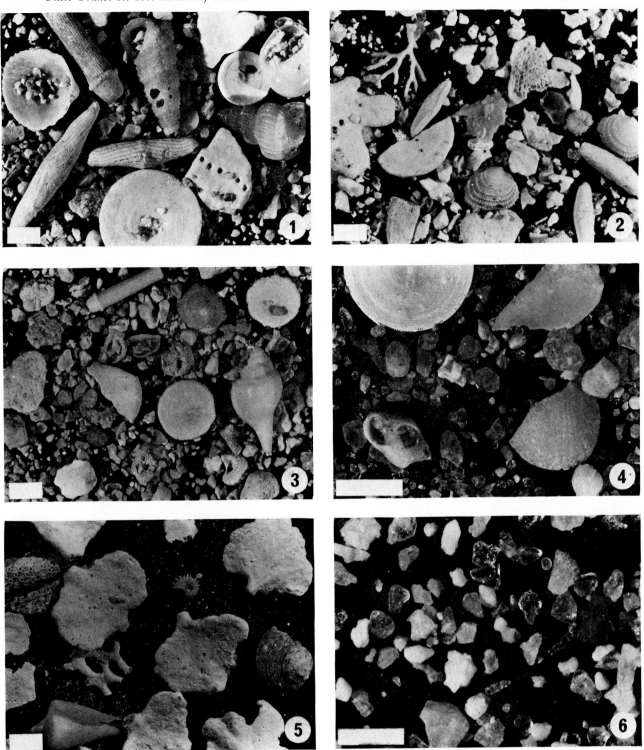

PLATE 1. For description see opposite.

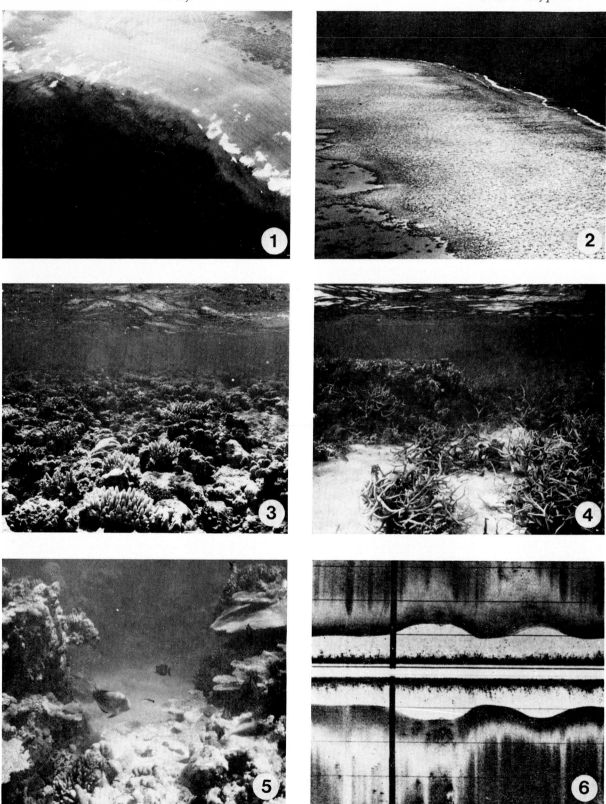

PLATE 2. For description see opposite.

Coral–algal gravels and sands. These represent the reef-top environment, particularly at the shelf edge, e.g. R163. This subfacies is restricted to reef-top situations. It is dominated by coral stick gravel with abundant *Halimeda*, coralline algal fragments, molluscs and foraminiferans. Worm tubes and bryozoans are among the minor components. Most samples of this subfacies are lag deposits but sample R163 (see figures 3 and 4) contains an extensive range of sand grade allochems, which washed into and were retained in the reef top pool from which the sample was collected. Of the sample, 71 % coarser than $-1\,\phi$ is mostly coral stick gravel.

Molluscan–foraminiferan gravels and sands. These represent a shallow lagoonal environment adjacent to reefs. Gastropods and pelecypods are dominant allochems with abundant and varied foraminiferans (compare R71, R75 and R148, figures 3 and 4, and plate 1, figures 1 and 2). Some coralline algal and coral fragments are present, and some samples also have a minor component of worn *Halimeda* fragments. Echinoid debris is conspicuous, and blue-grey mud is present. Some detrital quartz grains occur in this subfacies near to the continental islands. A conspicuous feature of the sediment adjacent to Lizard Island is the iron staining and partial solution of many contemporary allochems. This poorly sorted, polymodal sediment is characteristic near reefs and in Lizard Island Lagoon. The various modes of the frequency curve (figure 3) reflect variations in relative importance of the different allochems.

(ii) *Impure Carbonate Facies* (35 % *insolubles*)

Owing to an increase in the insoluble components, blue-grey mud and/or detrital quartz, the High Carbonate Facies passes to Impure Carbonate Facies near the southwestern margin of the region. Near Lizard Island, subfacies variations are complex.

(iii) *Transitional Facies*

Near the continental islands some sediments have intermediate proportion of insolubles due to mixing of terrigenous components (e.g. R1–58 % insolubles), mainly quartz and skeletal debris. This composition places such sediments in the Transitional Facies of Maxwell (1968 a).

Description of Plate 2

FIGURE 1. Oblique aerial view of part of Carter Reef from the northeast at half-tide, showing the fore-reef slope descending steeply to the Queensland Trough, and reef-top with 'sanded zone' at the top right hand corner of the photograph.

FIGURE 2. Oblique aerial view of the northern part of Carter Reef from the south showing the reef-top partly exposed at low tide, with algal zone, coral zone, and the leeward 'sanded zone' (bottom left). At the top right of the photograph is the deep water of the Queensland Trough, and the deep water to the left of this is Cook's Passage.

FIGURE 3. Underwater view of the coral zone at high tide: Carter Reef.

FIGURE 4. Underwater view of an *Acropora* thicket on the leeward side of Carter Reef near the margin of the 'sanded zone'.

FIGURE 5. A biogenic-sand covered channel in the 'Coral Zone' of Carter Reef.

FIGURE 6. Part of a side-scan sonar record of the seabed in an area of *Halimeda* covered banks to the southeast of Lizard Island (see figure 1). On each side of the centre line the record first shows a profile of the seabed, and thereafter extends continuously outwards as an oblique view of the seabed relief. The horizontal timing lines are spaced to show intervals of slant range of 25 m. Notable features are the uniform depth of the 'bank' tops, and their smooth profiles which contrast with the uneven 'rocky' surfaces of the hollows.

(iv) *Terrigenous Facies*

The insoluble end-members are of two types, namely blue-grey mud (mainland provenance) and coarse grained terrigenous end-members dominated by quartz (continental island provenance). The relative abundance of these two types of insolubles is the basis for subdivision of this facies.

Quartz-rich subfacies. The dominant grains are terrigenous, consisting of igneous quartz with feldspar, mica and tourmaline, together with some granite and beach-rock fragments. The biogenic carbonate allochems are fragmentary and worn, and consist of molluscs with a limited foraminiferan component which is chiefly represented by *Marginopora*. There is a lack of coral and coralline algal debris and only a limited number of *Halimeda* fragments. Many of the allochems are iron stained. It is a well sorted, unimodal sediment (see figures 3 and 4), with the mode in the fine sand range due to the dominant terrigenous end-members, particularly quartz. The inflexion in the cumulative curve at approximately 2 ϕ also reflects this characteristic. This subfacies is present, particularly to leeward of Lizard Island.

High mud subfacies. In the samples examined the blue-grey mud exceeds 50 %, and the quartz present is of a limited size range. The high gravel component (21 %) in the sample used to illustrate this subfacies (figures 3 and 4) comprises skeletal allochems dominated by molluscs, foraminiferans, and *Halimeda* fragments. Many of these are very worn and appear to have been transported. Coralline algae and corals are generally lacking.

(c) *Grain size distribution patterns*

In this outer shelf area there is a considerable diversity of bathymetric relief, sedimentary environments and ecological situations, in addition to complex provenance and dispersal factors, and owing to the varied interaction of such factors no simple interpretation of granulometric data will suffice. The cumulative curves (figure 4) illustrate the considerable granulometric differences which exist between sediments of this region. Marked points of inflexion on these curves separate discrete grain size populations, and indeed in R5, which is dominantly

---

DESCRIPTION OF FIGURE 5

FIGURE 5. Continuous seismic reflexion profile (high-resolution boomer) and interpretation along line 62, extending southwestward from Half-mile Opening, and viewed from the northwest. (Note that the top of the record has been removed so that the profile begins at the 30 ms level, which is 22.5 m below sea level.)

A figure of 1500 m/s, centrally bracketing the seismic velocity of water, has been used throughout this work. The reproduction of original seismic profiles have reference timing lines at 0.01 s intervals corresponding to an interpreted depth scale of 7.5 m per timing line. The velocity used is not substantiated by drilling or seismic measurements. However, on the basis of extensive (substantiated) experience this figure of 1500 m/s is reasonable, and follows convention under the circumstances. An error of up to 10 % *under-estimation* of *sub-bottom* thicknesses is the most that is envisaged in deeply buried sections of these records.

The profile extends from the shelf edge to the mid-shelf region between Lizard Island and North Direction Island (see figure 1). It shows fore-reef deposits at the top of the continental slope, a reef mass below Half-mile Opening, a back-reef zone of interdigitating reef-rock and bedded deposits, and an accumulation of lagoonal deposits with low angle cross bedding, which overlie the most prominent sub-bottom reflector, disconformity 'A'. The varied nature of the seabed, ranging from the smooth profiles of the bedded back-reef 'banks' to the irregular surfaces of reef masses, is clearly shown. A rugged seabed exists in some hollows in the 'bank' area, but some of the apparent irregularity shown on the profiles is due to side reflexions from adjacent objects. A subsidiary reflector 'S' is an erosional surface which is detectable in part of the 'bank' deposits. It is partially obscured by the apparent reflector 'E', which follows the profile of the seabed, and which is really an instrumental effect. The marked depositional reflector 'R' probably represents a change of sediment type, a nodular or concretionary layer, or a hiatus, during which lithification of a uniform layer occurred.

terrigenous, may relate to different transportation modes (Visher 1969). The coarse fraction of R133 and R154, *Halimeda* debris, is a reflexion of an ecological situation and the point of inflexion therefore may not imply a differentiation of grain size populations due to transportation modes, but may relate to grain size differences due to different grades of *in situ* post-mortem contributions, and to some extent represents a lag deposit. Similar factors may prevail in R163 where coral stick gravel is a lag deposit. However, some mixing has occurred in this particular case.

The points of inflexion on the curve representing sample R128 separate, on the one hand, carbonate gravels and carbonate and terrigenous sands, and on the other hand differentiate carbonate sands and terrigenous sands from the dominating terrigenous blue/green mud. In this case there may be, indeed, coincidence between these points of inflexion and the separation of discrete grain size populations moved by different transportation modes, but clearly there has been mixing of carbonate and terrigenous end-members in this inter-reef environment. The extreme skewness values (inclusive graphic skewness, Sk_I), see figure 3, suggest the mixing of different grain size populations under a variety of conditions.

5. RESULTS OF THE SEISMIC SURVEY

(a) Side-scan sonar records

Side-scan sonar revealed submarine banks at a fairly uniform depth of 25 m, with intervening troughs and hollows descending to − 37 m. The banks are not regular in form or size, and their smooth surfaces contrast with the rugged, rocky features of the hollows (see plate 1, figure 6). No small scale ripples were detected.

(b) High resolution profiles†

The most conspicuous feature is a prominent sub-bottom reflector 'A' which occurs over the entire area, extending from beyond the shoreward southwestern boundary eastwards to the shelf edge. It lies beneath bedded lagoonal deposits except near Cook's Passage where it is coincident with a current-swept rugged surface at a depth of 45 m (figure 6, line 54). In detail the surface is irregular, probably karstified, is especially uneven where it truncates reefal limestones, and it has been channelled to a depth of 24 m (69 m below present sea level) at the lagoonal approaches to Cook's Passage (see figure 6, line 54). Locally, there is discordance between bedding above with that below this surface (figure 5, pullout 1). Near Half-mile Opening (line 62) and on the leeward side of Carter Reef (line 61) surface 'A' loses (seismic) definition as it passes into a complex of reef-rock and bedded carbonate deposits. Reflector 'A' is clearly a major disconformity which records an important event in the geological history of this part of the continental shelf.

Another erosional seismic reflector ('B'), approximately 4 m below 'A', is evident locally, e.g. near Half-mile Opening and below the leeward side of Carter Reef.

A very striking feature recorded by all the seismic profiles is the 'bank' forming deposit which overlies reflector 'A' and extends from the leeward side of the shelf edge reefs almost to the mid-shelf line of continental islands (see figure 5). This accumulation reaches a maximum thickness

† In the discussion of the continuous reflexion seismic profiles it is (reasonably) assumed that the interpreted seismic reflectors are either coincident with significant geological interfaces referred to throughout as 'surfaces' or with a layering (producing seismically reflective velocity contrasts) within consanguineous deposits. Certain instrumental phenomena by which some seismic interfaces appear at computed depths, consistently slightly different from those revealed by confirmatory drilling, is known to the authors, but is not in this work considered to be of significance.

of 18 m near the outer reefs and thins toward the mid-shelf area, probably accompanied by a facies change. Low-angle cross bedding is here interpreted, and there is seismic evidence also of an erosion surface within the bank deposits (reflector 'S'). It is noticeable that the bank tops are at a fairly uniform depth of −25 m, beneath the side-scan sonar tracks (plate 2, figure 6); in profile 62 and 61 (figures 5 and 6) they occur at −24 to −26 m, occasionally descending to −30 m. It is also significant that the distribution of the 'banks' coincides with the area of luxuriant *Halimeda* growth, and the surface deposits of *Halimeda* gravels and sands.

It should be noted that the apparent seismic reflector ('E') which follows the outline of the profile (figure 5) at a uniform sub-bottom depth of approximately 4 m is considered to be an instrumental effect. This partially masks the erosion surface (seismic reflector 'S'), which may correlate with the local reflector detected within the reef mass near Decca Hi-fix position 1237. The depositional seismic reflector 'R' (see figure 5) is remarkably smooth and level in contrast to the irregular surface 'A', and the comparative intensity of seismic reflector 'R', corresponding to a significant velocity contrast is interpreted as due to advanced lithification of the materials underlying surface 'R', possibly composing a nodular or concretionary layer.

Other significant features shown by the profiles are well developed terraces at −22.5 and −30 m on the leeward side of Carter Reef (see line 61, figure 6), and evidence of ancient buried channels filled with acoustically more transparent deposits (probably poorly sorted with a high component of fines) than those of overlying 'banks' (e.g. line 69, figure 6). The similarity, in terms of internal structure and acoustic properties, between accumulations above and below reflector 'A' suggest that a similar environment and range of sedimentary facies prevailed before and after the formation of surface 'A'. Reef-rock extends beneath Half-mile Opening (figure 5), which indicates the former existence of a more continuous shelf edge reef at this site.

(c) The geological significance of the features recorded in the seismic profiles

The major event recorded in these profiles is represented by the seismic reflector interpreted as disconformity 'A'. This surface was produced when a marine regression exposed the entire continental shelf to subaerial weathering and erosion, allowing it to become karstified. A further lowering of the sea level resulted in the channelling of the shelf to the new base level, coincident with the formation of a continental slope terrace at −70 m. At this time the exposed part of the shelf edge reefs must have formed a limestone cliff which descended steeply to the Queensland Trough.

FIGURE 6. Interpretations of segments of high-resolution boomer profiles of the back-reef area near Carter Reef.
Line 61 extends from Carter Reef in a southwesterly direction, and is viewed from the northwest. It shows two terraces on the leeward side of Carter Reef, and two disconformities which can be traced into Carter Reef. Disconformity 'A' lies at approximately −45 m, and occurs over the entire area. Above this level the bedded deposits of the *Halimeda* covered 'banks' reach a thickness of 21 m. Disconformity 'B', lying 5 m below A, is of local occurrence. It crosses both bedded and unbedded (reefal) deposits.
Line 54 trends parallel with the shelf edge and crosses the approaches to Cook's Passage. This cross section is viewed from the southwest and shows disconformity 'A' exposed as a 'rocky', non-depositional seabed composed of both bedded deposits and reef-rock. The exposed part of 'A' lies at a depth of 45 m, and has been channelled to a depth of 69 m.
Line 69 shows a cross section of the back-reef area along a line running parallel with the shelf edge, and is viewed from the southwest. Disconformity 'A' occurs between the younger bedded deposits of the *Halimeda* covered banks, and the underlying bedded deposits and reef-rock. Where the older accumulations have been channelled, 'A' descends to a depth of 66 m below sea level. The channel was filled by acoustically more transparent sediments before the formation of the bank deposits which are continued over the buried channel as a layer which is approximately 9 m in thickness.

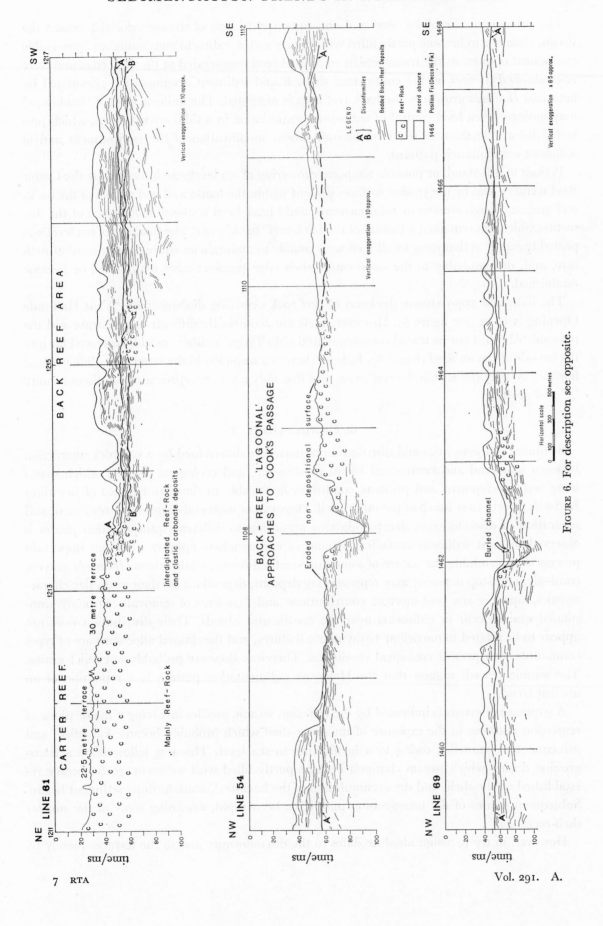

FIGURE 6. For description see opposite.

The subsequent rise of sea level and consequent reduction of stream velocities caused the stream channels to become partly filled with poorly sorted sediment containing an appreciable component of fines. As the transgression continued reefs regenerated at the shelf edge and were re-established at sites on the continental shelf. Rapid sediment accumulation stimulated by luxuriant *Halimeda* growth in the back-reef area is suggested. The uniform level of 'bank-tops' may indicate a sea level control of sediment accumulation in a tidal environment, which prevailed before elevation of sea level brought about modification of the environment and of sediment accumulation patterns.

Periods of still-stand, or possible temporary lowering of sea level, are indicated by the Carter Reef terraces, and by the erosion surfaces present within the banks and reef masses of the back-reef area, although erosion by tidal currents could have been responsible for some of the dissection which has remained a feature of the back-reef 'bank' area. The final rise of sea level was probably rapid, so that some small reefs were unable to maintain an equivalent upward growth rate, and, due probably to the same cause, shelf edge passages were maintained or became established.

The indicated approximate thickness of reef rock overlying disconformity 'A' at Half-mile Opening is 10 m (see figure 5). However, reefs are acoustically difficult to penetrate and the reflector 'A' could not be traced completely to the shelf edge, neither could it be traced far into the lee side of Carter Reef (figure 6). Indeed, there is a suspicion in the latter case (line 61) that horizon 'A' may rise within the reef mass, and that only a veneer represents post-disconformity 'A' reef growth.

6. Conclusions

Sediment characteristics and distribution patterns are determined by a complex interaction between mainland and continental island provenance, and ecological controls which determine reef development and promote conditions favourable to the maintenance of luxuriant *Halimeda* growth over much of the outer shelf. Dispersal of most end-members is restricted, and separation of discrete grain size populations according to different transportation modes is observed in most sediment subfacies. Mixing of end-members appears to be an important process in determining the nature of some inter-reef sediments, while others (*Halimeda* gravels, coral–algal reef-top deposits) may represent lag deposits, dependant for their grain size characteristics, upon *in situ* post-mortem contributions and 'currents of removal'. Strongly iron-stained grains occur in sediments near the continental islands. Their distribution does not appear to be related to particular bathymetric features, and the stained allochems are of types compatible with present ecological conditions. Therefore they are probably not relict grains. The seismic records suggest that the Holocene sedimentation pattern is a repetition of an ancient trend.

A sequence of events is indicated by the reflexion seismic profiles involving a major phase of regression resulting in the exposure of the outer shelf which probably became karstified, and subsequently channelled owing to a further fall in sea level. This was followed by a transgression during which stream channels became partly filled with sediment, reefs became re-established on the shelf, and the accumulation of the back-reef, bank-forming sediment began. Subsequent phases of the transgression may have been rapid, drowning some of the smaller shelf-reefs.

However, to try to assign absolute dates to the disconformity and to the terraces merely by comparison with eustatic curves would be presumptuous, especially since tectonic effects in this area have not been fully explored. Furthermore, it is possible that exhumed or relict surfaces which are apparent in the seismic profiles may be a source of confusion to researchers attempting to establish a chronology of sea level changes based on the recognition of a sequence of marine terraces.

An outline of the results of the investigations of the shelf in the Lizard Island area has been presented in this paper, but examination of the exact relation of surface 'A' to the continental islands, the mainland, and the large platform reefs, and an assessment of the significance of other disconformities, are continuing.

The detailed sequence of events and the full implications of the thickness and distribution of reefal and non-reefal facies in time and space may become apparent when detailed analyses of the seismic records from adjacent parts of the shelf investigated during the 1973 Expedition have been completed. Nevertheless, submarine drilling and isotope dating of features revealed by the seismic reflexion profiles presented here are desirable and would solve some of the problems concerning the geological history of the Great Barrier Reef Province.

The authors gratefully acknowledge the Royal Society of London, the University of Queensland, and the Australian Research Grants Committee (grant to Dr G. R. Orme for an investigation of sedimentary processes in the coral reef environment) for the support of this phase of the work of the 1973 Expedition. They are indebted to the Royal Australian Navy (Hydrographic Department) for the provision of Decca Hi-fix facilities, and to N.E.R.C. (U.K.) for the loan of supplementary seismic equipment, and record their gratitude to Mr E. Laundon and Mr A. G. Smith for invaluable technical assistance with the seismic operations and sediment sampling programmes respectively.

References (Orme *et al.*)

Fairbridge, R. W. 1950 *J. Geol.* **58**, 330–401.

Fairbridge, R. W. 1967 In *Landform studies from Australia and New Guinea* (eds J. N. Jennings & J. A. Mabbut), pp. 386–417. Canberra: A.N.U. Press.

Flood, P. G., Orme, G. R. & Scoffin, T. P. 1978 *Phil. Trans. R. Soc. Lond.* A **291**, 73–83 (this volume).

Flood, P. G. & Scoffin, T. P. 1978 *Phil. Trans. R. Soc. Lond.* A **291**, 55–71 (this volume).

Folk, R. L. & Robles, R. 1964 *J. Geol.* **72**, 255–292.

Frankel, E. 1974 In *Proc. 2nd Int. Symp. Coral Reefs, 1973*, vol. 2, pp. 355–369.

Maiklem, W. R. 1970 *J. sedim. Petrol.* **40**, 55–80.

Maxwell, W. G. H. 1968*a* *Atlas of the Great Barrier Reef*. Amsterdam: Elsevier.

Maxwell, W. G. H. 1968*b* *Aust. J. Sci.* **31**, 85–86.

Maxwell, W. G. H. 1973 *Biology and geology of coral reefs* (eds O. A. Jones & R. Endean), vol. 1 (Geology 1), pp. 299–345. New York: Academic Press.

Maxwell, W. G. H., Jell, J. S. & McKellar, R. G. 1964 *J. sedim. Petrol.* **34**, 294–308.

Maxwell, W. G. H. & Swinchatt, J. P. 1970 *Bull. geol. Soc. Am.* **81**, 681–724.

Orme, G. R., Flood, P. G. & Ewart, A. 1974 *Proc. 2nd Int. Symp. Coral Reefs, 1973*, vol. 2, pp. 371–386.

Purdy, E. G. 1974*a* *Bull. Am. Ass. Petrol. Geol.* **58**, 825–855.

Purdy, E. G. 1974*b* *Soc. econ. Palaeont. Miner. spec. Publ.* **18**, 9–76.

Sargent, G. E. G. 1969 *9th Commonwealth Mining and Metallurgical Congress*, paper 28. London: Institution of Mining and Metallurgy.

Visher, G. S. 1969 *J. sedim. Petrol.* **39**, 1074–1106.

comparison with relative curves would be presumptuous, especially since tectonic effects in this area have not been fully explored. Furthermore, it is possible that estimated or relief surfaces which are apparent in the seismic profiles may be a source of confusion to researchers attempting to establish a chronology of sea level change based on the recognition of a sequence of marine terraces.

An outline of the results of the investigations of the shelf in the Lizard Island area has been presented in this paper, but examination of the inter-relation of surface 'A' to the continental islands, the mainland, and the large platform reefs, and an assessment of the significance of other discontinuities, are continuing.

The detailed sequence of events and the full implications of the thicknesses and distribution of reefal and non-reefal facies in time and space may become apparent when detailed analyses of the seismic records from adjacent parts of the shelf investigated during the 1973 Expedition have been completed. Nevertheless, submarine drilling and isotope dating of features revealed by the seismic reflection profiles presented here are desirable and would solve some of the problems concerning the geological history of the Great Barrier Reef Province.

The authors gratefully acknowledge the Royal Society of London, the University of Queensland, and the Australian Research Grants Committee (grant to Dr G. R. Orme for an investigation of sedimentary processes in the coral reef environment) for the support of this phase of the work of the 1973 Expedition. They are indebted to the Royal Australian Navy (Hydrographic Department) for the provision of Decca Hi-fix facilities, and to N.E.R.C. (U.K.) for the loan of supplementary seismic equipment, and record their gratitude to Mr E. Lumsden and Mr A. G. Smith for invaluable technical assistance with the seismic operations and sediment sampling programmes respectively.

REFERENCES (Orme et al.)

Fairbridge, R. W., 1950 J. Geol. 58, 330-401.
Fairbridge, R. W., 1967 In Landscape studies from literature and art (ed. T. Y. Jennings & J. A. Mabbutt), pp. 386-417. Canberra: A.N.U. Press.
Flood, P. G., Orme, G. R. & Scoffin, T. P. 1978 Phil. Trans. R. Soc. Lond. A 291, 73-83 (this volume).
Flood, P. G. & Scoffin, T. P. 1978 Phil. Trans. R. Soc. Lond. A 291, 70-71 (this volume).
Jelly, R. T. & Robbie, R. 1960 J. Geol. 72, 283-292.
Kendall, T. 1972 In Proc. 24th Int. Geol. Congr., 1972, vol. 2, pp. 384-400.
Maxwell, W. R. 1970 J. sedim. Petrol. 40, 85-86.
Maxwell, W. G. H. 1968 Atlas of the Great Barrier Reef. Amsterdam: Elsevier.
Maxwell, W. G. H. 1968 a Submar. Geol. J. 34, 45-60.
Maxwell, W. G. H. 1973 Biology and ecology of coral reefs (ed. O. A. Jones & R. Endean), vol. 1, Geology, pp. 299-345. New York: Academic Press.
Maxwell, W. G. H., Jell, J. S. & McKellar, R. G. 1964 J. sedim. Petrol. 34, 294-308.
Maxwell, W. G. H. & Swinchatt, J. P. 1970 Bull. geol. Soc. Am. 81, 691-724.
Orme, G. R., Flood, P. G. & Ewart, A. E. 1974 Proc. 2nd Int. Symp. Coral Reefs, 1973, vol. 2, pp. 371-386.
Purdy, E. G. 1974 Bull. Am. Ass. Petrol. Geol. 58, 825-855.
Purdy, T. C. 1974 Sea-floor Relaxant. Mem. geol. Soc. Lond. 18, 9-76.
Sargent, G. E. G. 1960 Soil conservation. Mining and Metallurgy of Copper, part 28. London: Institution of Mining and Metallurgy, 36.
Veevers, O. S. 1969 J. sedim. Petrol. 39, 1074-1084.

Phil. Trans. R. Soc. Lond. A. **291**, 101–117 (1978)   [101]

Printed in Great Britain

Reef island sediments of the northern Great Barrier Reef

By R. F. McLean† and D. R. Stoddart‡

†*Department of Biogeography and Geomorphology, Australian National University,
Canberra, Australia* 2600

‡*Department of Geography, University of Cambridge, Downing Place, Cambridge, U.K.*

[Plate 1]

The reef islands are composed almost exclusively of bioclastic materials locally supplied from adjacent reef flats and reef crests. No sediment of terrigenous origin was encountered on the islands investigated (except drift pumice). Islands are built either of sand or of gravel, and only rarely mixtures of both sand and gravel. The *sands* are typically well sorted with a prominent size mode within the medium–coarse size range (0.25–1.00 mm). Major skeletal components include corals, Foraminifera, molluscs and crustose coralline algal fragments, either whole or broken. Most grains show signs of considerable abrasion. The *gravels* are more homogeneous in composition (mainly corals) but reveal a great range of size, shape and surface characteristics. Elongated clasts, derived from branching corals, provide the main components but corals of other growth forms are present. The sand deposits and gravel deposits are normally spatially discrete on any one reef but mixtures of these two size grades do occur, notably on the Turtles. Reasons for the presence or absence of spatial discrimination are discussed.

1. Introduction

Reef islands are morphologically coherent accumulations of bioclastic materials standing on reef tops and exposed above the level of the sea at high tide. During the 1973 Great Barrier Reef Expedition, island sediments were examined on 31 reefs between Low Isles in the south and Waterwitch in the north (figure 1). It was found that single islands were built (i) predominantly of sand, (ii) predominantly of gravel, or (iii) of mixtures of sand and gravel (either sandy-gravels or gravelly-sands depending on the dominant size grade), or (iv) parts of the same island comprised a combination of the above grades. In reef areas 'shingle' rather than 'gravel' is commonly used to describe the coarser sediments and this usage is adopted here. Thus, in terms of sediment calibre reef islands in the area can be classified as sand cays, shingle cays, mixed sand–shingle cays, composite cays.

Certain characteristics of the reefs and islands visited are summarized in table 1. On 21 of the reefs only one island was present although most possessed additional surficial sediment bodies which were not emergent at high water. The other ten reefs contained two or more islands which in the case of Ingram–Beanley and Sinclair–Morris are identified by different names. In general the reef islands occupy only a small proportion of the reef top on which they stand and rise but a few metres above low water level. In addition to unconsolidated sediments, many of the islands possess consolidated deposits, particularly beach-rock and rampart-rock. These exposed limestones are described elsewhere (Scoffin & McLean 1978, this volume).

The purpose of the present paper is to describe the nature of unconsolidated reef island sediments in terms of particle size, mineralogy and composition, and discuss the source and supply of sediments, local and regional distribution of sediment types and history of island sedimentation.

The account is a general one. Conclusions are based on data from some 200 sediment samples. Detailed results will be presented elsewhere. The foregoing division of island types based on sediment calibre will be followed.

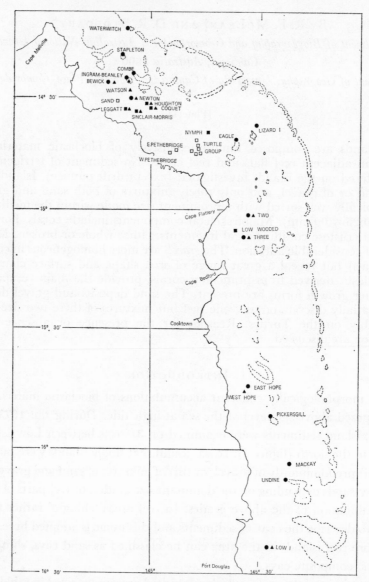

FIGURE 1. Location of reef islands on the northern Great Barrier Reef classified in terms of sediment type: ●, sand cay; ▲, shingle island; □, mixed sand–shingle island; ■, composite island.

2. SAND CAYS

The sand cays investigated can be divided into three types on the basis of the number of islands on the reef and the complexity of the sand cay itself.

(A) Single island on reef top

   I. Sand cay, unvegetated

      (i) Without beach-rock: Pickersgill, Undine, Mackay

      (ii) With beach-rock: Waterwitch

II. Sand cay, vegetated, with beach-rock

Combe, Stapleton, Eagle, East Hope

(B) More than one island on reef top

III. Sand cay of type II, plus shingle island and mangroves on same reef flat, i.e. the sand cay of a 'low wooded island' (Steers 1929): Ingram, Two Isles, Three Isles, Newton, Bewick

TABLE 1. REEF ISLAND CHARACTERISTICS

reef	percentage of reef occupied by island†	distance from coast km	distance from barrier km	reef type‡	island type sand	shingle	mixed	composite
Undine	<1	20	23	L	X	—	—	—
Mackay	2	22	23	L	X	—	—	—
Pickersgill	<1	20	28	L	X	—	—	—
East Hope	2	13	39	H	X	—	—	—
Eagle	1	28	29	L	X	—	—	—
Combe	<1	24	23	L	X	—	—	—
Stapleton	1	26	16	L	X	—	—	—
Waterwitch	2	35	1	L	X	—	—	—
Watson	9	18	31	H	—	X	—	—
West Hope	5	11	41	H	—	X	—	—
Turtle I	38	15	45	H	—	—	X	—
Turtle II	33	15	44	H	—	—	X	—
Turtle III	15	16	46	H	—	—	X	—
Turtle IV	4	17	43	H	—	—	X	—
Turtle V	32	17	42	H	—	—	X	—
Turtle VI	37	19	40	H	—	—	X	—
E. Pethebridge	6	9	48	H	—	—	X	—
W. Pethebridge	9	7	49	H	—	—	X	—
Sand	20	9	37	H	—	—	X	—
Nymph	52	22	33	H	—	—	—	X§
Low Wooded	59	15	36	H	—	—	—	X§
Low	4	18	36	H	X	X	—	—
Three	19	18	32	H	X	X	—	—
Two	18	17	29	H	X	X	—	—
Newton	8	15	34	H	X	X	—	—
Ingram	3	22	26	H	X	X	—	—
Bewick	16	18	29	H	X	X	—	—
Leggatt	8	7	41	H	—	X	—	X‖
Houghton	20	16	33	H	—	X	—	X‖
Sinclair	10	9	39	H	—	X	—	X‖
Coquet	32	15	34	H	—	X	—	X‖

† Includes island-enclosed ponds.
‡ Reef type after Maxwell (1968, fig. 72): L, lagoonal platform or platform reef; H, high reef.
§ Mainly shingle.
‖ Mainly sand.

The essential features of these sand cays have been well described by Steers (1929, 1937) and need not be repeated here. They are all located on the leeward (west) side of their respective reefs. In plan they possess flask, teardrop or ovate shapes with shape and long-axis orientation depending on the geometry of the surrounding reef in relation to prevailing southeasterly seas. Cays of type I and II are situated on large reefs with shallow lagoons. They are small islands with areas ranging from *ca.* 4000 m² (Undine and Pickersgill) to 45000 m² (Combe and Stapleton), which cover no more than 2 % of the total reef area (table 1). In contrast, cays of type III are located on high lagoonless reef tops. Excluding Ingram, the reefs are small and the cays cover a proportionally greater area of the reef top, up to 15 % in the case of Two Isles. Two Isles, with an area of some 195000 m², is the largest sand cay in the whole region. However, in spite of differences in island size, reef size and type, as well as the presence or absence of vegetation, beach-rock or other islands on the same reef, the sand cays possess a surprisingly homogeneous sediment population. Data from over 100 samples from these cays indicate that variation between cay sediments on different reefs is less than variation between reef flat and cay sediments on the same reef. Nevertheless, differences do exist and these are elucidated below.

2.1. *Constituent composition, mineralogy and morphology*

Apart from drift pumice, sand cay sediments are made up almost exclusively of skeletal reef materials. In order of abundance, Foraminifera, coral, *Halimeda*, molluscs and coralline algae are the most important constituents. Bryozoa, crustacea, echinoid and other fragments are rare and together account for less than 5 % of components. Typically, taxonomically unrecognizable coral fragments provide 20–25 % of the sands, Foraminifera 25–30 % and molluscs 10–15 %. Values for coralline algae and *Halimeda* are more variable, although the latter makes up more than 10 % of each sample. Proportions of the various skeletal types differ on different islands suggesting that each reef possesses a unique biota and set of environments. Sediments from Low Isles and Ingram are particularly rich in coral (30 %) and coralline algae (25 %) and poor in Foraminifera (8 %), while Bewick, Newton and Two Isles have a high percentage of forams (30 %) and are low in coralline algae (5 %). Sediments from Waterwitch cay on the outer ribbon reef are unusual in that they have both the highest percentage of molluscs and lowest percentage of forams in the sample suite.

There are also indications, as yet unquantified (1) that the relative abundance of constituents changes from reef flat to reef island, e.g. the proportion of *Halimeda* declines; (2) that the cays of high reefs possess a different suite of components than those of reefs with lagoons, e.g. the latter are especially poor in coralline algae; and (3) that contemporary beach materials differ in composition from older adjacent sands on the same cay, e.g. there is more coralline algae and less *Halimeda* in the modern sands. These suggest that the nature of skeletal production and reef-top environments have changed through time.

Bulk mineralogical determinations indicate the sands contain mixtures of aragonite and high magnesium calcite, although low magnesium calcite was present in a number of samples. Percentages of aragonite range from 20 to 70 % and calcite from 20 to 80 % depending on the relative proportions of the various skeletal types.

Grain morphology is also dependent on constituents. Foraminifera are either discoidal (e.g. *Marginopora*) or spherical (e.g. *Calcarina*), *Halimeda* leaf-like, calcareous algae elongate while coral fragments and molluscs have highly irregular shapes. All grains show signs of wear and abrasion and many are broken fragments. Nevertheless there is great variation in grain surface texture depending on skeletal type and grain size, though most are edge rounded and polished.

2.2. *Grain size and sorting*

In spite of variations in constituent components, the entire suite of cay samples are surprisingly uniform in terms of size grading (figure 2) irrespective of whether they were collected from vegetated or unvegetated cays, windward or leeward beaches, berms, soil horizons or island surfaces or subsurfaces.

It became apparent during plotting of cumulative curves that the division between sand and silt sizes in the Wentworth scale (4 φ) was a less appropriate break between sand and fines than 3 φ. None of the beach, and very few of the cay sands, possessed material finer than 3 φ, although it was detected in some soils. This implies that fines are rarely transported to and deposited on sand cays. Their presence is indicative of post-accumulation weathering and soil development. Also, plots of individual samples showed that grain size distributions are typically unimodal. Those possessing bimodal or polymodal distributions were few in number and commonly resulted from the occurrence of a few coarser coral or shell fragments. Mean sizes all fall within a range of less than 2 units (−0.3 to 1.6 φ) (figure 3a). Within this range, 75 % have means in the coarse sand category (0–1 φ), 11 % in the very coarse sand and 14 % in the medium sand grades. In terms of sorting, the majority of samples 54 % fall into Folk's moderately well sorted category (0.5–0.71 φ) with 17 % and 19 % being better or poorer sorted respectively. Thus, as a generalization, the cay sediments can be classified as moderately well sorted coarse sand.

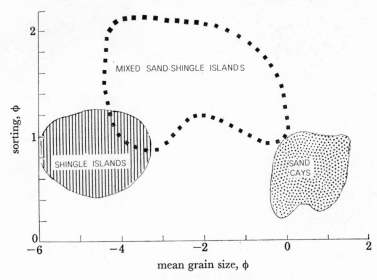

FIGURE 2. Fields of sorting versus mean grain size for three sediment types.

There are, however, fairly systematic differences in texture depending on whether samples were collected from contemporary beach, cay surface and subsurface or soil horizons (figure 3). Beach sands include the coarsest cay sediments. Figure 3b shows a sinusoidal trend which is accounted for by the difference between windward reef flat facing beaches of vegetated cays which possess larger sizes than sands of lee beaches and unvegetated cays. Unmodified clean cay sands, that is, those that do not have clear indications of soil development, commonly possess means between 0.3 and 1.0. Presently developing soils and buried soil horizons provide the finest cay sediments (figure 3d). Sample means show a prominent mode around 1 φ and a minor mode at 1.5 φ, the first being slightly offset from that of parent cay sands. This suggests that the soils are composed basically of parent sand with the addition of finer organic matter. This addition of fines accounts for the poorer sorting in those samples.

Earlier it was noted that the size–sorting distributions for the suite of basic cay sands were relatively uniform. However, it is clear from figure 3c that there are subtle differences between cays such that each cay has its own distinctive sediment population. For instance, Bewick sands are fine and moderately sorted whereas Ingram sands, although equally fine, are somewhat better sorted. Low Isles sands are generally coarser than both Bewick and Ingram samples and possess a greater range of sorting values. Three Isles sands, although within the combined size range of Low, Bewick and Ingram samples are well sorted and this better sorting distinguishes them from the three others. Sands from the other cays all fall within the fields set by Low, Bewick, Ingram and Three Isles but again subtle differences can be recognized (figure 3c). Reasons postulated to account for the differences in sediment size and sorting between cays include: (1) variations in proportions of different constituent components in areas of sediment production; (2) differences in distances, modes and rates of transport from source area to cay sinks and the degree of sorting and abrasion during transport; and (3) variations in residence-time since deposition.

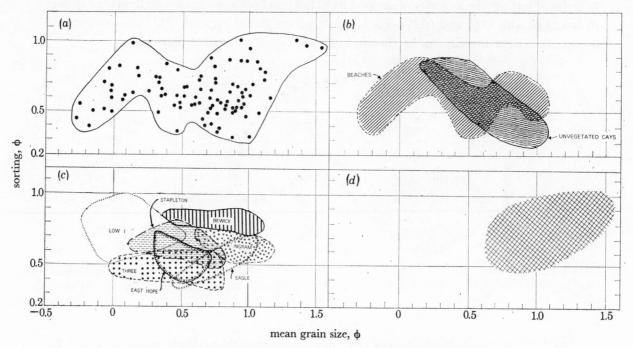

FIGURE 3. Sorting versus mean size of sand cay sediments. (a) All samples; (b) fields of beach and unvegetated cay samples; (c) fields of particular sand cays; (d) fields of samples from contemporary and buried soil horizons.

2.3. Source of sand cay sediments

The sediments which ultimately go into cay building originate mainly from the reef flat with a small proportion coming from biota (mainly molluscs) living within the beaches and beach-rock around the cay's margin. Sand cays develop on the lee corner of the reef platform at the zone of convergence of refracted southeast swell that wraps around the reef and crosses the reef flat at high tide. Their location is a relatively stable one and is dependent on reef configuration and orientation in relation to predominant swell direction. Ramparts, shingle islands, man-groves and other features on reef tops cause obstructions to wave fronts and create secondary zones of wave convergence and divergence which may influence the incidence and direction of wave attack on the cay. Thus cay shapes may change in detail through time depending on the number, size, location and time of development of other features on the reef top.

The nature of reef flat sediments for a number of the reefs in this area are described by Flood & Scoffin (1978, this volume). A comparison between reef flat and cay sediments for Three

Isles, Stapleton, Ingram and Low Isles samples is made in terms of size–sorting plots (figure 4). It is clear from these plots, for the first three reefs at least, that there are two sediment populations. However, the populations are not discriminated on grounds of mean size, but on the basis of their sorting values: the cays sands are uniformly better sorted than reef flat materials. Thus in the interval between production of skeletal debris on the reef flat and its accumulation on the cay, changes in one but not both sediment parameters takes place. The high sorting values in the

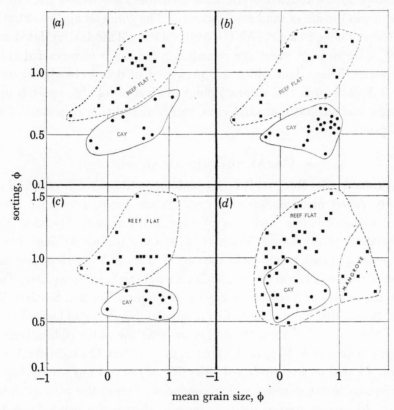

FIGURE 4. Comparison of reef flat and cay sands in terms of sorting versus mean size for four reefs: (a) Stapleton; (b) Three; (c) Ingram–Beanley; (d) Low Is.

reef flat sediments is the result of the presence of both coarse and fine tails in the size distributions: their frequency curves are lower and more spread out than the cay samples which possess much steeper unimodal curves. The finest reef flat material does not reach the sand cays and the coarsest is broken down during transport. Thus wave agitation during passage from reef flat to the cay selectively sorts the sediments, bringing the basic array of reef flat sizes (excluding finest and coarsest) to the cay in a graded state. The greater the distance between reef flat source and cay sink, the greater the degree of grading. Thus, for the Low Isles samples the cay sands fall within the field of the reef flat sediments presumably because the distance of travel and/or frequency of particle movements is less than on the other three reefs. Another reason for the good discrimination between the reef flat and cay sediments of Three Isles, Stapleton and Ingram reefs could be that the older cay samples were derived from reef flats in the past which possessed material different from that currently forming.

2.4. *Age of sand cay sediments*

Sixteen radiocarbon dates of bulk sand samples from ten sand cays which included a small unvegetated cay (Pickersgill), simple cays (Eagle, Stapleton, East Hope), and sand cays of low wooded islands (Bewick, Ingram, Leggatt, Low Isles, Two Isles and Three Isles) were obtained. In spite of this range of cay types and in spite of the great range of sample locations and elevations nine of the samples fell within a 500-year time span from 2900–3400 a B.P. which indicates considerable clustering in the age of sand cay bioclastics. The youngest age was 2190 ± 70 a B.P. (ANU-1641) and oldest 4380 ± 80 a B.P. (ANU-1559). Pickersgill sands were dated at 2230 ± 70 a B.P. (ANU-1606), a surprisingly great age considering the cay is unvegetated and only just emergent at high water. These data which are presented in detail elsewhere (see McLean, Stoddart, Hopley & Polach 1978, this volume) point to a period of high reef-top productivity some 3000 years ago, and suggest that sand cays, rather than being ephemeral features, are relatively stable deposits.

3. CORAL SHINGLE ISLANDS

In the northern Great Barrier Reef, reef islands made up predominantly of gravel sized clasts have been variously called mangrove–shingle cays (Steers 1929), mangrove islets, shingle islands, shingle and mangrove islets (Steers 1937), vegetated ramparts (Spender 1930) and at Low Isles rampart systems (Fairbridge & Teichert 1948). On a reef top a shingle island may be the only deposit emergent at high water (as in the case of Watson, West Hope and Sand reefs), or there may be more than one island present, either a sand cay (such as at Low, Two, Three, Bewick, Newton and Ingram) or a composite cay (as at Coquet, Leggatt, Sinclair–Morris and Houghton), in addition to the shingle island. Coral shingle islands are all located on high reefs (Maxwell 1968) of the inner shelf some 30–40 km in from the outer ribbon reefs (figure 1, table 1). They typically occupy 5–10 % of the area of the reef top. On individual reefs shingle islands are generally located on the windward (SE) side of the reef top where they form continuous or discontinuous linear deposits which commonly mimic the plan geometry of the peripheral reef edge. Islands are frequently quite narrow, their width being dependent on the number and nature of ramparts, ridges and swales. Mangroves colonize the sheltered leeside reef top while to windward the exposed island beach frequently overlies or is fronted by a consolidated platform or shallow moat. In some cases, such as Watson, the loose shingle deposits which make up the island surmount cemented platforms rather than directly overlie the reef flat. The variable morphology of shingle islands and rampart systems in this area have been described by Steers (1929, 1937), Fairbridge & Teichert (1948), Stoddart, McLean & Hopley (1978, part B of this Discussion) and others. However, sediments making up the islands have not been described in such detail, although most authors have commented on the fact that fragments of branching corals (*Acropora*) make up a very large percentage of the shingle.

3.1. *Particle size and sorting*

Because coarse materials cannot be easily measured by using sieving techniques, tri-axes measurements were made on 100 clasts from each sample. Mean size and sorting were therefore computed on a number rather than mass frequency basis (Folk 1962). Field inspection suggested that the shingle deposits were relatively uniform in terms of size grading (figure 5, plate 1) and results from five samples indicate that this is the case (table 2).

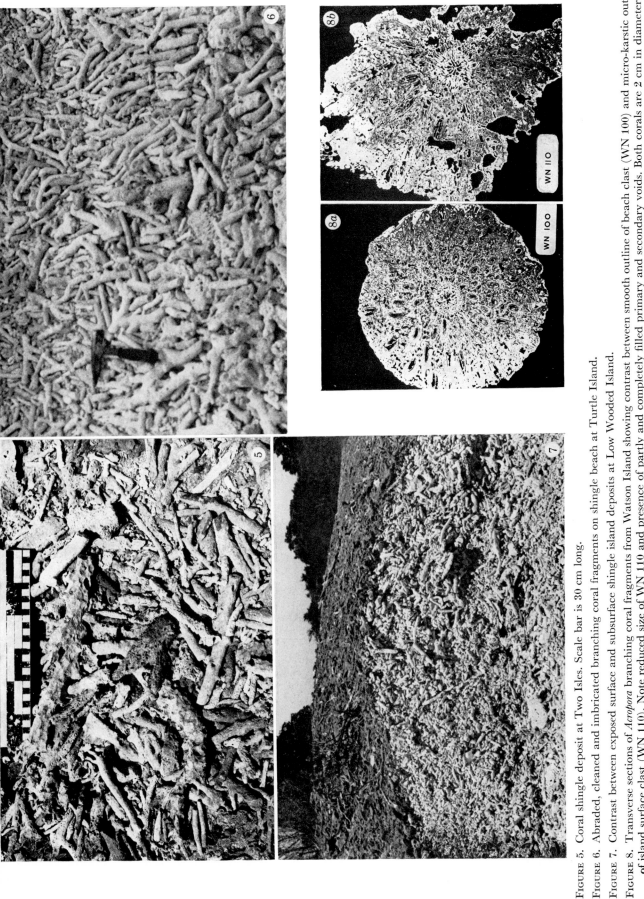

FIGURE 5. Coral shingle deposit at Two Isles. Scale bar is 30 cm long.

FIGURE 6. Abraded, cleaned and imbricated branching coral fragments on shingle beach at Turtle Island.

FIGURE 7. Contrast between exposed surface and subsurface shingle island deposits at Low Wooded Island.

FIGURE 8. Transverse sections of *Acropora* branching coral fragments from Watson Island showing contrast between smooth outline of beach clast (WN 100) and micro-karstic outline of island surface clast (WN 110). Note reduced size of WN 110 and presence of partly and completely filled primary and secondary voids. Both corals are 2 cm in diameter.

Mean size for the intermediate b-axis falls around $-4\,\phi$ (16 mm), which is the typical diameter of living reef edge corals of branching growth form. While the extreme range of clast diameters covered up to $8\,\phi$ units (2–512 mm), all samples possessed relatively normal frequency distributions. Particle length of a-axis medians registered about $1.5\,\phi$ greater than b-axis means. Cumulative curves of particle diameter and length indicate a regular relation such that the length of smaller particles is about $0.7\,\phi$ greater than their diameter, while for larger particles it reaches over $2\,\phi$. In terms of size grading, island deposits can be described as moderate to moderately well sorted medium sized pebbles. Typically, shingle clasts have lengths of some 50–60 mm and diameters of 15–20 mm. These sizes result from breakage of branching corals in the original coral thickets.

TABLE 2. SHINGLE SIZE AND SORTING (ϕ UNITS)

reef	mean size, b-axis	sorting	maximum size	minimum size	median size, a-axis	remarks
Watson	−4.3	0.55	−6.0	−2.0	−5.5	beach
Watson	−4.3	0.61	−6.0	−1.0	−5.6	third rampart
Watson	−4.3	0.61	−9.0	−2.0	−5.8	innermost ridge
Two	−3.8	1.06	−9.0	−1.0	−5.2	outer rampart
Low	−4.1	0.96	−6.0	−2.0	−5.7	inner rampart

3.2 Particle shape

Particle shapes derived from the three-axes measurements clearly show the dominance of rod-shaped clasts. Plots on Folk form triangular graphs indicate that 65 % of clasts possess elongate (20 %) or very elongate (45 %) shapes. Blades are the next most abundant group followed by those of compact (spherical) shapes. Less than 5 % of particles fit into the platy (discoidal) class, which is perhaps the most common particle form on shingle islands of Pacific atolls.

The visual ubiquity of elongate pebbles on shingle islands of the northern Great Barrier Reef is confirmed, the obvious ultimate source of this sediment being the thickets of branching corals so characteristic of the windward periphery of reefs in the area. During storms, colonies and individual branching corals are broken down to finger-like or Y-shaped segments (coral sticks) which are readily transported across the reef top, built into ramparts and ultimately islands.

Nevertheless, while stick corals of pebble size do make up the major proportion of shingle island deposits, the contribution of both larger and smaller particles, as well as fragments possessing non-rod-like shapes, may be underestimated. First, the use of mean size values tends to obscure the fact that maximum dimensions of measured individual particles ranged from 5 to 500 mm. Secondly, 35 % of particle shapes are not rod-like, and these invariably comprise the largest and smallest particles. Thirdly, our observations have been predominantly of surface deposits rather than sections. They are thus biased to the zone in which sediments have accumulated at maximum swash limit and storm washover levels, that is, above the reach of normal high water. Thus, where sections are available on eroding island shores or beaches, stratified deposits are frequently observed. In some, the surface deposit continues down to a distinctive basal layer, while in others it is separated from the basal layer by a zone of smaller coral fragments and sand which accumulate in the swash zone under normal wave conditions. The basal unit is most distinctive. Typically it consists of larger cobble-sized hemispherical and highly irregularly shaped corals, plus occasional massive clam valves. This lag gravel rarely reaches above mean sea level, although storm-tossed individual components are found scattered on island surfaces. Sections also reveal the loose packing of constituents which results in highly unstable erosional cliff faces, except in cases where stick corals are preferentially orientated and imbricated. Interclast voids are rarely occupied by sand or mud sized sediment above mean sea level.

3.3. *Constituent composition*

In common with sand cays, the shingle islands are made up exclusively of bioclastic carbonates (excluding drift pumice). However, fewer taxa are represented in the shingle deposits. Scleractinian corals are clearly dominant and perhaps account for 95% of the constituent components. Mollusca, hydrozoan corals (*Millepora*), Octocorallia (*Heliopora* and *Tubipora*) and crustose coralline algae are subordinate contributors. Lithoskels reworked from cemented rampart-rocks are also locally present.

Of the corals, the prolific genus *Acropora* dominates the shingle rampart and island deposits as it does most living reefs in the area. Staghorn and bushy forms are the most abundant of the various acroporid growth forms. Other branching forms include *Porites*, *Seriatopora* and particularly *Pocillopora*. Rounded and hemispherical corals, especially various members of the family Faviidae also occur, notably in the basal layer. Fragments of tabular, foliaceous, and encrusting corals as well as solitary free corals are also found in island deposits.

The richness of the coral suite of the shingle deposits is illustrated by the fact that J. E. N. Veron (personal communication) observed a similar number of coral genera in the windward shingle deposits of Bewick reef as were presently living on the reef edge between depths of 0 and 12 m. Moreover, surface collections of easily recognizable clasts of different growth forms were made at about 40 localities on all the shingle islands visited and it was possible to collect quickly at least two branching corals (*Acropora*, *Pocillopora*), one foliaceous or encrusting coral (commonly *Turbinaria* or *Echinopora*), one solitary coral (*Fungia*), one hemispherical coral (Faviidae) and the rounded spiky *Galaxea*, as well as one clam, either *Tridacna* or *Hippopus*, from each site. Thus, despite the visual dominance of branching *Acropora*, the suite of corals (and probably molluscs) present in shingle deposits appears to include most windward reef flat and reef edge species.

3.4. *Particle surface characteristics*

In the field a major visual contrast is between the dull grey of island surface shingle and white brilliance of windward beach shingle. This contrast differentiates stabilized from active deposits. Wave action in the harsh windward beach zone abrades particles and ensures that most surface epibionts, obtained during post-death reef flat residence, are cleaned from the clasts before they reach their final sites of deposition. Windward beach clast surfaces are typically smooth and edge rounded (figure 6, plate 1), although original skeletal architecture is frequently maintained in taxonomically recognizable, although subdued, form. Particle surface textures obtained in the beach environment appear relatively unchanged in subsurface island deposits, but those exposed on the surface obtain a micro-phytokarstic architecture of tiny jagged spongy pinnacles and pits (figure 7, plate 1), which result from colonization by boring filamentous algae. The extent and intensity of pitting varies systematically across a shingle ridge or rampart system, being dependent principally on the age of the deposit and length of time of exposure. Old surfaces are more intensely pitted than young surfaces. Porous corals and massive clam valves are equally subjected to post-depositional pitting in the subaerial environment.

Thus, as a result of both abrasion and corrosion, shingle island surface clasts are smaller than their living or broken reef flat counterparts. Reconstruction of original skeleton circumferences from transverse thin sections of branching *Acropora* clasts allows the magnitude of change from source area to final site of deposition to be evaluated. Clasts from shingle beaches at Low Isles, Bewick and Watson possessed 90–98 % of their original cross sectional areas, while those from the surfaces of adjacent islands were only some 51–67 % of their original size. These figures indicate the severity of *in situ* degradation of exposed island surface shingle (figure 8, plate 1).

3.5. *Mineralogy, porosity and diagenesis*

Unlike sand cays which are composed of bioclastic components of both calcitic and aragonitic organisms, shingle islands are made up almost exclusively of corals which build skeletons of aragonite. However, bulk mineralogical determinations of over 70 clasts from shingle islands on ten reefs in the area show that high magnesium calcite is present in almost all samples. Commonly the percentage is between 3 and 10 but values up to 52 % were registered. Large sample to sample variation exists and adjacent coral sticks from the same locality have given up to tenfold differences in calcite values. The presence of calcite in individual shingle clasts is indicative of early diagenesis, and can result either from recrystallization of the original skeleton or the presence of secondary calcitic cement or sediment. Thin section analysis favours the second alternative, with the percentage of high magnesium calcite being a function of the amount of adhering and intraskeletal void sediment and cement.

Corals are porous structures and porosity measurements of live branching *Acropora* from the area average 25 %. Post-death occupation by boring organisms creates secondary voids such that there is an increase in the total void space. For island shingle clasts this averages 45–65 %, an increase of no less than 20 % on primary pores. However, voids serve as sediment traps and loci for intraskeletal cementation and are partially filled or completely filled synchronously. Thus it is unlikely that the magnitude of primary porosity is greatly exceeded at any one time during the passage of an individual clast from live source area to its island site of final deposition. Measurements of thin sections of *Acropora* sticks from shingle islands and ramparts in the area show that on average between 25 and 40 % of clast volume is occupied by cement and sediment. The lack of a significant difference between beach or rampart clasts and those in island deposits suggests that both intraskeletal void creation and void filling is accomplished primarily during the particle's residence in the reef flat environment. Rapid diagenesis of particles is indicated in the intertidal zone. Subsequent changes which take place on the beach and post-island deposition (abrasion and corrosion) are essentially size reducing and void enlarging.

3.6. *Age of shingle clasts*

Components of shingle islands, either corals or tridacnids from both consolidated and unconsolidated shingle deposits, have been radiometrically dated. The oldest dated bioclasts are from the cemented rampart-rocks of the upper platform and these give a mean age, for eight determinations, of 3500 a B.P. (Scoffin & McLean 1978, this volume). Bioclasts from the consolidated lower platforms and unconsolidated shingle rampart and island deposits are significantly younger, with one group (three determinations) averaging 1500 a B.P. and another (five determinations) averaging 750 a B.P., there being no difference between loose and cemented shingle. These data which are presented elsewhere in detail (McLean *et al.* 1978, this volume) suggest two major phases of shingle island development, one pre-dating 3000 a B.P., the second post-dating 1500 a B.P., with the latter being subdivisible into earlier and later minor phases.

While the difference of approximately 2000 years between the major episodes may be an artefact of sampling, the general consistency of ages of equivalent deposits between different reefs is striking and suggests that the development of shingle islands has been episodic with periods of accumulation being separated by phases of stability or erosion. These episodes may reflect variations in relative sea level, storminess or reef productivity, or possibly a combination of all three.

TABLE 3. PARTICLE SIZE AND SORTING OF SAMPLES FROM TURTLE I ISLAND (ϕ units)

sample no.	mean size	sorting	sand (%)	gravel (%)	sand fraction median	gravel fraction median
TON-104	−1.69	1.37	21	79	+0.6	−2.4
TON-105	−3.26	1.36	9	91	+0.5	−3.5
TON-106	−3.33	1.73	10	90	+1.0	−3.7
TON-107	−1.48	1.93	41	59	+1.0	−3.2
TON-110	−0.30	1.57	75	25	+0.3	−2.5
TON-111	−4.16	2.02	13	87	+0.9	−4.7

4. MIXED SAND–SHINGLE ISLANDS

Sand cays and gravel islands when they occur on the same reef are generally geographically discrete sediment bodies. While small quantities of shingle are found on sand cays and sand on shingle islands, proportions of the secondary size components are insignificant and may be ignored. However, on some islands sediments consist of sandy-gravel or gravelly-sand such that the two size components are thoroughly mixed. Islands of the Turtle Group particularly fall into this category, as do the two Pethebridge islands and Sand Island (figure 1). It is notable that in all cases these reefs possess only one island, which in sediment terms can be described as a mixed sand–shingle cay. Reefs occupied by mixed sand–shingle cays are generally small high reefs located close to the mainland and the cay covers a large proportion of the reef top, up to 40 % (table 1). These features distinguish mixed sand–shingle cays from other reef islands in the area. Steers (1937) recognized that islands of the Turtle Group 'are examples of a kind of inter-mediate stage between the simple sand cay and the complex cay' but he did not explicitly relate this to differences in sediment type. While island surface sediments appear to be either sandy or shingly, and recently formed ramparts possess sediment characteristics similar to the shingle islands described above, pits dug in the older parts of the islands show that the two size components are mixed and have been deposited together.

4.1. Mean size and sorting

Sediment samples were sieved and weighed and grain size curves plotted by using mass frequency. Because the sediments comprise mixtures of sand and shingle in variable propor-tions, the range of sample mean sizes has a wide ϕ spread, from 0 to −5 ϕ, and sorting values are high, commonly between 1 and 2 ϕ units (figure 2). In textural terms the sediments can be classified as poorly sorted sandy-shingle. Some poorly sorted shingly-sands are also present. Results from analyses of six soil pit samples from Turtle I Island cover the range of values found on mixed sand–shingle islands (table 3).

Grain size curves are typically bimodal, the strength of the sand and shingle modes being variable. Mean sizes fall between the sand and shingle end-members with some overlap into the latter's range. Each of the two modal fractions is relatively well sorted. Size parameters calculated for the sand fraction and shingle fraction of the mixed sediments are quite similar to those described earlier for sand cay and shingle cay sediments respectively (table 3). Thus, the unique granulometric feature of these deposits is that the two sizes are thoroughly mixed and not geographically separated or sorted as elsewhere.

4.2. *Particle shape and surface features*

Axes measurements of shingle sized clasts from the mixed deposits of Turtle I Island plotted on Folk form triangular graphs show that blade shapes are the dominant group (43%) followed by platy (23%) and elongate (22%) particles. Compact shapes account for 12% of the clasts. Only the last value is similar to those from pure shingle islands. These data show that the shingle component of the mixed deposits is not dominated by elongate stick coral fragments and that there is a more equable distribution of the various shape categories. Part of this is the result of the presence of a large number of *Acropora* joints as well as the presence of corals of different growth forms. Moreover, clasts are more thoroughly edge rounded and polished such that distinct primary skeletal surface markings of corals are only recognizable on about 10% of the particles. While the shape characteristics do reflect different proportions of constituents, particularly growth form constituents, surface features also suggest that the clasts of mixed deposits have been more severely abraded than their shingle island counterparts.

4.3. *Constituent composition and mineralogy*

The sand fraction of the mixed deposits contains a similar suite of skeletal components as the pure sand cays, but the proportions of the various constituents are markedly different. Sands from the Turtle islands are particularly rich in coral fragments, which account for over 50% of constituents, and also molluscan fragments (20–25%). Equivalent values for sand cay sediments are 20–25% and 10–15%. Coralline algae is the third most important component (10%) while Foraminifera and *Halimeda* make up 7% and 5% of the sands respectively, compared with 25–30% and 10% for cay sediments. Thus, the sand fraction of mixed sand–shingle island sediments is clearly distinguished from those of pure sand cays.

Similarly the shingle fraction of mixed deposits can be distinguished from sediments of pure shingle islands. While *Acropora* continues to dominate the coarse fraction, the proportion of encrusting or foliaceous corals such as *Turbinaria*, stalky corals such as *Lobophyllia*, and smaller hemispherical faviid corals appear more abundant in the Turtle Island deposits. Two other features can be noted. First, there is a large number of lithoskels present. These clasts, which were noted only rarely in pure shingle island deposits, are characterized by a hard brown micrite streaky coating on the skeletal surface, the micrite also being present in intraskeletal voids. For the most part the original coral skeletons are taxonomically unrecognizable. The similarity between the adhering micritic cement of lithoskels and the high magnesium calcite cements of rampart-rocks (Scoffin & McLean 1978, this volume) strongly suggests that they have been reworked from these exposed limestones. Secondly, both the range of species and absolute numbers of molluscs appears to be greater. While clams and larger molluscs are present in roughly similar proportions to pure shingle deposits, smaller molluscs, particularly gastropods (e.g. *Nerita*, *Melaraphe*) and the oyster *Crassostrea*, are present in significantly greater numbers. Both whole and broken shell fragments are evident.

Bulk mineralogy of the sand and coral shingle fractions of the mixed deposits fall within the ranges of the sand cay sediments and shingle island sediments respectively. However, the sands possess a greater than average proportion of aragonite, and stick corals a greater than average proportion of high magnesium calcite, the former resulting from the larger number of aragonite building organisms and the latter from the presence of void infills and adhering surface cement.

4.4. *Age of sand–shingle deposits*

The only radiocarbon dates available for materials from mixed sand–shingle islands relate to one island of the Turtle Group, Turtle I. Coral from a loose coral and shell deposit beneath mangrove mud in the centre of the island dated 4910 ± 90 a B.P. (ANU-1479) while a *Tridacna* from rampart-rock of the upper platform gave an age of 4420 ± 90 a B.P. (ANU-1478). These dates suggest that island sediments began accumulating at least 4000 years ago. Bulk determinations of unconsolidated sandy-gravels from soil pits on the upper and lower levels towards the northwestern end of the island range over 3320 ± 80 a B.P. (ANU-1388) for the upper level to 2760 ± 80 a B.P. (ANU-1598) and 2480 ± 70 a B.P. (ANU-1597) for the lower level. The presence of lithoskels from rampart-rocks, intertidal rock-dwelling molluscs and highly abraded clasts in the shingle, plus the high coral and mollusc and low *Halimeda* and Foraminifera content of the sands, suggests that there was continual reworking and local redistribution of reef top materials in the interval 4000–2000 years ago. The mixed nature of the sediments comprising these deposits also supports this view. These data and geomorphological evidence additionally suggest that the island has been basically stable during the last 2000 years, although fresh material in the form of narrow shingle ramparts has been added to the island's periphery in this period. A *Tridacna* from one of these deposits dated 1430 ± 70 a B.P. (ANU-1477).

Regrettably, there are no radiocarbon dates from other mixed sand–shingle islands upon which to base an absolute chronology for comparison with Turtle I. Nevertheless, the relative size (in relation to reef-top area), location and nature of deposits of some other islands in the Turtle Group, notably Turtle II, V and VI, suggests that these at least may have had a similar history of sedimentation. Other islands in this group, Turtle III and IV, together with Sand Island, East Pethebridge and West Pethebridge which occupy less than 10% of the available reef top on elongate reefs, may well have a different history.

5. COMPOSITE ISLANDS

Composite islands are those that contain areas of at least two of the three sediment types described above (table 1). None of the islands possess large areas of mixed deposits, the main components being either sand or shingle. Frequently these sediment types are zonally arranged and although there may be a narrow swale between the two, more usually they abut or overlap one another with little obvious morphological break.

The simplest composite islands consist of a large sand deposit with a narrow band of shingle (frequently cemented) on one side. Examples of this type are the leeward islands on Leggatt, Sinclair–Morris and Houghton reefs, while Coquet has a broader loose shingle deposit backed against the 'sand cay' in addition to a cemented shingle shore. These reefs also have another island (shingle) on the windward side, or, in the case of Leggatt, residuals of cemented shingle, rampart-rock. Nymph and Low Wooded Island are more complex islands. They consist of a single island which encircles a large central pond or mangrove swamp. The islands are highly variable in width and their morphology suggests that there were formerly a number of discontinuous deposits which have since become united. On the leeside both islands possess what Steers called a 'sandy cay-like area' which passes rather indefinitely into mixed sand–shingle or shingle ridges and ramparts which make up the bulk of the islands. Nymph and Low Wooded Island are the largest reef islands in the region. They both cover an area of some 450 000 m² (including enclosed ponds and mangrove swamp), and occupy 50–60 % of the available reef top space.

5.1. *Sediment characteristics*

Composite islands have been distinguished as a separate category in this paper only because the geographical arrangement of sediment types, but not the nature of those sediments, is different from other islands. Analysis of seven sand samples from the main island of Leggatt and the cay-like area of Nymph show they fall within the size-sorting and compositional fields of the other sand cays. The sediments from both sites can be described as Foraminifera and coral-rich moderately well sorted coarse sands. Likewise, the shingle deposits of composite islands are similar to those of pure shingle islands, being composed predominantly of moderately sorted medium pebble sized stick coral fragments.

5.2. *Age of composite island sediments*

It is believed that the majority of the sand and shingle deposits of composite islands accumulated around 3000 a B.P., though there are insufficient radiocarbon dates to substantiate this conclusion. A bulk calcarenite from Houghton Island was dated 2670 ± 70 a B.P. (ANU-1596). This sample included secondary aragonite as a cementing medium, the presence of which post-dates the age of the bioclastic components. Overlying the calcarenite was a coral shingle veneer, one component of which dated 3550 ± 80 a B.P. (ANU-1413). At Nymph and Low Wooded Island tridacnids from rampart-rock of the upper platform aged 3540 ± 80 (ANU-1383) and 3320 ± 70 (ANU-1604) a B.P. respectively. These data suggest an early phase of island building and that island cores were established by about 3000 a B.P. More recent peripheral accumulation is indicated by samples of loose shingle and cemented shingle from Coquet and Nymph which have been dated at 1070 ± 60 (ANU-1411) and 520 ± 70 (ANU-1476) a B.P.

All these ages are similar to those from equivalent deposits on pure sand cays and shingle islands in the region and it is likely that composite islands have similar chronologies. The essential difference is that the windward shingle deposit has extended leeward to encompass, partly or wholly, the sand cay to form what here is called a composite island. In other words spatially disjunct islands on the same reef have joined to form a single island processing two or more contrasting sediment populations.

6. Conclusions

1. Three distinct sediment types exists on the islands of the northern Great Barrier Reef. These sediment types are discrete populations, distinctive in textural properties, composition and geographic occurrence, and allow single reef islands in the region to be classified as either sand or shingle or mixed sand–shingle islands. A fourth type, composite islands, is included to cater for those instances where two of the foregoing sediment types have been morphologically united. Size and geometry of the reef, relative exposure to prevailing and catastrophic waves, as well as reef-top morphology govern in large measure the nature and distribution of reef organisms and local sediment types.

8-2

2. Sand cay sediments can be described as moderately well sorted coarse sands composed predominantly of worn foraminiferal, coral and molluscan fragments. Subtle differences in texture and constituent composition exist between cays and between beach, basic cay and soil horizons on the same cay. Cay sands are derived from reef flat bioclasts. Sediments of the two environments are discriminated on textural grounds, particularly sorting.

3. Shingle island sediments can be described as moderately sorted medium sized pebbles composed predominantly of elongate shaped corals. Fragments of branching corals, particularly *Acropora*, account for the major constituents though deposits include the range of coral growth forms and species present on windward reef flats and edges. Coral thickets and colonies are broken down during storms and built into shingle ramparts, and ultimately islands. Boring organisms increase skeletal porosity, and voids are partly or completely filled with sediment or cement while particles are on the reef flat. Wave action on beaches abrades, cleans and smooths bioclasts. Beach shingle surface textures are unchanged in subsurface island deposits, but exposed bioclasts develop a micro-phytokarstic architecture.

4. Mixed sand–shingle island sediments can be described as poorly sorted sandy-gravels or gravelly-sands. The two modal populations, sand and shingle, are moderately well sorted but vary considerably in relative abundance. Both fractions possess similar textural characteristics as the pure sand and shingle deposits, though there are some distinguishing features. Mixed sands have a significantly higher percentage of coral and molluscan constituents and lower percentage of Foraminifera and *Halimeda* than pure sand deposits. Likewise, the shingle components differ in that the mixed shingle has a more equable distribution of particle shapes, contains a greater percentage of molluscan fragments and lithoskels, and all clasts are well worn and some polished. The two size populations are thoroughly mixed indicating contemporaneous deposition, and considerable reworking of original materials.

5. Composite islands do not possess a unique sediment type. Instead they are made up of two or more of the foregoing types, commonly pure sand and pure shingle.

6. An essentially zonal pattern of island sediment types exists across the shelf from the mainland to outer ribbon reefs. Sand cays (types I and II) occur in the outer zone where they are the only islands on the reefs. In the central zone, both sand cays and shingle islands are present on the same reef. Composite islands also occur in this region. Nearer the mainland, sole shingle islands and mixed sand–shingle islands are present. This zonal pattern mainly reflects variations in energy conditions in the manner outlined by Stoddart (1965).

7. At the local reef level, sand deposits occur on the leeward side and shingle deposits on the windward side of the reef top. Occasionally the latter extend leeward to partly or completely surround the sand cay. Mixed sand–shingle deposits are more centrally located on a reef. An idealized distribution of sediment types on reefs of the area has been described and explained by Steers (1929, 1930).

8. Radiocarbon dates indicate that some reef islands in the Northern Province of the Great Barrier Reef were formed at least 4000 years ago. However, the major period of both sand cay and shingle island building took place around 3000 a B.P. The basic outlines of islands established at this time have since become modified to a greater or lesser degree through subsequent erosion and redistribution of island materials and additions of fresh reef flat and reef edge detritus. Considerable enlargement, particularly of shingle islands, has taken place in the last 1500 years and some new islands may have been created. Nevertheless, the basic pattern of islands, both at regional and local scales, pre-dates this more recent period. As a consequence of

reef tops becoming more packed with emergent sedimentary deposits, the available space for production of primary sediment has been reduced such that some islands, for example Turtle I, V, VI, Nymph and Low Wooded Island, may well have now reached their maximum size.

We wish to acknowledge the assistance of Expedition colleagues, particularly Mr P. G. Flood, Mr H. Polach, Dr T. P. Scoffin and Dr J. E. Veron who provided results of some of their work for this paper. We also thank Mr K. Fitchett, Mr M. Campion and Mr J. Caldwell of the Australian National University for carrying out much of the laboratory work associated with the sediment samples.

REFERENCES (McLean & Stoddart)

Fairbridge, R. W. & Teichert, C. 1948 *Geogrl J.* **111**, 67–88.

Flood, P. G. & Scoffin, T. P. 1978 *Phil. Trans. R. Soc. Lond.* A **291**, 55–71 (this volume).

Folk, R. L. 1962 *N.Y. Acad. Sci. Trans.* **25**, 222–244.

Maxwell, W. G. H. 1968 *Atlas of the Great Barrier Reef.* Amsterdam: Elsevier.

McLean, R. F., Stoddart, D. R., Hopley, D. & Polach, H. 1978 *Phil. Trans. R. Soc. Lond.* A **291**, 167–186 (this volume).

Scoffin, T. P. & McLean, R. F. 1978 *Phil. Trans. R. Soc. Lond.* A **291**, 119–138 (this volume).

Spender, M. A. 1930 *Geogrl J.* **76**, 194–214 and 273–297.

Steers, J. A. 1929 *Geogrl J.* **74**, 232–257 and 341–370.

Steers, J. A. 1930 *Scient. Rep. Gt Barrier Reef Exped. 1928–29* **3**, 1–15.

Steers, J. A. 1937 *Geogrl J.* **89**, 1–28 and 119–146.

Stoddart, D. R. 1965 *Trans. Pap. Inst. Br. Geographers* **36**, 131–147.

Stoddart, D. R., McLean, R. F. & Hopley, D. 1978 *Phil. Trans. R. Soc. Lond.* B **284**, 39–61 (part B of this Discussion).

Phil. Trans. R. Soc. Lond. A. **291**, 119–138 (1978)   [119]

Printed in Great Britain

Exposed limestones of the Northern Province of the Great Barrier Reef

By T. P. Scoffin† and R. F. McLean‡

†*Grant Institute of Geology, University of Edinburgh, West Mains Road, Edinburgh, EH9 3JW, U.K.*

‡*Department of Biogeography and Geomorphology, Australian National University, Canberra, Australia* 2600

[Plates 1–4]

The exposed reef limestones occur principally on the inner-shelf reefs and can be separated into two groups – organically cemented (reef-rock) and inorganically cemented (beach-rock, rampart-rock, boulder-rock and phosphate-rock). No examples were found of exposed subtidal reef framework; the reef-rock exposed is entirely of intertidal origin resulting from incipient encrustation by intertidal corals and coralline algae. Most of the beach-rock, rampart-rock and boulder-rock exposures are intertidal and many show vadose cement fabrics. The cements, chiefly aragonite needles in beach-rock and cryptocrystalline high Mg calcite in rampart and boulder-rocks, are thought to be derived from seawater, though the environments of precipitation on windward sides of reefs where rampart-rocks form are quite different from those on the leeward sides where beach-rocks form. Phosphate-rock develops supratidally on the surface of some sand cays. Solutions derived from guano precipitate thin layers of phosphatic cement which bring about the centripetal replacement of carbonate grains.

1. Introduction

The reefs examined in the Northern Province of the Great Barrier Reef lie on the Queensland continental shelf between Cairns and Cape Melville (about 300 km) to a distance from shore of about 40 km. The consolidated carbonate deposits that were found exposed above low water mark on these reefs were subdivided into the following types: beach-rock, rampart-rock, boulder-rock, phosphate-rock, and reef-rock.

The first three listed are essentially intertidal and are quantitatively the most important. Phosphate-rock was found on a number of sand cays. Reef-rock, although common on the reefs, is exceptional in that it is the only consolidated deposit with a predominantly organic cement. Table 1 lists the reefs' associated rock types and summarizes other relevant data concerning the reefs visited. The criteria used in subdividing these five limestone types include field occurrence, grain composition and matrix composition.

1.1. *Reef-top sediments*

It is necessary to consider briefly the sedimentary deposits above low water spring tide level on the reefs. The composition and form of the deposits relate very closely to the ecological distribution of the dominant carbonate producers and to the depositional processes operating.

1.1.1. *Windward side*

In the shallow-water zone of the reef front on the windward (southeast) side, branching and platy corals of *Acropora* are particularly abundant (figure 1)§. These are readily broken by

§ Figures 1–8 appear on plate 1, figures 9–16 on plate 2, figures 18–25 on plate 3, and figures 26–31 on plate 4.

TABLE 1

				rampart-rock									
reef†	sand cay	beach-rock	ramparts	low bassett edge	high platform	low platform	boulder tract	boulder-rock	mangroves	freshwater vegetation	distance from shore/km	Maxwell's (1968) zone‡	Latitude S
Pipon	×	×	×	×	—	×	×	—	×	×	6	4	14° 07′
Waterwitch	×	×	—	—	—	—	—	—	—	—	35	5	14° 11′
Stapleton	×	×	—	—	—	—	—	—	—	×	26	5	14° 18′
Combe	×	×	—	—	—	—	—	—	—	×	24	5	14° 24′
Ingram–Beanley	×	×	×	×	—	×	×	—	×	×	22	5	14° 25′
Bewick	×	×	×	×	×	×	×	—	×	×	18	4	14° 27′
Watson	×	×	×	×	×	×	×	×	×	×	18	4	14° 28′
Howick	×	×	×	×	—	×	×	×	×	×	15	4	14° 30′
Newton	×	×	×	×	—	×	×	×	×	×	15	3	14° 31′
Sand	×	×	—	×	—	—	×	—	—	—	9	2	14° 31′
Houghton	×	×	×	×	×	×	×	—	×	×	15	4	14° 32′
Coquet	×	×	×	×	×	×	×	—	×	×	15	3	14° 33′
Leggatt	×	×	×	×	×	×	—	—	×	×	7	2	14° 33′
Sinclair–Morris	×	×	×	×	×	×	—	—	×	×	9	2	14° 33′
Hampton	×	—	×	×	—	—	—	—	×	×	7	2	14° 36′
Nymph	×	×	×	×	×	×	×	—	×	×	22	4	14° 39′
Eagle	×	×	—	—	—	—	—	—	—	×	28	4	14° 42′
Turtle V	×	×	×	×	—	×	×	—	×	×	17	4	14° 42′
Turtle VI	×	×	×	×	—	—	—	—	×	×	17	4	14° 42′
Turtle IV	×	×	×	×	—	—	—	—	×	×	17	4	14° 43′
Turtle III	×	×	×	×	—	×	×	—	×	×	15	4	14° 43′
Turtle II	×	—	×	×	×	×	—	—	×	×	15	4	14° 44′
Turtle I	×	×	×	×	×	×	×	—	×	×	15	4	14° 44′
E. Pethebridge	×	—	×	×	—	×	—	—	×	×	9	3	14° 44′
W. Pethebridge	×	×	×	×	—	×	—	—	×	×	7	3	14° 44′
Two Isles	×	×	×	×	—	×	×	—	×	×	17	4	15° 01′
Low Wooded	×	—	×	×	×	×	×	—	×	×	15	4	15° 06′
Three Isles	×	×	×	×	×	×	×	—	×	×	18	4	15° 07′
East Hope	×	×	×	×	—	—	—	—	×	×	13	3	15° 43′
West Hope	×	—	×	×	—	×	—	—	×	×	11	3	15° 44′
Pickersgill	×	—	—	—	—	—	—	—	—	—	20	4	15° 53′
Mackay	×	—	—	—	—	—	—	—	—	—	22	5	16° 03′
Undine	×	—	—	—	—	—	—	—	—	—	20	4	16° 08′
Low Isles	×	×	×	×	—	—	×	—	×	×	18	3	16° 23′
Michaelmas	×	×	—	—	—	—	—	—	—	×	39	5	16° 37′
Arlington	×	—	—	—	—	—	—	—	—	—	37	5	16° 40′
Upolu	×	×	—	—	—	—	—	—	—	×	28	4	16° 40′
Green	×	×	—	—	—	—	×	—	—	×	28	4	16° 46′

† All of these reefs were visually examined during the expedition. Many other reefs occur between latitudes 14° 07′ S and 16° 46′ S but other than the high (continental) islands, nearly all of those not included in this list have no surface exposure at low spring tide.

‡ Maxwell's (1968) zones are: 1, high terrigenous; 2, terrigenous; 3, transitional; 4, impure carbonate; 5, high carbonate.

strong wave action into platy and stick branches (figure 2) which, on near-mainland reefs, pile up to form ridges or ramparts (figure 3). These ramparts have a characteristic plan and profile form. In plan, their outer margins are convex and parallel to the reef edge. Their inner margins are cuspate and in places develop tongues (up to 100 m long) over the reef flat (figure 3). In profile these ramparts appear as large, asymmetric ripples with a steep foreset (60°) and a shallowly inclined (5°) stoss slope. The amplitude is normally 1–2 m and wavelength about 30 m. The existence of several discrete loose ramparts on one reef supports the theory that they are not being continually built, that is with new material added as they migrate to lee, but rather that each forms, in the main, during one depositional event and may later be moved, added to or eroded. Comparisons of rampart positions on reefs over several years suggest that the movement is not regular (Stoddart, McLean, Scoffin & Gibbs 1978, part B of this Discussion; Fairbridge & Teichert 1948). Coarse debris of coral plates, fungiid corals and *Tridacna* valves accumulate at the foot of the foreset slopes; the crests and stoss slopes consist predominantly of stick-like branches of corals. On lithification, these deposits become the *rampart-rocks* (figure 4). The overall distribution and plan morphology of the rampart-rocks are similar to that of the ramparts, though in detail, because of irregular cementation and also subsequent erosion or addition of deposits, the profiles are dissimilar. The highest shingle rampart had an elevation of 3.1 m above low water datum at Cairns (mean high water springs is 2.3 m) but loose ramparts rarely accumulate above high water mark. Their structures are normally sufficiently impervious to pond seawater into moats on the reef flat during low tide. Mangroves commonly colonize loose and cemented ramparts.

1.1.2. *Leeward side and flanks*

Waves are refracted over and around each reef such that at the confluence of the opposing sets, sediment accumulates as a leeward sand deposit which develops a cay with beaches and spits (figure 5). Fine sand is transported by wind from the beach and together with storm-wave swash sediment the cay can build higher than high water mark and support freshwater vegetation. Intertidal lithification of the beach material results in the formation of *beach-rock* (figure 6). Beaches and associated beach-rock are also occasionally found on the windward sides of reefs, usually in small embayments in the cliffs of rampart-rock (figure 7). Lithification of cay sediments above high water mark can result from phosphatic mineralization forming *phosphate-rock*.

In the shallow water on the flanking margins of the reef, massive corals, principally *Porites*, take over in abundance from the branching *Acropora* corals. These spherical or dome-shaped corals grow up to several metres in diameter. Movement of these corals, presumably by large waves during storms, produces either scattered isolated reef-blocks (figure 8) or a linear pile of contiguous boulders (figure 9) at the leeward flanks of the reef edge: the boulder tract. A large majority of the boulders (90 %) are single colonies of massive corals, suggesting that at the time of their erosion, these corals were not part of a solid reef framework of interconnected corals. The lithification of boulders results in the formation of *boulder-rock* (figure 10).

1.1.3. *Reef flats*

The reef flats that are exposed at low water are predominantly sandy with coral microatolls and *Tridacna* clams being the major macroorganisms that grow where sea water is ponded. In places of permanent water, such as well established ponds or even the outer edges of ramparts through which water is constantly seeping during low tide, corals (dominantly *Porites* and

Montipora) grow in sufficient abundance to build a thin organic framework. Encrustation by coralline algal growth on exposed shingle at the windward margin of the reef produces another form of organic cementation. These two are the major examples of organic cementation producing a consolidated *in situ* skeletal rock above low water mark – here termed *reef-rock*.

Although gradations do exist, in the main, each limestone type has its own characteristic location, gross morphology, constituent grain types and matrix. These characteristics will be discussed for each limestone type.

2. Beach-rock

2.1. *Location*

Beach-rock was exposed on sand cays on the leeward side of most of the reefs in the Northern Province (table 1). Steers (1929, 1937) noted that beach-rock was found only on vegetated cays. Although this is the general rule we also found extensive outcrops bordering the unvegetated cay on Waterwitch reef. However, it was absent from other unvegetated cays, such as those on Pickersgill, Undine and Mackay reefs, that are just awash at high tide.

Beach-rock was rarely seen completely encompassing a sand cay, though obviously the present distribution of loose sand can obscure its true extent. Nevertheless, on some of the simpler cays like those on Eagle and Combe reefs, as well as those of 'low wooded islands' like Low, Two and Three Isles, the cays were bordered on two or three sides by bands of beach-rock. Normally the broadest expanse was found on the windward side of cays, suggesting a leeward migration of loose cay sands.

2.2. *Gross morphology*

Beach-rock takes on the form and disposition of the parent beach. In the field two types were recognized, here called inclined and horizontal beach-rock. The first and most common type occurs as linear or arcuate strips of thinly bedded units dipping seawards at 10–15°, and is similar to that described from beaches in many tropical areas. In places, discordant overlapping sequences occur which reflect local changes in beach position, for example on Newton cay five differently dipping sets of beach-rock are found superimposed. Inclined beach-rock was limited to the contemporary intertidal zone, commonly between heights of 0.8–2.3 m above datum, the latter being the level of mean high water spring tide. Frequently loose sand obscures the lower or upper portion of an outcrop. The friable nature of some beach-rocks suggest that they are forming at present. The second type, horizontal beach-rock, is characterized by an upper surface that is nearly horizontal. The outer edge is commonly eroded into a steep scarp, frequently notched or undercut at the base, which rises directly from the reef flat. The upper surface is typically at an altitude between 2.5 and 3.0 m above datum and is exposed for 3–5 m before passing inland beneath a veneer of cay sand. A bulk beach-rock sample from Houghton Island was dated at 2670 ± 70 a b.p. (ANU-1596).

A zonal arrangement of algae and animals was frequently observed across exposed beach-rock. Bioerosion was most active near to low water mark and locally beach-rock showed extensive encrustation by the oyster *Crassostrea* at the level of high water neaps. In some instances, surfaces were clean and showed signs of active wave abrasion. Elsewhere jagged irregular spray-pitted surfaces had developed around high water mark. However, on horizontal beach-rock this micro-morphology was replaced by smoother shallow pools and pits on the upper surface which frequently had a discontinuous cover of the succulent plant *Sesuvium*.

TABLE 2. GRAIN AND CEMENT COMPOSITION OF BEACH-ROCKS

sample no.	island	coral[†] (%)	mollusc (%)	coralline algae (%)	Halimeda (%)	echinoderm (%)	benthonic Foraminifera (%)	unknown (%)	mean grain size/mm	insoluble[‡] residue (%)	cements§
30	Two	18	14	5	6	1	52	4	1.0	3.3	aragonite needles + calcite mud
43	Three	28	10	7	26	0	29	8	1.4	2.9	aragonite needles
19	Low	35	28	22	14	0	7	1	1.5	3.2	aragonite needles
3	Eagle	19	9	3	36	1	32	0	0.9	2.5	aragonite needles
10	Turtle I	48	23	10	7	0	13	0	1.1	2.7	aragonite needles
4	Ingram	28	16	23	21	1	9	3	1.3	0.8	aragonite needles
9	Turtle I	58	22	10	3	1	2	3	2.0	6.3	calcite mud + aragonite needles
7	Newton	14	16	4	34	1	26	4	1.0	2.5	aragonite needles
5	Bewick	24	13	8	19	0	32	3	1.5	3.1	aragonite needles
2	Stapleton	27	18	9	12	2	23	9	1.4	3.0	aragonite needles
35	Howick	17	10	22	32	2	10	5	0.7	3.6	aragonite needles
1	Waterwitch	20	40	7	24	0	3	6	1.5	2.9	aragonite needles
54	Nymph	23	13	15	9	0	29	11	0.8	4.4	calcite mud

† Percentage of total grains present in thin section: average number of grains of each thin section = 212.
‡ Insoluble residue as mass percentage of bulk sample after treatment in 10% HCl.
§ Cement mineralogy determined by staining with Feigl's solution.

2.3. *Grain composition*

Irrespective of morphological type, the constituent grains of beach-rock are principally sand-sized and show a high degree of rounding and polishing. The grain composition of several beach-rocks was determined from grain-count analysis of thin sections and the results shown in table 2. The skeletal remains of corals, molluscs, benthonic Foraminifera and algae are the dominant grains. Two of the major benthonic foraminiferal components (together they constitute 17 % of the total grains) are *Baculogypsina* and *Calcarina* which live in abundance attached to short tufts of *Laurencia* algae on the windward lip of reefs. These foraminiferans are detached by wave action and then swept across and round the flanks of the reefs by long-shore drift to accumulate finally with other fine reef debris on the leeward sand cay.

Two types of reworked limestone material were found in beach-rocks: one was where pieces of earlier beach-rock had been recemented with no evidence of much movement of the cobble-sized fragments; the other was on those reefs where the windward and leeward intertidal deposits and rocks were adjacent (for example Turtle I). Here the beach-rock contained polished grains consisting of pieces of stick coral with a thin layer of brown micrite. As these stick corals and the brown micrite matrix are characteristic of the windward intertidal deposits (see later) it is assumed that these coated coral fragments have been eroded from rampart-rocks.

Little evidence of post-depositional solution was found in the beach-rock constituents and most grains appeared in a fresh state of preservation.

2.4. *Cements*

The most common cement fabric observed in the leeward cay beach-rocks was that of a thin fringe of acicular crystals around grains (figure 11). The individual crystals have a length: breadth ratio of about 15:1 (figure 12). These needles were identified as aragonite by staining (with Feigl's solution) and the needles grew in the interparticle pore spaces with the same fabric from substrates of all compositions – aragonite, calcite and non-carbonate rock fragments. Some beach-rock samples showed an even isopachous fringe of needles surrounding grains, but the majority had cement concentrated at grain contacts in the 'meniscus' position (Dunham 1971). As a result, the small interparticle pore spaces were commonly totally occluded by cement whereas the large pores were mostly vacant. Needles of a similar fabric to those in interparticle pores also occurred lining intraparticle pores, though it was noticed that the chambers of aragonite skeletons (*Halimeda*, gastropods, corals) commonly showed a better development of cement than the chambers of calcite skeletons (Foraminifera). It is quite common to see two or three generations of acicular cement in one rock. The fibrous layers are then normally separated by a very thin dark line.

Several examples of beach-rock have a micritic cement, and some such as the horizontal beach-rock at Houghton cay, contain both aragonitic and micritic cements. The latter takes various forms. It occurs normally at grain contacts and there may show flat-floored internal sediment characteristics indicating its sedimented origin, though it also occurs as an even layer around grains and in such cases it is commonly seen, under high power on the microscope, to be fibrous in detail. One example of micritic beach-rock cement revealed a vaguely pelleted botryoidal texture.

The beach-rocks with micritic cement were found only on those sand cays that were connected with the windward shingle deposits, and normally formed close to mangroves. More details of this micrite cement are given later for it is this same type of cement fabric which characterizes the rampart-rocks. Those beach-rocks with a micritic cement normally have a higher percentage of insoluble residue than the beach-rocks with acicular cements (table 2) and also they commonly show signs of a degree of post-depositional solution of grains, which was not apparent in the purer beach-rocks.

3. RAMPART-ROCK

3.1. *Location*

Coarse coral fragments accumulate as shingle ramparts or ridges only on those reefs close to the mainland and it is only these reefs that have rampart-rock (table 1). Steers (1929, 1937) and Spender (1930) called these exposed limestones coral or shingle conglomerates or conglomerate platforms, but the term 'rampart-rock' is favoured because, as Steers (1937) noted, they derive originally from ramparts or shingle ridges. Rampart-rocks occur on the windward side of the reefs usually about 50–100 m in from the perimeter at low water.

3.2 *Gross morphology*

Rampart-rock retains the main elements of the form of ramparts. In plan they commonly show a broad crescent shaped seaward margin and cuspate inner margin. In section the general shape and internal structure of a rampart is apparent but commonly the detail of the profile of an asymmetrical ripple is lost; for example, normally there is no obvious crest as in most ramparts.

Rampart-rock occurs either with a planar, essentially horizontal surface (figure 13), 'the upper and lower platforms, pavements or promenades' of Steers (1929, 1937) and Spender (1930), or with a jagged saw-tooth profile of inclined beds 20 cm thick projecting a fairly uniform distance above the reef flat and dipping steeply (20–70°) to lee (figure 14). The latter are the 'bassett edges'† of Steers (1929). Limestones with platform and bassett edge surface morphologies have essentially similar composition and both occur on the windward perimeter of reefs.

3.2.1. *Platforms*

Each continuous platform of rampart-rock has a fairly constant height. On several reefs (e.g. Three Isles, Nymph, Low Wooded Island) two platforms occur: an upper and a lower. The upper is normally found to the leeward of the lower and the two may be separated by a narrow shallow moat. The heights of the platform surfaces of several outcrops of rampart-rock on the reefs of the northern Great Barrier Reef were surveyed. Results revealed considerable variation in the altitudes of the upper and lower platforms from reef to reef, but where continuous traverses were run across both platforms on the same reef a difference in level of 1.0–1.2 m was found. In these instances the maximum elevation of the upper platform ranged from 3.10 to 3.53 m and lower platform 1.96–2.49 m above datum. (See McLean, Stoddart, Hopley & Polach 1978, this volume.)

3.2.2. *Bassett edges*

The inclined bedding of bassett edges represents cemented foresets of ramparts. The irregular projections result from differential cementation and weathering of the bedding with the layers more resistant to erosion, having more cement and also generally finer constituents than the less

† Basset or bassett: 'The edge of a stratum showing at the surface of the ground; an outcrop' [O.E.D.].

resistant layers. The bedding occurs as steeply dipping foresets (40–70°) on the tongue shapes, like anticlines plunging to leeward (figure 4) but as shallowly dipping (20–40°) arcuate bands between. Presumably at least the inner buried portions of ramparts have to remain stationary for some time to allow lithification. It is perhaps for this reason that those parts of the shingle tongues more removed from wave action at the central parts of reef flats are the better cemented and preserved.

3.2.3. *Surfaces of rampart-rock platforms*

Although rampart-rock platforms appear at first sight to be remarkably level, height measurements indicate that they do fluctuate in elevation along their length, typically by about 0.4 m. The leeward margins of platforms are normally obscured by mangroves or superficial deposits and this masks their rampart profile. It should be pointed out that along their strike, loose ramparts too have a fairly constant elevation. The flat surface of platforms could be explained by either erosion down to a level or deposition and lithification up to a level.

If marine truncation, such as that at the reef flat rim, is invoked to account for the flat surface of platforms, one would expect a bevelling down of the upper parts of those coral boulders that project well above the general level of the platform in which they are seated, but this is not seen (figure 15). Also, it was noted that where a lower platform abuts an upper platform the two rocks are of different compositions, the contained coral debris are of different

---

DESCRIPTION OF PLATE 1

FIGURE 1. Windward coral assemblage of branching and platy forms of *Acropora*. 5 m depth. Lizard Group.

FIGURE 2. Windward beach of broken *Acropora* corals. Hammer 80 cm long. Turtle III Reef.

FIGURE 3. Loose shingle rampart on the windward side of Three Isles Reef. Rampart amplitude 1 m.

FIGURE 4. Arcuate outcrop of cemented foresets of ramparts giving bassett edge morphology. Beds 20 cm thick. Low Isles Reef.

FIGURE 5. Oblique aerial photograph of Sinclair–Morris Reef, showing leeward sand cay with spit. The sand cay has beach-rock around part of its rim and a cover of freshwater vegetation. Mangroves occur on the windward side of the reef. Altitude 200 m.

FIGURE 6. Intertidal beach-rock exposed on the leeward sand cay of Two Isles Reef.

FIGURE 7. Cemented beach in a small gap in the rampart-rock on the windward side of Nymph Reef.

FIGURE 8. Scattered coral boulders partly buried by sand on the leeward margin of Mid Reef.

DESCRIPTION OF PLATE 2

FIGURE 9. Boulder tract exposed at low water on the leeward flank of Low Wooded Island. Boulders 30–100 cm in diameter.

FIGURE 10. Boulders cemented in a sandy matrix at the leeward margin of the sand cay on Howick Reef. Hammer 30 cm long.

FIGURE 11. Photomicrograph of rounded coral, algal and foraminiferal grains cemented by a fringe of aragonite needles. Plane polarized light. Beach-rock, Bewick Reef. Scale bar = 0.5 mm.

FIGURE 12. Scanning electron micrograph of a broken surface of beach-rock showing needles of aragonite cement on a sand grain. Bewick Reef. Scale bar = 20 μm.

FIGURE 13. Horizontal surface of rampart-rock, the lower platform, on the windward margin of Ingram–Beanley Reef.

FIGURE 14. Bassett edge surface morphology of rampart-rock. Beds 20 cm thick. Watson Reef.

FIGURE 15. Coral boulder cemented in rampart-rock projecting above platform surface. Ingram–Beanley Reef.

FIGURE 16. Rampart-rock coated on its leeward side by a thin friable veneer of shingle. Ingram–Beanley Reef.

FIGURES 1–8. For description see opposite.

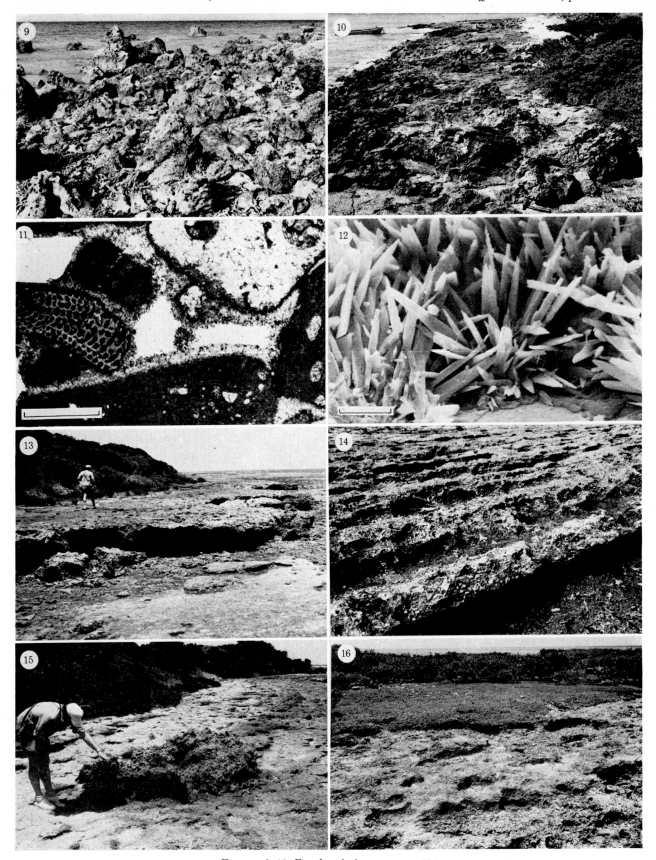

FIGURES 9–16. For description see page 126.

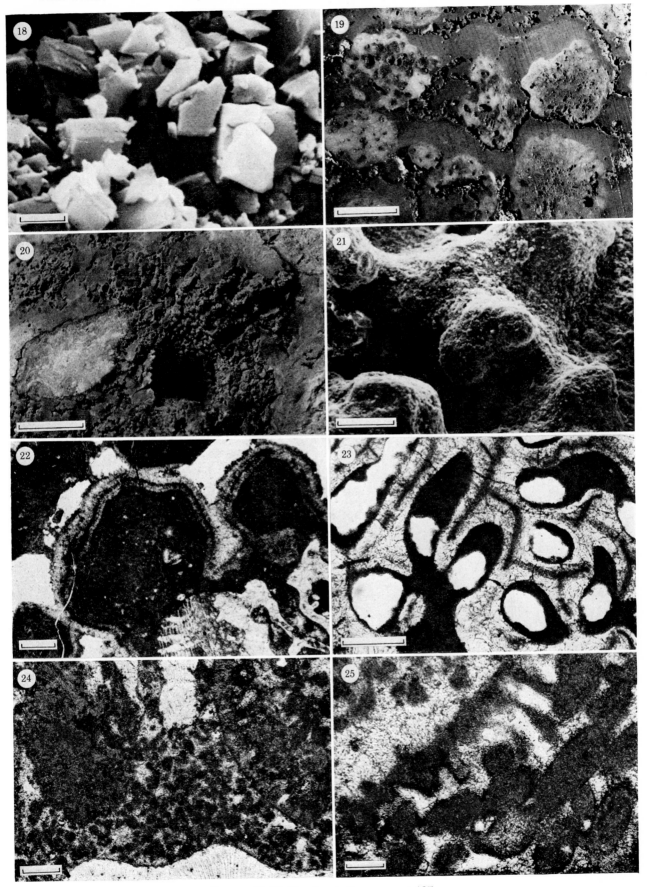

Figures 18–25. For description see page 127.

FIGURES 26–31. For description see opposite.

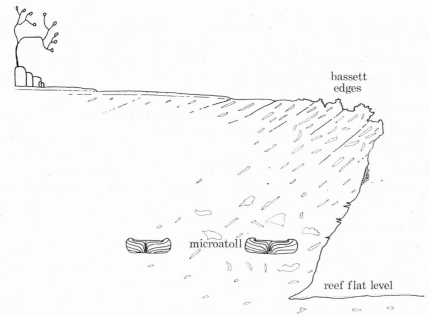

FIGURE 17. Sketch of a cross section of rampart-rock showing the lower facies with *in situ* microatolls and coarse coral debris and an upper facies of coral branches. The surface at the seaward edge, to the right, shows a bassett edge morphology with phytokarst, and to the leeward a veneer of shingle with mangroves.

DESCRIPTION OF PLATE 3

FIGURE 18. Scanning electron micrograph of the surface of an untreated sample of rampart-rock micrite matrix. Turtle VI Reef. Scale bar = 2 μm.

FIGURE 19. Cut cross section of rampart-rock showing micrite matrix draped over coral branches. Watson Reef. Scale bar = 1 cm.

FIGURE 20. Broken surface of rampart rock revealing in cross section the botryoidal texture of the micrite matrix in partly filled cavities between coral fragments. Nymph Reef. Scale bar = 1 cm.

FIGURE 21. Scanning electron micrograph of broken surface of micrite matrix in partly filled cavities in rampart-rock. Houghton Reef. Scale bar = 0.5 mm.

FIGURE 22. Thin section of fibrous cement coating the micrite matrix of rampart-rock. Howick Reef. Plane polarized light. Scale bar = 0.3 mm.

FIGURE 23. Thin section of micrite lining the walls of coral chambers in rampart-rock. Bewick Reef. Plane polarized light. Scale bar = 0.2 mm.

FIGURE 24. Thin section of pelleted matrix of rampart-rock. The pellets are in a calcite cement and fill an interstice between two altered *Halimeda* grains and, at the base, a mollusc fragment. Houghton Reef. Plane polarized light. Scale bar = 0.1 mm.

FIGURE 25. Thin section of partially dissolved *Halimeda* fragment which contains micrite-filled utricles. The top left corner of the photograph is pellet- and calcite cement-filled interstice. Houghton Reef. Plane polarized light. Scale bar = 0.1 mm.

DESCRIPTION OF PLATE 4

FIGURE 26. Sheet of cobble-sized equant fragments of corals on the flanks of Low Isles Reef.

FIGURE 27. Low coral conglomerate. Windward side of Low Isles Reef.

FIGURE 28. Framework of branching corals encrusted by coralline algae surrounding a giant clam (50 cm diam.) growing in ponded water on Turtle III Reef.

FIGURE 29. Cross section of coral shingle with a very thin veneer on the surface and around some grains of crustose coralline algae. Pipon Reef. Scale bar = 1 cm.

FIGURE 30. Thin section of thinly laminated phosphatic cement on algal and foraminiferal grains. Green Island. Plane polarized light. Scale bar = 0.1 mm.

FIGURE 31. Loosely consolidated shingle platform with scattered vegetation. Nymph Reef. Hammer 30 cm long.

ages and excavation shows that the deposits of the lower actually bank against a buried cliff of the upper platform deposits. Therefore, if marine planation did produce the flat surface of the lower platform, it is a remarkable coincidence that this cutting back ended exactly at the junction of the two rampart-rock deposits. These observations argue against a wave-cut terrace origin.

So an origin by deposition and/or lithification up to a level remains. An upper limit to marine lithification for one still-stand of the sea would be expected to give a fairly uniform upper level of rock formation but is unlikely to account for the smoothness or flatness of local surfaces; this feature is more probably the result of deposition up to a level.

Commonly there is a noticeable change in the character of the surface of platform rocks from their seaward edge to the lee. At their seaward edge the surface is hard and jagged with a bassett edge morphology but this is normally lost to the lee where the surface is smooth and consists of less tightly cemented shingle (figure 16). Cross sections of some platform rocks show an indistinct bedding with a steep leeward dip (the subsurface extension of bassett edges) near the seaward edge, whereas to the leeward commonly the upper few centimetres show horizontal bedding or else layers with a shallow seaward dip similar to that characteristic of beach-rock (figure 17).

As it is not uncommon to find thin deposits of loose sand or shingle capping, sometimes being bound by vegetation to the surfaces of rampart-rocks, then it is reasonable to assume that where cementation conditions are favourable a cemented veneer on top of exposed limestones will result. At the southeast end of the rampart-rock at Watson Island, cemented coral shingle could be seen to be partly filling the potholed relief of an older rampart-rock. The distinction between veneer and earlier limestone is more obvious on some reefs than others. Where the veneer is similar in composition to the rock it blankets, and where no obvious bedding features are present in either deposit and no corrosion surfaces occur, then field evidence alone is inadequate to distinguish separate increments of sediment. It is concluded that generally a bassett edge relief characterizes the irregularly eroded surface of cemented ramparts and that this morphology may be masked by the subsequent lithification of thin blanketing deposits of sand or shingle to give the even surface of a platform (figure 16). The location, internal structure and surface detail of these flat topped deposits suggest that they accumulated on top of cemented ramparts by a ponding to the leeward. The stabilization of this veneer may have been greatly assisted by the vegetation common in this position such as mangroves, *Sesuvium*, grasses and filamentous algae.

3.3. *Petrography of rampart-rocks*

3.3.1. *Grain composition*

The constituent grains of most rampart-rocks are fundamentally the same as those of the present loose ramparts and unconsolidated shingle island deposits, that is over 90 % of the grains are broken clasts of branching corals, notably *Acropora*. Typically the clasts are about 10 cm long and 1–2 cm in diameter. Coral fragments of other growth forms, as well as molluscs such as tridacnids, are also present, but in subordinate numbers. In thin section some of the coral fragments show patchy encrustation by coralline algae, foraminiferans and bryozoans and many possess intragranular sediment or cement as void fill, the presence of which, based on analysis of unconsolidated shingle deposits from the same reefs, may pre-date intergranular lithification. (See McLean & Stoddart 1978, this volume.)

When examined closely in cliff section, rampart-rocks of the upper platform show two, and locally three, vertical facies. The lowest part contains *in situ* microatolls, coarse coral debris and sand sized particles with a whitish chalky cement that grades into an upper zone of smaller, predominantly stick coral fragments set in a splintery hard, brown micritic cement (figure 17). This succession represents one complete cycle of advance of a rampart across a reef flat, or moat, that contained microatoll corals. The sediment at the foot of the foresets of loose ramparts is normally coarser than that towards the crest producing the differentiation into lower and upper facies. This stratigraphy is also observed on shingle island beaches and has been described by McLean & Stoddart (1978, this volume). A thin, commonly friable, sandy or stick coral zone is often found at the surface of platforms to leeward of the exposed cliff. This represents a later addition to the rampart deposits. Exposed in some cliff sections are two distinct layers of the coral stick deposits and at low tide seaward-draining water seeps out of the cliff along the junction of the two layers. In such a case the lower layer has a less permeable matrix than the upper and commonly a brown film marks the boundary of a corrosion surface on the lower unit.

While the matrix between the stick corals is normally very fine grained, sand-sized particles are not uncommon. Locally cemented cross-bedded sands occur within sequences of rampart-rock, presumably as a result of beach-rock formation in gaps in the cliffs of rampart-rocks.

TABLE 3. ANALYSES OF THE MICRITE MATRIX OF RAMPART-ROCKS

	aragonite (%)	calcite (%)	$MgCo_3$ in calcite (mol %)	non-carbonate (%)
Brown hard micrite				
Low I.	16	80	13	4
Bewick I.	10	83	n.d.	7
Bewick I.	12	80	n.d.	8
Bewick I.	15	77	n.d.	8
Nymph I.	16	80	13	4
Houghton I.	10	82	14	8
Houghton I.	10	82	14	8
Low Wooded I.	10	86	13	4
White chalky micrite				
Three I.	29	68	8	3
Low I.	15	80	25	5

n.d., not determined.

3.3.2. *Cements*

Examination of over 50 thin sections and s.e.m. samples showed the cementing medium of all shingle rocks to be micrite which varied in colour from white to rusty brown. This micrite consists of roughly equant grains of 0.5–6 μm diameter (figure 18) and was found by X-ray diffraction to be 68–95 % high magnesium calcite (14 % $MgCO_3$) and 0–25 % aragonite (table 3). Clay minerals constituted 5–8 % of the matrix with kaolinite being the most abundant (65 % of clay mineral fraction) and lesser amounts of illite (32 %) and montmorillonite (3 %) being present.

There is a consistent difference between the textures of the lower and upper facies of platform rocks. The lower parts, where water saturation is more persistent, have a relatively soft chalky micrite matrix which normally completely fills interparticle pores. The upper part which suffers greater exposure has a splintery hard brown matrix that only partially occludes interparticle pores. Both matrices have the same mineralogical composition. The splintery hard brown micrite has a most unusual texture for it does not always floor interparticle cavities producing a flat-floored geopetal fabric in a manner typical of normal internal sediments. In some cases it drapes over coral fragments and shows vague banding (figure 19), in others it accumulates in a pendant attitude rather like miniature stalactites (figure 20) of speleothem deposits. In most rocks both the 'drape' and 'drip' fabrics occur together commonly with the micrite at the inner margin of incompletely filled cavities having a botryoidal texture (figure 21).

This final pore lining is soft mud in some samples, brittle micrite in others. Although a flat-floored geopetal fabric is rarely seen, the presence of soft mud, the laminated texture in places, and the presence in the micrite of clay minerals and coccoliths (seen by using a scanning electron microscope) point to a sedimentary origin for at least part of the matrix.

Commonly, loose corals on the surface of rampart-rocks show an undersurface that is coated with a film of soft mud whose surface texture is similar to that of the splintery hard botryoidal micrite. Close inspection shows the pattern of the surface of the soft mud to be a replica of that of the skeletal architecture of the coral to which it sticks. Mud occurs in only the lower chambers of such corals; therefore the fine particles cannot have trickled through but must have adhered to the skeleton from beneath, perhaps during the evaporation of water draining from the skeleton. The dripstone texture of the brown micrite matrix of the upper facies of the rampart-rocks strongly suggests that lithification took place in a vadose zone of impermanent saturation by seawater.

The texture of the micrite matrices was commonly blotchy and when viewed with a high powered microscope some areas contained what appeared to be very small comminuted skeletal remains. The incompletely filled cavities normally showed a scalloped margin (without a flat floor) and in a few samples radiating clusters of fibrous cement could be seen coating parts of the surfaces of the cavities (figure 22). This fibrous cement was shown by staining to be calcite. Most coral fragments contained micrite in their outer chambers. In partly filled pores this micrite did not show sedimented characteristics such as a flat-floored geopetal fabric but instead was commonly seen lining the walls of the skeletal cavities (figure 23). At the centre of some corals, fibrous aragonite cement crystals formed a lining or a total filling of the pores.

Several samples of rampart-rock have matrices with a markedly pelleted texture (figure 24). The pelleted matrix commonly shows flat-floored geopetal fabrics and is found in rocks containing abundant *Halimeda* remains. Many of these *Halimeda* grains showed extensive solution and it is considered possible that the matrix pellets represent, at least in part, the deposition of the utricle contents (micrite) after solution of the *Halimeda* plates (figure 25). Schlanger (1964) noted marked solution of *Halimeda* particles in some fossil reef limestones in Guam. The mechanism of skeletal solution and chamber-fill stability was proposed for pelleted fabrics in internal sediments associated with bryozoans in Silurian reef limestones (Scoffin 1972).

3.3.3. *Alteration of grains*

Mineralogical analyses show coral and *Tridacna* skeletons from some of the oldest rampart-rocks to be still aragonitic. No evidence of recrystallization from aragonite to calcite in these skeletons was found.

However, there is marked evidence of skeletal and matrix solution in the rampart-rocks though the style of this solution varies from lower to upper facies. In the upper facies the skeletons (principally corals and *Halimeda*) and layers of the hard micrite matrix are seen to have been irregularly truncated before later matrix deposition though the remaining skeleton structure is well preserved. The brown colouring in the micrite is commonly most intense in a layer at the truncation surface and also coats those surfaces presently in contact with circulating water. Analyses of these brown surface films showed them to be relatively rich in iron and manganese. The suggestion is that those surfaces coated with such a film experienced prolonged contact with circulating seawater during which time iron and manganese salts precipitated. The corals of the lower facies with the chalky matrices showed a degree of peripheral truncation but also pervasive internal solution, giving all the internal surfaces a markedly etched appearance.

Though peripheral solution of grains was not uncommon, recrystallization was rare for all but coralline algal grains. The fine cellular texture of these algae is commonly partly or totally lost and many grains now appear as homogeneous micrite with just a vestige of cellular texture in places.

TABLE 4. RADIOCARBON AGES OF SOME RAMPART ROCKS

code no.	reef	age/a B.P.	mineralogy aragonite (%)	calcite (%)	material
		upper platform			
ANU-1604	Low Wooded	3320 ± 70	97	3	*Tridacna* upper facies
ANU-1595	Houghton	3330 ± 80	99	0	coral basal facies
ANU-1413	Houghton	3550 ± 80	98	2	*Tridacna* surface
ANU-1592	Nymph	3420 ± 75	100	0	*Tridacna* surface
ANU-1383	Nymph	3540 ± 80	98	2	*Tridacna* middle facies
ANU-1380	Three	3750 ± 110	95	5	*in situ* coral basal facies
ANU-1382	Three	3050 ± 70	99	1	*Tridacna* surface
ANU-1478	Turtle I	4420 ± 90	90	10	*Tridacna* upper facies
		upper platform cements			
ANU-1602	Nymph	2350 ± 70	16	80	middle facies
ANU-1381	Three	2260 ± 80	29	68	basal facies
		lower platform			
ANU-1385	Bewick	640 ± 70	99	1	coral surface
ANU-1390	Watson	810 ± 70	98	2	coral surface
ANU-1477	Turtle I	1430 ± 70	100	0	*Tridacna* beneath surface
ANU-1475	Three	1460 ± 70	99	1	*Tridacna* beneath surface
ANU-1476	Nymph	520 ± 70	100	0	*Tridacna* beneath surface
		bassett edges			
ANU-1607A	Low	740 ± 70	n.d.	n.d.	coral
ANU-1601	Low	380 ± 80	15	80	cement

Ages determined by Radiocarbon Laboratory, Australian National University (see Polach *et al.* 1978, this volume).

3.4. *Age of rampart-rocks*

Radiocarbon dates of corals and molluscs from the upper and lower platforms and bassett edges provide evidence for the age of the rampart-rocks (table 4). Eight determinations on constituents of the upper platform from locations where the surface equals or exceeds 3.0 m above datum give an average of 3547 a B.P. with a range from 3050 ± 70 to 4420 ± 90 a B.P. (Polach, McLean, Caldwell & Thom 1978, this volume). Ages of high magnesium calcite matrices are some 1200–1500 radiocarbon years younger than adjacent skeletal components. These dates strongly suggest that loose shingle ramparts accumulated before 3000 a B.P. and became cemented during the following thousand years to form rampart-rocks which are now in a sense fossil forms. Lower platforms and bassett edges that occur above the reef flat level contain corals and shells all dated at younger than 1500 a B.P., and the friable nature of parts of these deposits suggests that lithification is continuing.

4. BOULDER-ROCK

Many reefs contain near the margins of the reef flat a few scattered boulders; likewise in some, rampart-rocks and beach-rocks are cemented scattered isolated boulders (figure 15). However, the term boulder-rock is here intended to refer to those rocks where the grains of the rock are generally greater than 25 cm median diameter and are in mutual contact.

4.1. *Location*

The reefs with loose or cemented boulder tracts are shown in table 1. Stephenson, Stephenson, Tandy & Spender (1931) have described boulder tracts from Low and Three Isles reefs. Although most of the reefs near to the mainland have beach-rock and rampart-rock, only a few reefs presently have exposures of boulder-rock. One reef may have a boulder tract and an exposure of boulder-rock while its neighbour has neither. Those reefs showing good developments of loose and cemented boulder tracts occur in the Howick Group.

Boulder tracts normally have a curved trend at the perimeter of one flank of the reef and roughly link the windward intertidal deposits (ramparts) with the leeward sand cay. Boulder-rocks occupy the same general position but lie a short distance (30–100 m) in from the reef perimeter. In several examples the boulder-rocks parallel the loose boulder tract and are separated from it by a narrow moat.

4.2. *Gross morphology*

Boulder tracts, and the stretches of rocks they become once lithified, are about 100–200 m in length and 20–50 m in width. The indication from cliff sections is that they attain a maximum thickness of about 3 m. The density of boulders increases towards the centre of the deposit. Normally all except the very large boulders of loose and cemented boulder tracts are immersed at high spring tides. The surface of a cemented boulder tract may be irregular with protruding boulders or else essentially flat like a platform deposit; the configuration of the surface depends upon the extent of matrix infilling.

4.3. *Petrography*

Spherical, hemispherical or mushroom shaped colonies are the major forms of boulders (figure 9). The dominant corals are *Porites* (70%) with *Goniastrea* and *Leptoria* making up the bulk of the remainder. A critical observation was that less than 5% of the boulders were made of more than one coral colony. It thus suggests that these boulders were not plucked out of a structure of coral intergrowth but rather that they were seated individually on a surface from which they could easily be dislocated, perhaps even a sandy bottom. Kornicker & Squires (1962) have shown how buoyant many of the massive corals are when in water, so it may not have taken the great force that one initially envisages to carry them up on to the reef. Nevertheless the existence on several reefs of a discrete boulder tract isolated from an earlier boulder tract that is now cemented, suggests that each tract deposit resulted from one major catastrophic event, or at least from a period of boulder-producing conditions which was ephemeral and that little subsequent dispersal of boulders has occurred. Many of the coral boulders are riddled with animal boreholes and burrows, with boring barnacles, bivalves, worms, sponges and echinoids being presently active. The consistent positions of bioerosion structures in the very large boulders suggest that once settled the boulders do not move.

The matrix of boulder-rock commonly varies from the windward end, where it is dominantly coral shingle with a micritic calcite cement, to the leeward end, where it is dominantly sand grains with a fibrous aragonite cement. The matrix of boulder-rock is therefore similar to that of rampart-rock on the windward side and beach-rock on the leeward.

Locally, especially at the windward flanks of reefs, low-profile ridges or sheets occur consisting of cobble-sized equant fragments of corals (figure 26). Both in grain size and in gross morphology these deposits represent an intermediate between shingle ramparts and boulder tracts. Several examples of weakly cemented (by micrite) cobble-sized corals are seen forming patchy outcrops of conglomerate (figure 27), with a low relief normally on the windward margins of reefs.

5. REEF-ROCK

This *in situ* rock results from organic cementation. Currently, around all the reefs examined, encrusting calcareous organisms build a rigid structure up to sea level. This subtidal growing structure is built chiefly of large branching corals and where growth is sufficiently sturdy or else cemented by crustose coralline algae, a characteristic framework is produced which eventually develops a flat upper surface just below low water level of spring tides. No exposures of fossil subtidal reef framework were found on any of the reefs visited. This absence is one of the strongest lines of evidence against a significant drop in sea level over recent time. It is possible that *in situ* deposits of fossil subtidal reef framework lie just below the loose or cemented sand and shingle on the reef flat – though shallow excavations did not reveal any and none were reported from the boreholes on Low Isles by Marshall & Orr (1931). Alternatively, any former subtidal reef framework may have been eroded by wave-transported coral debris moving across the reef flat, though the preservation of fossil microatolls, boulders and rampart-rocks on the present reef flats argues against this.

The exposures of *in situ* skeletal framework above low water mark are all the result of growth that was above low water mark in areas of permanently ponded, draining or splashing water.

Massive (chiefly *Porites*) and branching (chiefly *Montipora*) corals grow and coalesce in reef flat moats and on the seaward slopes of those ramparts where tidal-flat water seeps through during low tide (figure 28). The anastomosing skeletons provide a degree of cohesion, though the deposit normally crumbles underfoot. Crustose coralline algae cement coral shingle into an algal rim at the windward margins of many, particularly outer-shelf reefs. The algal coat is normally quite thin (figure 29) and the rim surface, which is a few metres wide, is only a few centimetres above the level of the neighbouring leeward moat. The depression of the moat represents a zone where normal intertidal erosion (physical and biological) is relatively more effective than at the rim, where construction is dominant. The algal rim deposits are therefore dynamic features that are simultaneously built on their outer edge and destroyed on their inner edge as the reef expands laterally.

The intertidal exposures of *in situ* skeletal reef-rock are normally patchily distributed near the rim of all reefs but the limited degree of cohesion resulting from this incipient organic binding does not produce rocks of the solidarity and physical prominence of the beach-rocks, rampart-rocks and boulder-rocks.

6. Phosphate-rock

Several sand cays on the reefs of the Northern Province have large colonies of migratory birds and some show consolidation of the cay sands by phosphate mineralization by guano. Such deposits are located on the highest parts of the sand cays. At Bewick and Three Isles the phosphatized sand occurs in a distinct horizon at depths between 20 and 60 cm with loose clean sand both above and below, while at Ingram the surface sands have been removed by deflation giving patchy exposures. On Stapleton cay the upper 10 cm of carbonate sand is bound by guano and organic debris, indicating the first stage in the development of phosphate rock. Green Island has a dense wooded vegetation and though at present it does not have an exceptionally large bird population, it is perhaps because it has a permanent one over a fixed location that results in the formation of phosphorites here. The consolidated grains form a veneer over loose sand and the field occurrence suggests a downward mineralization by phosphate solutions. The cemented sands contain 23 % P_2O_5 and the phosphates (X-ray diffraction indicates chiefly hydroxy-apatite) occur as either a thinly laminated wavy layer (0.02 mm thick) around carbonate grains (figure 30) or as a brown structureless matrix. The phosphatic coating brings about a centripetal replacement of the grains to phosphate. This replacement affects *Halimeda* grains more readily than corals and the foraminiferal grains are the least altered.

It is significant that the only high supratidal cementation on sand cays was by phosphate. No occurrences were found of calcarenite cemented by sparry calcite on the reefs.

7. Origin of intertidal rocks with inorganic cements

7.1. *Beach-rock*

Numerous workers have suggested that the aragonite cement of beach-rock is derived directly from seawater (Stoddart & Cann 1965, Ginsburg 1953, see also Bricker 1971). Bathurst (1971) pointed out that in some areas of the world, calcium carbonate cements beaches made entirely of non-carbonate grains, endorsing the seawater origin. On the sand cays of the Northern Province of the Great Barrier Reef, the carbonate grains showed no post-depositional solution effects in the aragonite cemented beach-rocks so the grains themselves were not the source of the cement. The fact that the vertical span of modern beach-rock over the beach relates closely to the tidal range shows a direct control by daily soaking and evaporation. A further critical control indicated by this study is that those beaches without supratidal exposure do not normally develop beach-rock. This could simply be a connection with the stability of the sand cay, i.e. the beaches that are built above high water mark are more stable than those that are not (also these high cays normally develop vegetation which further stabilizes the cay), or alternatively it could be more complex, relating to either exposure to meteoric waters or the degree of exposure to evaporation.

7.2. *Rampart-rock*

Most rampart-rocks occur intertidally. The only exceptions are parts of the upper platforms which locally reach more than 1 m above present mean high water spring tide level. The cementation of the upper part of this deposit by high Mg calcite could be explained by one of three processes: (1) being on the windward side of reefs the upper surface receives a high quantity of spray, thus raising the effective level of lithification; (2) capillary forces within the muddy matrix draw evaporating seawater above h.w.s.t. level; (3) during its formation some

3000 years ago, the level of high spring tides was higher than it is today. Evidence from fossil microatolls (Scoffin & Stoddart 1978, this volume) contained in the upper platform does support the last theory. The highest fossil *in situ* corals contained in the upper platform are 0.7 m higher than the highest living corals found on the reefs at present. This strongly suggests some relative lowering in either mean sea level or tidal range since the upper platform formed. The exact amount of this fall is not possible to assess from lithification criteria alone for the surfaces of platforms need not necessarily represent the upper limits of lithification. The general elevations of platforms may even relate more closely to the size of the original ramparts, and it can be seen today that there is a considerable range in the scale of loose ramparts and in the elevations of the loosely consolidated lower platforms (figure 31). Nevertheless it still remains that some 3000 years ago, ramparts were built and lithified to heights of at least 3 m, and during the last 1500 years ramparts were built and lithified to heights of about 2 m.

The lower limit of rampart-rock formation is not precisely known though it is thought to be near low water mark. Excavation into the bassett edge structure of low rampart-rocks revealed the absence of cement at a shallow depth (30 cm).

Shingle ramparts occur on only those reefs on the inner portion of the shelf close to the mainland (whereas beach rocks occur across the shelf). The critical distance from shore appears to be about 20–22 km (table 1); beyond this distance ramparts do not currently build intertidally and rampart-rocks do not form. The majority of the reefs within the 20 km zone have intertidal ramparts.

To facilitate lithification, ramparts have to be stationary for some time and also they probably require a degree of interstitial fine sediment to aid in the retention of saturated waters at shingle grain contacts. Satisfactory stabilization may be brought about by a decline in the frequency and intensity of shore-face erosion and washover as the rampart migrates across the reef flat and its distance from the reef edge increases. Alternatively it can be achieved through the development of another rampart to windward, thus excluding wave action from the older deposit. Sequences of shingle ramparts and ridges, frequently separated by moats, are not an uncommon feature of inner-shelf reefs of the Northern Province. Mangrove colonization of rampart surrounds and surfaces may also assist in stabilization. In the sheltered area leeward of the ramparts, quiet water conditions allow the accumulation of fine sediment and the extension of mangrove swamps. Mud from these swamps later finds its way to rampart interstices. It is clear that the areal distribution of rampart-rocks relates very closely to that of mangroves, both at a regional scale (table 1) and local reef scale. Furthermore, most loose ramparts without associated mangroves have no muddy matrix, whereas most cemented ramparts have large quantities of interstitial micrite.

Rampart-rocks have an average mass percentage insoluble residue of 6.7, whereas sand cay beach-rocks have a much lower insoluble residue of 2.7 % (table 5). This variation is noted even on one reef. Analyses of sea floor sediments from just off the reefs show that similar quantities of clay minerals occur on the windward and leeward sides of reefs, so it appears that it is the special local reef top conditions governing deposition and preservation of fine sediment that control the amount of insolubles of these rocks.

An obvious contrast exists between the sand grains on the leeward sand cay beaches that are constantly being agitated by the lapping, small amplitude waves and the shingle grains of the ramparts that provide a semi-rigid network into which mud can percolate especially during low tide when reef-flat water drains seawards through the ramparts. It is something of a paradox that on the windward side of reefs in the zone of heavy surf are built rocks with a large quantity of fine interstitial sediment, whereas on the sheltered leeward sides of reefs the common rock type has clean, well washed grains free from mud.

The most striking contrast between the leeward and windward limestones is the composition of the cements. Characteristically, beach-rocks of leeward sand cays are lithified by fibrous aragonite crystals, whereas the rampart-rocks are cemented by cryptocrystalline high magnesium calcite. Regrettably, chemical analysis of interstitial waters was not feasible during the

TABLE 5. MASS PERCENTAGE INSOLUBLE RESIDUES OF BULK SAMPLES OF
EXPOSED LIMESTONES (IN 10% HCl)

rock type/island	insoluble residue (%)	average of similar types	location on reef
beach-rock			
Pipon	2.0		leeward – sand cay
Waterwitch	2.9		leeward – sand cay
Stapleton	3.0		leeward – sand cay
Ingram	0.8		leeward – sand cay
Howick	3.6		leeward – sand cay
Bewick	3.1		leeward – sand cay
Newton	2.5	2.7	leeward – sand cay
Eagle	2.5		leeward – sand cay
Turtle I.	2.7		leeward – sand cay
Two Isles	3.3		leeward – sand cay
Three Isles	2.9		leeward – sand cay
Low Isles	3.2		leeward – sand cay
beach-rock			
Houghton	2.9	3.7	leeward flanks – in platform
Nymph	4.4		windward – in gap in platform
rampart-rock			
Beanley	7.0		windward – lower platform
Houghton	8.1		windward flank – upper platform
W. Pethebridge	8.2		leeward spit
Three	3.0	6.7	windward – lower platform
West Hope	7.5		windward – lower platform
Low	7.0		windward – low bassett edge
Three	7.2		windward – upper platform (up. facies)
Three	5.3		windward – upper platform (low facies)
boulder-rock (matrix only)			
Howick	4.1	4.1	leeward flank
beach-rock (consisting of grains of reworked cemented rampart-rock)			
Turtle I.	6.3	6.3	leeward – sand cay

expedition. However, some important differences between the two environments of cement precipitation were recognized and we believe the cause of the contrast in the petrography of the two deposits to lie in these differences. Beaches on sand cays are kept free from fines and are bathed in open-shelf seawater throughout the tide cycle. Ramparts, on the other hand, are not cleaned of trapped fines and are immersed in a range of water types: normal open-shelf seawater at high tide, and at low tide ponded reef-flat water seeps through the rampart during its seaward drainage. This reef-flat water may have been (a) concentrated by prolonged evaporation, (b) diluted by freshwater after heavy rain, or (c) influenced by passage through mangrove swamps.

Apart from cement mineralogies the occurrence of solution fabrics in the rampart rocks and their absence in beach-rocks suggest conflicting compositions of the interstitial waters. The beach-rock interstitial water is seemingly permanently saturated or supersaturated while the rampart-rock water undergoes periods of undersaturation.

It was noted that where small beaches occur in gaps in rampart-rocks on the windward sides of reefs, the cement is similar to that of the neighbouring rampart-rocks, indicating the control this windward-mangrove environment has on the nature of the cement. Aragonite can precipitate on the windward side shown by its occurrence in the inner chambers of coral fragments in shingle rocks, but its presence here likely pre-dates rampart-rock formation. On the other hand, high magnesium calcite was only found in those beach-rocks that either contained grains reworked from rampart-rocks or were associated with mangroves and windward type environments. Recently, Morita (1976) has suggested that the microflora that is associated with coral debris that becomes anaerobic a few centimetres below the surface is responsible for the precipitation of a calcitic cement.

The main differences between the windward rampart-rocks and the leeward sand cay beach-rocks are summarized in table 6.

TABLE 6. COMPARISON OF WINDWARD AND LEEWARD INTERTIDAL ROCKS

	beach-rock	rampart-rock
location on shelf	across entire shelf	inner-shelf position less than 22 km from mainland
location on reef	leeward edge	windward edge
bed thickness	7 cm	20 cm
bed attitude	15° away from cay	60° to reef centre
grain composition	corals, molluscs, *Halimeda*, coralline algae, benthonic Foraminifera	*Acropora* coral branches
grain size	equant, 1 mm diameter	rods 100 mm long, 15 mm in cross section
mass percentage insoluble residue in 10 % HCl	2.7	6.7
colour	white to fawn	white to rusty brown
cement mineralogy	aragonite	calcite (14% $MgCO_3$) (with local aragonite)
cement fabric	radiating fringes of needles most abundant in meniscus position at grain contacts (though isopachous fringes occur)	multiple generations of micrite commonly developing a dripstone fabric
post-depositional solution effects	absent	present

The suggestion is that the characteristics of rampart-rocks (the inner-shelf location, the occurrence of mangroves, the presence of fine sediment, the abundance of clay minerals and the high magnesium calcite cement) are interrelated. It is concluded that the hierarchy of dependency is as follows: the position on the shelf determines which reefs develop stable ramparts, and these ramparts govern the distribution of mangroves which trap fine sediment including abundant clay minerals. The ramparts pond water during low tide and this water drains through mangrove humus carrying fine sediment into the ramparts. This water, perhaps in conjunction with the special anaerobic surfaces of the shingle, permits calcite to precipitate here, whereas on the leeward sand cay aerated beaches, evaporation of open seawater causes aragonite to precipitate.

References (Scoffin & McLean)

Bathurst, R. G. C. 1971 *Carbonate sediments and their diagenesis (Developments in sedimentology*, vol. 12). Amsterdam: Elsevier, 620 pages.

Bricker, O. P. (ed.) 1971 *Carbonate cements (Johns Hopkins University Studies in Geology*, No. 19). Baltimore: Johns Hopkins Press, 376 pages.

Dunham, R. J. 1971 Meniscus cement. In Bricker (1971), pp. 297–300.

Fairbridge, R. W. & Teichert, C. 1948 *Geogrl J.* **111**, 67–88.

Ginsburg, R. N. 1953 *J. sedim. Petrol.* **23**, 85–92.

Kornicker, L. S. & Squires, D. F. 1962 *Limnol. Oceanogr.* **7**, 447–452.

Marshall, S. M. & Orr, A. P. 1931 *Scient. Rep. Gt Barrier Reef Exped. 1928–29* **1**, 93–133.

Maxwell, W. G. H. 1968 *Atlas of the Great Barrier Reef.* Amsterdam: Elsevier, 258 pages.

McLean, R. F. & Stoddart, D. R. 1978 *Phil. Trans. R. Soc. Lond.* A **291**, 101–117 (this volume).

McLean, R. F., Stoddart, D. R., Hopley, D. & Polach, H. A. 1978 *Phil. Trans. R. Soc. Lond.* A **291**, 167–186 (this volume).

Morita, R. Y. 1976 *Joint Oceanographic Assembly, Edinburgh 1976*, p. 117. Rome: Food and Agriculture Organisation of the United Nations.

Polach, H. A., McLean, R. F., Caldwell, J. R. & Thom, B. G. 1978 *Phil. Trans. R. Soc. Lond.* A **291**, 139–158 (this volume).

Schlanger, S. O. 1964 *U.S. geol. Surv., Prof. Pap.* **403-D**, 1-52.

Scoffin, T. P. 1972 *Geol. Rdsch.* **61**, 565–578.

Scoffin, T. P. & Stoddart, D. R. 1978 *Phil. Trans. R. Soc. Lond.* B **284**, 99–122 (part B of this Discussion).

Spender, M. A. 1930 *Geogrl J.* **76**, 193–214 and 273–297.

Steers, J. A. 1929 *Geogrl J.* **74**, 232–257 and 341–370.

Steers, J. A. 1937 *Geogrl J.* **89**, 1–28 and 119–146.

Stephenson, T. A., Stephenson, A., Tandy, G. & Spender, M. A. 1931 *Scient. Rep. Gt Barrier Reef Exped. 1928–29* **3**, 17–112.

Stoddart, D. R. & Cann, J. R. 1965 *J. sedim. Petrol.* **35**, 243–247.

Stoddart, D. R., McLean, R. F., Scoffin, T. P. & Gibbs, P. E. 1978 *Phil. Trans. R. Soc. Lond.* B **284**, 63-80 (part B of this Discussion).

Phil. Trans. R. Soc. Lond. A. **291**, 139–158 (1978)   [139]

Printed in Great Britain

Radiocarbon ages from the northern Great Barrier Reef

By H. A. Polach,† R. F. McLean,‡ J. R. Caldwell§ and B. G. Thom‖

†*Radiocarbon Laboratory, Australian National University, Canberra, Australia* 2600

‡*Department of Geography, University of Auckland, Auckland, New Zealand*

§*Department of Biogeography and Geomorphology, Australian National University, Canberra, Australia* 2600

‖*Department of Geography, Faculty of Military Studies, University of New South Wales, Duntroon, Australia* 2600

1. Introduction

The aims of the 1973 Great Barrier Reef Expedition's radiocarbon dating programmes were: (i) to collect live specimens from various reef environments to serve as modern reference standards; (ii) to evaluate the suitability of materials from the drilling, geomorphic and sediment programmes for dating purposes; and (iii) to date appropriate samples related to those programmes. The radiometric ages provide a time scale for the evolution of reefs and reef islands, and the history of sea level in the area.

The purpose of this paper is to report results of all ages determined to date, to describe field and laboratory methods used and to assess the reliability of the ages in terms of (i) the actual materials dated and (ii) the geomorphic, ecological or stratigraphic units from which the samples were obtained. Seventy-nine determinations based on 74 samples are reported. No interpretation of the results is attempted here.

2. Field collection

Samples were obtained from various reef environments within a 5° latitudinal range between Raine Island (11° 36′ S) and Low Isles (16° 23′ S) on the northern Great Barrier Reef. However, nearly 90% of the samples were collected within a 1° band between Stapleton Island (14° 19′ S) and Three Isles (15° 07′ S). Samples can be grouped into the following categories: (1) live specimens, (2) fossils, rocks and sediments, and (3) samples from drill cores.

2.1. *Live specimens*

Specimens of living corals were collected from reef flats and reef slopes at Houghton and Magaera reefs; these included species of *Symphyllia, Favia, Turbinaria, Physogyra, Porites, Goniastrea, Acropora, Fungia, Astreopora* and *Platygyra*. Other living material collected included *Nerita, Lithophaga, Trochus, Tridacna, Cypraea* and other molluscs, and *Acanthaster* and *Halimeda*. Of this collection only the last is reported here. All living specimens were preserved with flesh in salt water under vacuum in Agee jars to check ^{14}C distribution and $^{13}C/^{12}C$ isotopic compositions. Polach was assisted by Thom, J. E. N. Veron and T. P. Scoffin in collecting this living material; Veron identified the corals.

2.2. *Fossils, rocks and sediments*

A large number of fossils, limestone rocks and sediments were collected during the Expedition, but only a small proportion of these have been dated. About half of the total samples reported here were collected with dating explicitly in mind at the time of collection.

All dated samples in this group were collected from above low water level and many were located on transects surveyed by McLean, D. R. Stoddart, D. Hopley, and A. L. Bloom. Of the samples reported 51 were collected by McLean, 10 by Hopley, 3 by Polach and 10 by Thom.

2.2.1. *Fossils*

Parts of entire specimens of individual fossil corals and molluscs were taken from a variety of reefs and situations, including coral colonies, boulder tracts, reef flats, shingle ramparts and ridges. Some specimens were in position of growth, others were not. Commonly, a number of different specimens were taken from the one locality, stratigraphic or geomorphic unit. Individual samples normally weighed 0.5–5.0 kg and were carried untreated in plastic bags to the laboratory. Massive and branching corals and tridacnids were most frequently obtained.

2.2.2. *Limestone rocks*

Rock types exhibiting some degree of consolidation included rampart-rock, beach-rock, boulder-tract-rock, reef-rock, island-rock and phosphorites. Descriptions of these rock types are given in Scoffin & McLean (1978, this volume). For dating purposes most attention was paid to rampart-rocks, which based on morphological and altitudinal criteria were called bassett edges, and lower and upper platform in the field.

Both large and small samples were broken off with geological or sledgehammer and crowbar; the samples normally contained both skeletal material and matrix. In some instances, particularly in the loosely consolidated deposits, it was possible to prise individual corals and clams from the matrix, although this was only rarely possible in the tougher rocks. Notes were made on position, facies, geometric form and relations of the samples and whether or not skeletal materials were in growth position.

2.3.3. *Sediments*

These comprised either carbonate sands of mixed skeletal composition or loose coral gravels from soil pits, scarp sections or surface deposits of sand and shingle cays. Details of the nature of these sediments are given by McLean & Stoddart (1978, this volume). Individual horizons or sediment units were sampled and bulk samples collected by hand, spade or trowel. Samples were packaged in plastic bags without treatment.

2.3. *Drill cores*

The drilling programme yielded both consolidated and unconsolidated material suitable for dating. The technique of drilling and method of sampling has been described by Thom (1978). Sample size was limited by the 35.5 mm internal diameter core barrel and highly variable lithologies with depth. Quantity of material submitted for dating was barely adequate, inhibiting further treatment (see Thom, Orme & Polach 1978, this volume).

3. LABORATORY EXAMINATION

Potential samples were carefully examined before submission to the Radiocarbon Laboratory. Modes of examination were determined by the nature of the samples and the submitter's reasons for obtaining a radiocarbon age. The validity of ages was regarded as of utmost importance and special attention was paid to removing post-death contaminants, in those samples which consisted of single organisms.

3.1. *Corals and molluscs*

Samples in individual isolated corals, and molluscs from colonies in growth position, boulder tracts and shingle ramparts, were sawn into cubes. Where possible, massive corals possessing thick walled corallites and large valves of *Tridacna* clams were selected. Interior portions of cut samples were checked initially in hand specimen and then under the binocular microscope for the presence of loose sediment of cement, filled or unfilled boreholes and bore-linings and areas of discoloration, which if present were cut out. The resulting cuttings were split into different portions for mineralogical determination, thin section preparation, radiocarbon determination, sample identification and archives. Thin sections of corals were studied to ascertain the degree of secondary carbonate deposition, recrystallization and foreign organism intrusion.

Contamination of massive corals in growth position and of reef blocks was found to be minimal. Thin sections were not prepared of tridacnids and reliance was placed on aragonite–calcite determinations (see below) for evaluation of the degree of contamination (which also was proved to be minimal). Of the 19 samples in this group submitted for radiocarbon dating, 11 were corals and 8 tridacnids.

3.2. *Limestone rocks*

These included beach-rocks, and the various categories of rampart-rocks which varied from loosely cemented bassett edges, through more firmly cemented lower platform rocks to the tough, firmly lithified rocks of the higher platform. With the rampart-rocks, ages were desired on both skeletal materials and some cements. Where massive corals and tridacnids were present in rampart-rocks these were cut out or broken from the matrix and prepared and examined in the manner indicated above (§ 3.1) with similar results.

Separation of smaller branching corals and molluscs from the cementing matrix, and removal and isolation of cement from within coral pores, proved much more difficult. Tiny slabs of coral and cement were cut and examined microscopically, and contaminants (either coral or cement) removed by saw and/or dentist's drill. Examination of thin sections of these samples suggests that it is unlikely that all contamination was removed. Thus the ages for this suite of samples, notably ANU-1208 (coral), ANU-1380 (coral), ANU-1381 (matrix), ANU-1601 (matrix), ANU-1602 (matrix), ANU-1607A (coral), and ANU-1607 (matrix) must be interpreted accordingly.

Only three beach-rocks were dated. Two contained *Tridacna* valves (ANU-1386, ANU-1581) on the surface, which were prised out, cut and examined in the usual manner. The third (ANU-1596) was a bulk sample; thin section examination indicated the presence of acicular aragonite cement. In this case no attempt was made to separate allochems and cement.

Dated samples from limestone rocks included 6 corals, 12 tridacnids, 4 matrix cements and 1 bulk calcarenite.

3.3. *Sediments*

Samples of loose sediments from individual horizons and sediment units comprised three types: sand (17 samples), coral fragments (4 samples) and mud (1 sample). The last, a fibrous mud, was submitted in bulk to the Radiocarbon Laboratory (ANU-1480) where rootlets and fibres were separated from the organic mud. Age determinations were made on the organic carbon of the fibrous material (ANU-1480A) and the mud fines (ANU-1480B). Cay sands were sieved at $\frac{1}{2}$ φ intervals and a split of the total detritus falling between medium to very coarse sand (2.0 to −1.0 φ units) was submitted for bulk age determination. Thin section examination indicated a variety of skeletal constituents, mainly Foraminifera, *Halimeda*, other calcareous algae, and molluscan and coral fragments, the proportion of each constituent varying between samples. In a number of instances secondary material was present in intraskeletal voids and chambers, but not in sufficient quantities to affect significantly the validity of ages.

Four samples of sandy-gravels from reef islands were dated. Between 5 and 15 stick coral fragments in the -1.0 to $-4.0\,\phi$ size range were selected from each sample and physically cleaned of surface contaminants. Interior void sediment and borehole linings were not removed; but such contaminants were estimated to compose less than 3 % of total sample mass.

3.4. *Drill core samples*

Nine samples recovered from the shallow coring on Bewick and Stapleton were radiocarbon dated. Of the six samples from Bewick, five were coral fragments and one was a biomicrite. ANU-1283, 1284 and 1395 were non-recrystallized corals with a high aragonite content. ANU-1281 and 1282 were clearly recrystallized samples in which the primary aragonite of the coral specimens had been converted to calcite. Similar modification is suggested by the low Mg calcite content of the biomicrite (ANU-1280).

Only three dates were obtained from the Stapleton core. Two of these were on samples of high aragonite content, while one, a reef calcerenite fragment possibly containing coralline algae, had a high magnesium calcite content (ANU-1664). The other samples were coral fragments (ANU-1721) and loose carbonate sand (ANU-1663).

3.5. *Living specimens*

At this stage dating has been restricted to *Halimeda*, a calcified green algae growing on reef flats of many islands. Both the organic carbon fraction (living plant tissue) and the inorganic carbonate plates have been dated (ANU-1272A, 1272B, ANU-1273A, 1273B).

4. MINERALOGICAL EXAMINATION

Corals and tridacnids build skeletons of aragonite. Thus detection of calcite serves as a guide to the degree of post-death contamination resulting from either recrystallization of the original skeleton, or the presence of calcitic cement or sediment. However, contamination by secondary aragonite cannot be gauged by this technique, hence the need for microscopic examination. Percentages of aragonite and calcite were determined for most of the samples submitted for radiocarbon dating. In the case of bulk carbonate sands, all of which possessed a mixed constituent composition, the proportion of aragonite and calcite reflects the relative abundance of aragonitic and calcitic organisms and the presence of calcite is not diagnostic of contamination. The calcite phase was also determined, because separation of low and high magnesium calcite is important in evaluating environments in which cementation and recrystallization occur. Sample validity is not affected by the presence of non-carbonate minerals, but these were determined as a matter of routine.

4.1. *Infrared quantification of calcite and aragonite*

Infrared spectrophotometry was used to analyse carbonate minerals present in a sample. The following methods were adopted to calculate quantities of calcite and aragonite.

A carbonate aliquot of 10–50 g was crushed by hand to pass 35-mesh. A 0.25 g aliquot if this was reduced in a rotomill for 8 min with one agate ball and 2 ml isopropanol, the latter acting as a mechanical buffer to reduce the likelihood of any aragonite being converted to calcite during grinding. The resulting 'milk' was diluted with ethanol and vacuum-sucked through a 5 μm sieve and desiccated with phosphorus pentoxide for 24 h. After this, 20.0 mg of desiccated carbonate was mixed with 1.0 g desiccated potassium bromide for 1 min in the rotamill and pressed into a 30 mm disk in a vacuum die taken quickly up to 196 MPa, and retained at that pressure for a minute.

Up to this point, standards and unknown samples must be treated in exactly the same manner, including grinding technique to ensure duplication of artificially prepared grain sizes. This must be done because the behaviour of the peaks is dependent on grain size duplication. Without grain size similarity, results will be erroneous. Using the above methods, no mineral segregation due to differences in hardness between calcite and aragonite was detected, nor was there a grinding or pressure induced conversion of aragonite into calcite.

Measurements were taken with a Unicam SP1000 at 4 cm/min scan, using pure potassium bromide as reference. Aragonite absorbs at wavenumbers of 694, 709, 860, 1078 and calcite at 709 and 880. By using pure standard minerals, the unknowns are quantified by measuring the peak height at 694 and adjusting that linearly to a predetermined constant value for 860/880 twins. That is to say, if the carbonate mixture is low in one component (up to 20%), either calcite or aragonite, then one of these peaks is zero. For low calcite values the 880 peak is zero; for low aragonite values the 860 peak is zero. For all other combinations, both peaks will be measurable and must be summed to account for the total energy emanation due to both carbonates. It is this summed height that becomes the common reference between all samples. This also corrects for carbonate concentration variations due to non-carbonate impurities which, if present in the sample, reduce the effective sample mass. The accuracy which is $\pm 2\%$ is unaffected by variations in magnesium contents and crystallinity.

4.2. *X-ray diffraction determination of* Mg *calcite phase*

The calcite phase quantified by infrared spectrophotometry was identified by X-ray diffraction. The following techniques, which basically follow standard procedures, were adopted.

A cleaned carbonate aliquot of 70–50 g was crushed by hand to pass 35-mesh; a 250 mg aliquot of crushed carbonate was mixed with 25 mg either NaCl or CaF_2 and reduced in a rotomill (RM 100) for 8 min with one agate ball and 2 ml isopropanol. The resulting 'milk' was air-dried; a 50 mg portion of the dry product was mixed with six drops of acetone and placed on a 4 cm² etched circle on a glass slide and allowed to dry while being gently tapped.

The X-ray work was done with the use of a Philips PM 8000 diffractometer fitted with a Siemens horizontal goniometer having a spinning and oscillation attachment. Nickel-filtered copper radiation and tube constants of 50 ku/35 mA were used throughout. Chart speed was 20 mm/min with scanning speed $\frac{1}{4}°$ 20 P.M., T/C 4 s, and range variable 200–400 Hz depending on calcite content.

Fluorite of halite internal references provided a constant check on the accuracy of angle measurement as well as providing correction factors. Fluorite *hkl* (111) reflexion at 3.15 Å† and halite *hkl* (200) at 2.821 Å are conveniently close to pure calcite *hkl* (104) 3.035 Å and its Mg-phases up to dolomite *hkl* (104) at 2.886 Å. One or more sharp peaks are produced depending on how many calcite phases are present. Provided the peaks were large enough to locate

† 1 Å $= 10^{-10}$ m $= 10^{-1}$ nm.

accurately, these were converted to atomic percentages Mg in the calcite structure based on the curve of Chave (1952, fig. 1, p. 192).

4.3. *Results*

A split of most samples selected for radiocarbon dating was analysed by the i.r. and X-ray diffraction methods outlined above. All analyses were carried out by Caldwell. Results are presented in table 1.

TABLE 1. MINERALOGICAL RESULTS, GREAT BARRIER REEF EXPEDITION, 1973

A.N.U. code	aragonite (%)	calcite (%)	MgCO$_3$ (mol %)	non-carbonates (%)	sample type
1207	96	4	n.a.	t	C
1208	92	8	n.a.	t	C
1280	2	94	2	6	BM
1281	0	100	0	0	C
1282	0	99	n.a.	1	C
1283	90	10	n.a.	0	C
1284	99	1	n.a.	0	C
1285	99	0	0	1	C
1286	98	2	0	0	T
1287	100	0	0	0	C
1380	95	5	1	0	C
1381	29	68	8	3	M
1382	99	1	0	0	T
1383	98	2	0	0	T
1384	99	0	0	1	C
1385	99	1	0	0	C
1386	96	4	0	0	T
1387	50	49	n.a.	1	S
1388	100	0	0	0	C
1389	98	2	0	0	T
1390	98	2	1	0	C
1391	99	1	0	0	T
1392	98	2	0	0	T
1393	97	3	0	0	C
1394	100	0	0	0	C
1395	37	56	3	7	C
1410	62	38	n.a.	0	S
1411	98	2	0	0	T
1412	59	41	n.a.	0	S
1413	98	2	0	0	T
1414	49	51	n.a.	0	S
1475	99	1	n.a.	t	T
1476	100	0	0	0	T
1477	100	0	0	0	T
1478	90	10	n.a.	0	T
1479	100	0	0	0	C
1554	53	47	13	t	S
1555	58	40	3	2	S
1556	62	35	3	3	S
1557	51	48	13	1	S
1558	57	42	9	1	S
1559	50	50	13	t	S
1560	63	36	13	t	S
1591	97	3	0	0	T
1592	100	0	0	0	T
1593	92	8	9	t	C
1594	94	4	9	2	C
1595	99	0	0	1	C

TABLE 1 (*cont.*)

A.N.U. code	aragonite (%)	calcite (%)	MgCO$_3$ (mol %)	non-carbonates (%)	sample type
1596	36	63	13	1	B
1597	59	41	15	0	S
1598	92	8	11	0	C
1599	96	2	13	2	C
1600	1	95	3	4	C
1601	15	80	25	5	M
1602	16	80	13	4	M
1603	97	3	3	t	C
1604	97	3	3	0	T
1605	98	2	13	t	C
1606	68	32	9	t	S
1607	38	53	13	9	C/M
1608	97	3	3	0	T
1609	98	2	0	0	T
1639	96	3	13	1	C
1640	95	2	0	3	T
1641	32	68	15	0	S
1642	52	48	13	0	S
1643	21	79	13	0	S
1663	48	52	7	0	S
1664	5	95	13	0	R
1721	70	30	13	0	C

Sample types: C, coral; T, *Tridacna*; M, matrix cement; B, beach-rock; S, sand; BM, biomicrite; R, reef calcarenite; n.a., not available; t, trace.

These results, together with thin section analysis, show that post-depositional contamination of dated samples was minimal, except in a few instances. Of the 19 tridacnid samples, calcite ranged from 0 to 4% and magnesium calcite (low) was detected in only two samples (ANU-1604 and 1608). Minimal post-death contamination was indicated, likewise for the reef and island corals which had calcite values up to 5%, being zero in seven samples. Calcite was generally pure; the low magnesium phase was identified in three samples and high magnesium phase in another three.

However, a number of samples possessed calcite values in excess of 5% (e.g. ANU-1208, 1280, 1281, 1282, 1283, 1395, 1593, 1598, 1600, 1607, 1721) and this must be taken into account in interpreting the dates. Many of these samples contained aggregates of a number of coral specimens either from shingle ramparts (ANU-1593, 1598) or rampart-rocks (ANU-1208, 1607) or in drill core (ANU-1281, 1282). The presence of high magnesium calcite in some cases suggests contamination. Hand specimens and thin section analysis shows that contamination in rampart environment results from the presence of matrix cements in rock corals and secondary carbonate deposition or intrusion in the interior of rampart corals, and not recrystallization of the original skeletons, except in the case of ANU-1600 and deeper drill core specimens from Bewick (e.g. ANU-1281). Sample ANU-1600 comprised a number of coral fragments collected from the surface of an old coral shingle ridge at West Hope Island. Although in hand specimen the corals (mainly *Acropora*) appeared similar to those on adjacent ridges, calcite–aragonite determinations showed the presence of an abnormally large amount of calcite (33% in a split of the dated sample and 95% in another sample from the same locality) of the low magnesium phase. Thin section and s.e.m. examination of stick corals from the same unit indicated post-death contamination by both the presence of intraskeletal void deposits (minor) and recrystallization (major). The radiocarbon age reported here is therefore not considered a valid age for the death of coral.

Separation of matrix and skeletal components proved difficult in some of the rampart-rocks (see § 3.2). The presence of aragonite (15–29%) in high magnesium cement samples (ANU-1381, 1601, 1602) may reflect the occurrence of precipitated aragonite. However, it is unlikely that all coral and other aragonitic skeletal particles incorporated in the matrix were removed.

Regrettably, the effect the observed degree of contamination has on the validity of the radiocarbon ages cannot be quantified. Nevertheless, mineralogical and microscope examination suggests that the great bulk of skeletal, rock and sediment sample ages are valid. Of the groups of samples listed in table 1 it is believed that of the surface samples only ANU-1600 gives an unreliable age, although some others should be used with caution. Drill core dates beyond 30000 a B.P. on recrystallized material should be considered as minimum ages only. The last interglacial (*ca.* 120000 a B.P.) is a more likely age for coral growth (Thom *et al.* 1978, this volume), but until unrecrystallized corals can be found this assumption cannot be tested, for example by uranium series dating.

5. RADIOCARBON AGE DETERMINATIONS

5.1. *Pretreatment in the Radiocarbon Laboratory*

Selection of samples, based on stratigraphic and geomorphologic criteria and physical pretreatment (see § 3) was carried out by the collector in consultation with the ^{14}C laboratory. Hence samples considered for ^{14}C age determinations were dated by the A.N.U. Radiocarbon Laboratory *as submitted*, without any further physical or chemical pretreatment.

5.2. *Method of radiocarbon age determination*

All results listed here were produced at the A.N.U. Radiocarbon Dating Laboratory. The low level liquid scintillation counting techniques used were described by Polach & Stipp (1967) and Polach (1969, 1974). However, it cannot and should not be taken for granted that errors of measurement, as expressed by the statistical error term associated with the results, fully expresses the range of possible laboratory errors. Indeed, proof of correlation and cross-checking on an international basis as well as duplicate pair agreement analysis and long term stability, such as given by Polach (1972), ought to be sought by the discriminating user wishing to evaluate the validity of the radiocarbon ages produced. The standard used for dating is the A.N.U. sucrose contemporary radiocarbon dating standard, a secondary standard whose ^{14}C activity with respect to the international standard N.B.S. oxalic acid, has been carefully evaluated. All ages are reported with respect to 0.95 oxalic acid as reference standard and are in conventional radiocarbon years B.P. (see below).

5.3. *Calculation of ^{14}C ages for the Great Barrier Reef samples*

In this paper all radiocarbon ages are expressed as Conventional Radiocarbon Ages B.P. (Olsson 1970). This means that the age was calculated relative to 95% of the activity of the National Bureau of Standards (N.B.S.) oxalic acid normalized for isotopic fractionation to δ^{13}C $= -19.0‰$ (editorial statement in *Radiocarbon* **3**, 1961) with respect to the ^{13}C/^{12}C ratio of a marine standard, PDB, a Cretaceous belemnite, *Belemnitella americana* from the Peedee formation of South Carolina (Craig 1957). Further, these conventional radiocarbon ages were calculated using Libby ^{14}C half-life of 5568; they were not corrected for secular variations of the ^{14}C/^{12}C ratio, but were corrected for ^{13}C/^{12}C isotopic fractionation of the sample.

The normal procedures for calculating ^{14}C ages have been published by Callow, Baker & Hassall (1965) and Polach (1969, 1976) and can be presented, somewhat simplified, as follows:

The activity of the sample is expressed as a per mille (‰) ratio of the standard:

$$d_{^{14}\text{C}} = 10^3 \left(\frac{A_{\text{sample}}}{0.95 A_{\text{oxal}}} - 1 \right), \tag{1}$$

where A_{sample} is the statistically averaged activity of the sample from which the natural background count rate was subtracted, and A_{oxal} is the activity, in 1950, of N.B.S. oxalic acid corrected for background and isotopic fractionation; 95 % of A_{oxal} is assumed to be equal to the activity of terrestrial plants (more precisely wood) in 1950.

Variation in the isotopic composition of the samples, caused by observed natural ^{13}C/^{12}C isotopic fractionation processes in nature (Rafter 1955) is corrected by normalizing the $d_{^{14}\text{C}}$ activity to the value which it would have been if the sample measured had been wood. The average values for terrestrial plants (wood) is taken to be $\delta^{13}\text{C} = -25‰$ with respect to PDB (cf. tabulation of mean δ^{13}C, relative to PDB, of materials used for radiocarbon dating in Polach (1969, fig. 5, p. 6; 1976, fig. 4, p. 268) and Olsson & Osadebe (1974, fig. 1, p. 141)).

$$D_{^{14}\text{C}} = d_{^{14}\text{C}} - 2(\delta^{13}\text{C} + 25)(1 + 10^{-3} d_{^{14}\text{C}}). \tag{2}$$

(The annotation $D_{^{14}\text{C}}$ has been used to denote the relative ^{14}C activity corrected for ^{13}C/^{12}C isotopic fractionation (δ^{13}C), but not age corrected, secular variation corrected or industrial effect corrected.)

Then, by using the half-life of 5568 years the conventional ^{14}C age B.P. (a_c) is obtained from:

$$a_c = -\{8033 \ln(1 + 10^{-3} D_{^{14}\text{C}})\}, \tag{3}$$

where 8033 is the *mean life* derived from 5568/ln 2.

6. Validity of ^{14}C age determinations

In order to evaluate further the validity of the ^{14}C ages, the following effects have to be considered.

6.1. *Isotopic fractionation effect*

The isotopic composition of carbon found in nature is *ca.* 98.9 % ^{12}C, 1.1 % ^{13}C and 1×10^{-10} % ^{14}C. Of these, ^{12}C and ^{13}C are non-radioactive. The isotopic composition of carbon-bearing compounds in nature varies, and isotopic fractionation is one of the factors responsible for this variation. While all carbon isotopes follow the same chemical reaction pathways, the rate at which they do so is related to their respective mass differences. Craig (1953) was the first to point out that there is a need to allow for the effect of carbon isotopic fractionation in nature by applying a correction to radiocarbon dates based on mass spectrometric measurement of the stable ^{13}C and ^{12}C isotopes. Lerman (1972) and Olsson & Osadebe (1974) have discussed and presented the total evidence available including references to early work in great detail, and Polach (1969, 1976) has presented in graphical form the observed variations. The stable isotopic composition of ^{13}C and ^{12}C is expressed as a ratio, ^{13}C/^{12}C of the sample (R) and its millesimal (‰) deviation ($\delta^{13}C$) from the $^{13}C/^{12}C$ ratio of a standard (R_0)†, by following the equation

$$10^3 \left(\frac{R - R_0}{R_0}\right) = \delta^{13}C = 10^3 \left[\frac{R}{R_0} - 1\right]. \tag{4}$$

The way in which this $\delta^{13}C$ correction factor enters our $^{14}C/^{12}C$ ratio measurement, our enrichment or depletion with respect to oxalic standard, is shown in equation (2). However, a further explanation may be appropriate. According to Craig (1954), '...the enrichment of C^{14} in a given compound should be almost exactly twice that of C^{13} in both equilibrium and rate reaction isotopic effects'. This works out in radiocarbon dating in the following way: for a 10‰ change in the $\delta^{13}C$ value we must expect a 20‰ change in the d_{14C} value. Given that a 1% change in d_{14C} throughout the age range represents a change of *ca.* 80 years and taking the extremes of $\delta^{13}C$ variations in nature into account, a marine shell at $\delta^{13}C = +3‰$ and ter-restrial plant remains (peat) at $\delta^{13}C = -30‰$ with respect to PDB have a natural difference in ^{13}C of 33‰ and consequently in ^{14}C a difference of 66‰, giving an apparent age discrepancy of some 545 years for these materials living at the same time.

The importance of the dating error due to fractionation, which may be much larger than the ^{14}C counting standard deviation, is the primary reason why some radiocarbon dating labora-tories have made mass spectrometric measurements of $\delta^{13}C$ an integral part of their work (e.g. the British Museum, New Zealand, Gröningen, Uppsala; see *Radiocarbon*) and why we at A.N.U. have since 1968 (Polach, Golson, Lovering & Stipp 1968) either measured or estimated the $\delta^{13}C$ value and applied a fractionation correction accordingly. Thus, for the study reported here, a value of $\delta^{13}C = 0 \pm 2‰$ with respect to PDB has been assumed to represent the mean of the natural $^{13}C/^{12}C$ ratio variations of samples in this marine environment. This, as a first approxi-mation, is adequate considering the context in which the results are going to be used as the approximate age errors involved can be shown to be *ca.* 20 years; ± 20 years is then the expected order of magnitude of variation in age due to natural $\delta^{13}C$ variations of the Great Barrier Reef samples.

Nevertheless, it is our intention in the near future to determine the precise $^{13}C/^{12}C$ ratio of all samples reported here, not only to derive a more correct radiometric age, but also to evaluate some geochemical aspects of carbonate deposition, mobility and exchange.

6.2. *Environmental effects*

The ^{14}C concentration of a sample is affected by the environment in which the material to be dated lived or was deposited. The inference in radiocarbon dating is that during its biological life (or at time of deposition) the ^{14}C concentration of the sample was equal to that of the radiocarbon *modern* reference standard.

The ocean and marine species living in the ocean are unevenly depleted in ^{14}C with respect to the land environment. For example, Rafter, Jansen, Lockerbie & Trotter (1972) report a value for cockle shells (*Protothaca crassitesta* (Deshayes)) collected in October 1955 (I.N.S. R.42) of $D_{14C} = -54 \pm 5‰$‡, a value which appeared to be in equilibrium with the Makara (N.Z.)

---

† A few early $\delta^{13}C$ values may be found reported with respect to wood, air, limestone, etc. We at A.N.U., together with the majority of radiocarbon laboratories, report $\delta^{13}C$ variations exclusively with respect to PDB.

‡ Note that Rafter further corrects this value for industrially produced CO_2 dilution which, because it is ^{14}C free, caused a depletion of ^{14}C activity in both the terrestrial and marine environments. Thus Rafter *et al.* (1972) obtained an industrial effect corrected value, $D_{14C} = -40‰$ for the cockle shell, which gives thus an apparent industrial effect corrected age of 330 a. This is the value used by Rafter as his marine environment age correction.

surface seawater D_{14C} values at that time. This value gives an apparent age for the cockle shell of 446 ± 40 a B.P., which is in excellent agreement with an average apparent age of 450 ± 40 a for marine shells from the Norwegian coast (Mangerud 1972), and the average apparent age of 450 ± 35 a for marine shells from the Australian east and south coast (Gillespie & Polach 1976). Contrast this with the higher southern latitudes, particularly where there is a much greater variation due to deep water upwelling. Rafter (1968) reports $D_{14C} = -269\%_0$, which is equivalent to an apparent age of surface waters of 2520 a B.P. Nevertheless, based on the study by Gillespie & Polach (1976), one can assume a relatively stable surface layer of ocean waters between latitudes $40°$ N and $40°$ S, with an apparent ^{14}C age of 450 ± 35 a, a value which we suggest be subtracted from the conventional radiocarbon ages reported here. Should one wish or have need to establish a chronology relative to the Christian calendar, this environmental correction has to be applied before looking up the secular variation correction factors in one of the many published tables (see, for example, Damon, Long & Wallick 1972). We have not made any of these corrections in this paper.

The assumption of uniform mixing of the marine environment at the Great Barrier Reef will be further investigated by us, as we have collected a cross section of living coral and other marine life and plant species, of which the *Halimeda* (ANU-1272 and 1273, table 2) constitute a preview. These results confirm, considering the overall increase of ^{14}C due to atom bomb testing, that our first assumptions of uniform mixing are essentially correct.

TABLE 2. RADIOCARBON RESULTS, GREAT BARRIER REEF EXPEDITION, 1973

A.N.U. code	$\delta^{13}C \pm \sigma$ (‰)	$d_{14C} \pm \sigma$ (‰)	$D_{14C} \pm \sigma$ (‰)	% modern $\pm \sigma$	conventional radiocarbon years B.P. $\pm$ error
1207	0.0 ± 2.0	-426.0 ± 4.4	-454.7 ± 4.7	54.5 ± 0.5	4870 ± 70
1208	0.0 ± 2.0	-260.5 ± 5.8	-297.5 ± 6.2	70.3 ± 0.6	2840 ± 70
1272A	-4.0 ± 2.0	$+178.7 \pm 6.4$	$+129.2 \pm 7.7$	112.9 ± 0.8	$>$ modern
1272B	-24.0 ± 2.0	$+175.5 \pm 7.1$	$+173.1 \pm 8.5$	117.3 ± 0.9	$>$ modern
1273A	-4.0 ± 2.0	$+171.3 \pm 4.9$	$+122.2 \pm 6.6$	112.2 ± 0.7	$>$ modern
1273B	-24.0 ± 2.0	$+117.5 \pm 7.1$	$+115.3 \pm 8.4$	111.5 ± 0.8	$>$ modern
1280	0.0 ± 2.0	-975.9 ± 3.2	-977.1 ± 3.1	2.3 ± 0.3	$>30350 \pm 1150$
1281	0.0 ± 2.0	-985.5 ± 3.9	-986.2 ± 3.7	1.4 ± 0.4	$>34400^{+2500}_{-1900}$
1282	0.0 ± 2.0	-989.8 ± 2.4	-990.4 ± 2.3	1.0 ± 0.2	$>37300^{+2200}_{-1700}$
1283	0.0 ± 2.0	-537.5 ± 6.9	-560.6 ± 6.8	43.9 ± 0.7	6610 ± 130
1284	0.0 ± 2.0	-555.0 ± 7.1	-577.3 ± 6.9	42.3 ± 0.7	6920 ± 130
1285	0.0 ± 2.0	-335.7 ± 6.6	-368.9 ± 6.8	63.1 ± 0.7	3700 ± 90
1286	0.0 ± 2.0	-448.5 ± 7.7	-514.0 ± 7.6	48.6 ± 0.8	5800 ± 130
1287	0.0 ± 2.0	-492.0 ± 10.1	-517.4 ± 9.9	48.3 ± 1.0	5850 ± 170
1380	0.0 ± 2.0	-340.1 ± 8.6	-373.1 ± 8.6	62.7 ± 0.9	3750 ± 110
1381	0.0 ± 2.0	-205.6 ± 7.3	-245.4 ± 7.6	75.5 ± 0.8	2260 ± 80
1382	0.0 ± 2.0	-280.0 ± 5.1	-316.0 ± 5.6	68.4 ± 0.6	3050 ± 70
1383	0.0 ± 2.0	-322.1 ± 5.8	-356.0 ± 6.2	64.4 ± 0.6	3540 ± 80
1384	0.0 ± 2.0	-216.7 ± 6.2	-255.9 ± 6.7	74.4 ± 0.7	2370 ± 70
1385	0.0 ± 2.0	-28.4 ± 6.9	-77.0 ± 7.6	92.3 ± 0.8	640 ± 70
1386	0.0 ± 2.0	-182.6 ± 6.4	-223.5 ± 6.9	77.7 ± 0.7	2030 ± 70
1387	0.0 ± 2.0	-270.5 ± 6.1	-307.0 ± 6.5	69.3 ± 0.7	2950 ± 80
1388	0.0 ± 2.0	-303.8 ± 6.0	-338.6 ± 6.3	66.1 ± 0.6	3320 ± 80
1389	0.0 ± 2.0	-11.9 ± 7.1	-61.3 ± 7.9	93.9 ± 0.8	510 ± 70
1390	0.0 ± 2.0	-47.8 ± 6.8	-95.5 ± 7.5	90.5 ± 0.8	810 ± 70
1391	0.0 ± 2.0	-132.0 ± 6.6	-175.4 ± 7.1	82.5 ± 0.7	1550 ± 70
1392	0.0 ± 2.0	-124.4 ± 6.7	-168.2 ± 7.2	83.2 ± 0.7	1480 ± 70
1393	0.0 ± 2.0	-27.9 ± 6.9	-76.5 ± 7.6	92.4 ± 0.8	640 ± 70

TABLE 2 (cont.)

A.N.U. code	δ¹³C ± σ (‰)	$d_{14C} \pm \sigma$ (‰)	$D_{14C} \pm \sigma$ (‰)	% modern ± σ	conventional radiocarbon years B.P. ± error
1394	0.0 ± 2.0	− 384.5 ± 6.7	− 415.3 ± 6.8	58.5 ± 0.7	4 310 ± 100
1395	0.0 ± 2.0	− 524.0 ± 6.5	− 547.8 ± 6.5	45.2 ± 0.7	6 380 ± 120
1410	0.0 ± 2.0	− 295.9 ± 6.0	− 331.2 ± 6.3	66.9 ± 0.6	3 230 ± 80
1411	0.0 ± 2.0	− 79.1 ± 5.1	− 125.2 ± 6.1	87.5 ± 0.6	1 070 ± 60
1412	0.0 ± 2.0	− 277.1 ± 5.9	− 313.2 ± 6.3	68.7 ± 0.6	3 020 ± 70
1413	0.0 ± 2.0	− 323.6 ± 5.8	− 357.4 ± 6.1	64.3 ± 0.6	3 550 ± 80
1414	0.0 ± 2.0	− 295.0 ± 5.8	− 330.2 ± 6.2	67.0 ± 0.6	3 220 ± 80
1475	0.0 ± 2.0	− 122.3 ± 6.9	− 166.2 ± 7.4	83.4 ± 0.7	1 460 ± 70
1476	0.0 ± 2.0	− 13.1 ± 7.4	− 62.5 ± 8.1	93.8 ± 0.8	520 ± 70
1477	0.0 ± 2.0	− 118.5 ± 7.1	− 162.6 ± 7.6	83.7 ± 0.8	1 430 ± 70
1478	0.0 ± 2.0	− 392.9 ± 6.1	− 423.3 ± 6.3	57.7 ± 0.6	4 420 ± 90
1479	0.0 ± 2.0	− 428.4 ± 6.0	− 457.0 ± 6.1	54.3 ± 0.6	4 910 ± 90
1480A	− 24.0 ± 2.0	− 126.7 ± 7.7	− 128.4 ± 8.4	87.2 ± 0.8	1 100 ± 80
1480B	− 24.0 ± 2.0	− 239.4 ± 15.1	− 240.9 ± 15.4	75.9 ± 1.5	2 210 ± 170
1553	0.0 ± 2.0	− 305.9 ± 5.7	− 340.6 ± 6.2	65.9 ± 0.6	3 350 ± 80
1554	0.0 ± 2.0	− 331.1 ± 5.6	− 364.5 ± 6.0	63.5 ± 0.6	3 640 ± 70
1555	0.0 ± 2.0	− 296.8 ± 5.7	− 331.9 ± 6.2	66.8 ± 0.6	3 240 ± 70
1556	0.0 ± 2.0	− 212.5 ± 6.0	− 251.9 ± 6.5	74.8 ± 0.6	2 330 ± 70
1557	0.0 ± 2.0	− 233.6 ± 6.0	− 272.0 ± 6.4	72.8 ± 0.6	2 550 ± 70
1558	0.0 ± 2.0	− 352.1 ± 5.6	− 384.5 ± 6.0	61.6 ± 0.6	3 900 ± 80
1559	0.0 ± 2.0	− 389.5 ± 5.5	− 420.0 ± 5.7	58.0 ± 0.6	4 380 ± 80
1560	0.0 ± 2.0	− 271.3 ± 6.0	− 307.8 ± 6.3	69.2 ± 0.6	2 960 ± 70
1591	0.0 ± 2.0	− 90.9 ± 6.2	− 136.4 ± 7.0	86.4 ± 0.7	1 180 ± 65
1592	0.0 ± 2.0	− 311.9 ± 5.6	− 346.3 ± 6.0	65.4 ± 0.6	3 420 ± 75
1593	0.0 ± 2.0	− 47.4 ± 6.5	− 95.1 ± 7.2	90.5 ± 0.7	800 ± 70
1594	0.0 ± 2.0	− 47.3 ± 6.5	− 95.0 ± 7.2	90.5 ± 0.7	800 ± 60
1595	0.0 ± 2.0	− 304.7 ± 5.7	− 339.4 ± 6.1	66.1 ± 0.6	3 330 ± 80
1596	0.0 ± 2.0	− 244.8 ± 5.9	− 282.6 ± 6.3	71.8 ± 0.6	2 670 ± 70
1597	0.0 ± 2.0	− 227.4 ± 6.1	− 266.0 ± 6.5	73.4 ± 0.6	2 480 ± 70
1598	0.0 ± 2.0	− 253.7 ± 6.6	− 291.0 ± 6.9	70.9 ± 0.7	2 760 ± 80
1599	0.0 ± 2.0	− 94.7 ± 6.7	− 140.0 ± 7.4	86.0 ± 0.7	1 210 ± 70
1600	0.0 ± 2.0	− 52.5 ± 6.8	− 99.9 ± 7.5	90.0 ± 0.8	(?) > 850 ± 70
1601	0.0 ± 2.0	+ 3.7 ± 9.3	− 46.5 ± 9.7	95.4 ± 1.0	380 ± 80
1602	0.0 ± 2.0	− 214.4 ± 5.9	− 253.7 ± 6.5	74.6 ± 0.6	2 350 ± 70
1603	0.0 ± 2.0	− 506.1 ± 5.3	− 530.8 ± 5.4	46.9 ± 0.5	6 080 ± 90
1604	0.0 ± 2.0	− 304.1 ± 5.7	− 338.9 ± 6.1	66.1 ± 0.6	3 320 ± 70
1605	0.0 ± 2.0	− 221.2 ± 6.2	− 260.2 ± 6.7	74.0 ± 0.7	2 420 ± 70
1606	0.0 ± 2.0	− 212.8 ± 6.3	− 252.2 ± 6.7	74.8 ± 0.7	2 330 ± 70
1607A	0.0 ± 2.0	− 40.0 ± 6.7	− 88.0 ± 7.5	91.2 ± 0.8	740 ± 70
1607B	0.0 ± 2.0	− 18.3 ± 12.2	− 67.4 ± 12.2	93.3 ± 1.2	560 ± 110
1608	0.0 ± 2.0	− 42.2 ± 6.4	− 90.1 ± 7.2	91.0 ± 0.7	760 ± 65
1609	0.0 ± 2.0	− 183.3 ± 6.4	− 224.2 ± 6.9	77.6 ± 0.7	2 040 ± 70
1639	0.0 ± 2.0	− 433.8 ± 5.2	− 462.1 ± 5.5	53.8 ± 0.6	4 980 ± 80
1639R	0.0 ± 2.0	− 432.2 ± 4.9	− 460.6 ± 5.2	53.9 ± 0.5	4 960 ± 80
1640	0.0 ± 2.0	− 520.3 ± 5.0	− 544.3 ± 5.1	45.6 ± 0.5	6 310 ± 90
1641	0.0 ± 2.0	− 198.3 ± 6.2	− 238.4 ± 6.7	76.2 ± 0.7	2 190 ± 70
1642	0.0 ± 2.0	− 300.3 ± 6.0	− 335.3 ± 6.3	66.5 ± 0.6	3 280 ± 80
1643	0.0 ± 2.0	− 274.5 ± 6.0	− 310.8 ± 6.4	68.9 ± 0.6	2 990 ± 80
1663	0.0 ± 2.0	− 287.0 ± 6.1	− 322.6 ± 6.5	67.7 ± 0.7	3 130 ± 80
1664	0.0 ± 2.0	− 289.8 ± 7.5	− 325.3 ± 7.7	67.5 ± 0.8	3 160 ± 90
1721	0.0 ± 2.0	− 453.0 ± 8.3	− 480.3 ± 8.5	52.0 ± 0.8	5 260 ± 130

6.3. *Post-depositional and contamination effect*

It is also important to evaluate whether the sample submitted for radiocarbon dating remains unaffected by possible isotopic and other chemical processes causing an addition or exchange with extraneous carbon of different ¹⁴C activity.

The standard technique used in radiocarbon laboratories is to wash shell samples to be dated in a weak acid, thus attempting to remove the outer part, the one most likely to be contaminated by secondary carbonate or, in other circumstances, by solution and ionic exchange of carbonate forms. No one has shown that acid treatment removes the carbonate from shells in a definite time sequence. In coral, with its multifaceted internal labyrinth, such treatment is certainly not appropriate. Dating the inner and outer portions of shells, as suggested by Olsson & Blake (1962) with the assumption that agreement is proof of validity, is also often undertaken. However, inner and outer portions often yield no ^{14}C concentration differences, suggesting no isotope replacement (cf. Shotton, Blundell & Williams 1970); yet such agreement would also be expected even if a sample has fully recrystallized in an open system where exchange of secondary carbon of different ^{14}C activity is possible (Chappell & Polach 1972†). Thus dating of fractions, in absence of X-ray diffraction data, may not constitute proof of validity of the ^{14}C age determination.

6.4. *Summary and conclusions*

The foregoing multidisciplinary approach to the problem of obtaining valid radiocarbon dates can be summarized as follows:

(*a*) careful selection of dating material according to visual and stratigraphic criteria;

(*b*) removal of the part most likely to be contaminated; in molluscs, physical removal of the outer part and in corals physical removal of all parts showing secondary carbonate (calcite or aragonite) deposition or intrusion;

(*c*) mineralogical analysis to determine aragonite/calcite content and Mg calcite phase;

(*d*) thin section microscopic analysis;

(*e*) $^{13}C/^{12}C$ ratio determinations;

(*f*) $^{14}C/^{12}C$ ratio (age) determinations and evaluation of their validity.

Thus where it can be shown that post-depositional exchange and contamination of samples dated for the Great Barrier Reef project were minimal (or their magnitude could be evaluated) the conventional ^{14}C ages B.P. reported here may be considered as valid.

7. DESCRIPTION OF SAMPLES

The following list has been arranged on an island-by-island basis starting from Raine Island in the north and continuing southward to Low Isles. Each entry includes the island or reef name and location, Australian National University Radiocarbon Dating Laboratory code number, Expedition sample number, and radiocarbon date in conventional ^{14}C years. These data are followed by brief sample descriptions, locations and collector's name.

Raine Island (11° 35′ S:144° 02′ E)

ANU-1591   RAI-172   1180 ± 65 a B.P.

Tridacna shell from surface of beach-rock near beacon at eastern end of Raine Island. Beach-rock overlain by guano rock. Coll. D. Hopley.

† Chappell & Polach (1972) have shown that two modes of carbonate recrystallization are possible: the open system, allowing exchange with secondary carbon, and the closed system, where recrystallization occurs involving only the carbonate species already locked into the material. They have further shown that the $^{13}C/^{12}C$ isotopic ratio, even coupled with calcite/aragonite ratio determinations, is not a good guide towards establishing whether secondary contamination has occurred during recrystallization. Indeed, microscopic thin section examination of crystalline forms involved in the two processes appears to be a more reliable guide of evaluating the validity of ^{14}C ages. Thus demonstration of recrystallization does not prove that exchange of secondary carbonate has occurred unless it is accompanied by the coarse sparry calcite crystalline structure of the open system.

Fisher Island

| ANU-1640 | FIS-228 | 6310 ± 90 a B.P. |

Tridacna shell resting on fossil microatolls in growth position at base of cemented coral shingle platform on southern side of Fisher Island. Sample from 0.6 m below platform surface. Coll. D. Hopley.

Stainer Reef (13° 57′ S:143° 50′ E)

| ANU-1639 | STA-106 | 4980 ± 80 a B.P. |
| ANU-1639R | STA-106 | 4960 ± 80 a B.P. |

Coral *Favites abdita* from area of fossil microatolls in growth position which emerge 5 cm above sandy reef flat, some 250 m ESE of Stainer sand cay. ANU-1639R is a repeat determination of the same specimen. Coll. D. Hopley.

Stapleton Reef (14° 20′ S:144° 50′ E)

ANU-1663	CORE-1-1	3130 ± 80 a B.P.
ANU-1664	CORE-1-2-3	3160 ± 90 a B.P.
ANU-1721	CORE-1-2-3	5260 ± 130 a B.P.

Drill core series: ANU-1663 was collected from a depth of 8.5 m below h.w.s.t. and consists of loose calcareous sand; ANU-1664 is a reef calcarenite fragment from *ca.* 10 m below h.w.s.t.; ANU-1721 is an unidentified coral fragment (possibly *Galaxa* sp.) from 13 m below h.w.s.t. Coll. B. G. Thom.

| ANU-1555 | ST-6D | 3240 ± 70 a B.P. |

Moderately well sorted clean coarse calcareous sand. Bulk sample from depth of 55–80 cm beneath grassed surface of 2.3 m high sand cliff exposure on south side of Stapleton cay some 100 m from its western end. Coll. R. McLean.

Bewick Reef (14° 28′ S:144° 47′ E)

| ANU-1387 | BE-5B | 2950 ± 80 a B.P. |

Moderately sorted medium sized calcareous sand. Bulk sample of partly weathered creamy-brown sand from soil pit horizon 25–40 cm beneath grassed surface on eastern slope of easternmost ridge on Bewick sand cay, some 25 m west of cay–mangrove margin. Coll. R. McLean.

| ANU-1559 | BE-1A | 4380 ± 80 a B.P. |

Moderately sorted medium sized calcareous sand. Bulk sample of greyish weathered sand from depth of 15 cm beneath surface at crest of westernmost ridge on Bewick sand cay. Located some 25 m southeast of drill hole site. Coll. R. McLean.

| ANU-1386 | BE-D1 | 2030 ± 70 a B.P. |

Tridacna shell from surface of high beach-rock outcrop on north shore of Bewick sand cay close to drill site. Base of tridacnid valve was lightly cemented to beach-rock surface, which had *Sesuvium* and algae growing on it. Beach-rock outcrop was fronted by mangroves and backed by a sandy slope covered with grasses and *Pemphis*. Coll. R. McLean.

ANU-1280	CORE 2-15-4	> 30 350 ± 1150 a B.P.
ANU-1281	CORE 2-13-2	> 34 400$^{+2500}_{-1900}$ a B.P.
ANU-1282	CORE 2-9-1	> 37 300$^{+2200}_{-1700}$ a B.P.
ANU-1283	CORE 2-7-1	6610 ± 130 a B.P.

ANU-1284   CORE 2-3-5 6920 ± 130 a B.P.

ANU-1395   CORE 2-4-1 6380 ± 120 a B.P.

Drill core series: samples ANU-1280, 1281 and 1282 were recovered below a disconformity in the drill core, as indicated by recrystallization of biomicrite (1280) and corals (1281, 1282) to low magnesium calcite. The ^{14}C results must be considered as *minimum ages* only. Samples ANU-1283, 1284 and 1395 are aragonite rich coral fragments (possibly *Porites* sp.). Sample 1283 was recovered from below the disconformity. The sample's location is believed to be a product of cave-in during drilling. The disconformity occurs at *ca.* 4 m below l.w.s.t., with sample ANU-1395 located at the contact. Coll. B. G. Thom.

ANU-1272A   % modern $= 112.9 \pm 0.8$

ANU-1272B   % modern $= 117.3 \pm 0.9$

ANU-1273A   % modern $= 112.2 \pm 0.7$

ANU-1273B   % modern $= 111.5 \pm 0.8$

Contemporary environment ^{14}C levels check: two samples of calcium carbonate secreting algae ANU-1272, 1273 were collected to check contemporary ocean bicarbonate ^{14}C levels. Both the organic carbon (fraction B, assumed $\delta^{13}C = -24\%_0$ PDB) and the inorganic carbonate (fraction A, assumed $\delta^{13}C = -4\%_0$ PDB) were dated. Results indicate the predictable increase in ^{14}C activity due to atom bomb testing and confirm that it is appropriate to subtract 450 ± 35 a (Gillespie & Polach 1976) from the conventional radiocarbon years B.P. (as given with all results reported here) in order to allow for pre-atom bomb (i.e. pre-1950) depletion of ^{14}C in the Coral Sea water mass. Coll. H. Polach.

ANU-1385   BE-D10B 640 ± 70 a B.P.

Coral *Platygyra* from loosely cemented coral shingle deposit with bassett edge – lower platform morphology on northeastern side of Bewick reefs, traverse no. 1. Outcrop located 100 m landward of reef edge, between inner edge of reef flat and 1.5 m high scarp of high platform which forms seaward side of gravel cay. Coll. R. McLean.

ANU-1208   BE-D11 2840 ± 70 a B.P.

ANU-1609   BE-D111 2040 ± 70 a B.P.

Coral (ANU-1208) and *Tridacna* (ANU-1609) from strongly cemented rampart rock of high platform 20 m southwest of ANU-1385 on traverse no. 1, northeastern part of gravel cay on Bewick reef. Both samples were firmly embedded in the corroded platform surface. Coll. R. McLean.

ANU-1608   BE-115 760 ± 65 a B.P.

Tridacna shell from surface of the innermost of three shingle ramparts on traverse no. 2, on eastern side of Bewick reef. The sample was exposed on a grassy slope which abutted the mangrove swamp 100 m west of the gravel cay's active beach. Coll. R. McLean.

Ingram–Beanley Reef (14° 25′ S: 144° 55′ E)

ANU-1393   IN-D1 640 ± 70 a B.P.

ANU-1394   IN-D2 4310 ± 100 a B.P.

Coral *Porites* from separate reef blocks of boulder tract on northern side of reef, 300 m northeast of Ingram Island. ANU-1393 was collected from the largest single coral head of the boulder tract. Its top was 2.2 m above reef flat level and was located 80 m from the reef edge. ANU-1394 was the innermost solitary coral head, 1.5 m high, surrounded by sandy reef flat 130 m from the reef edge. Coll. R. McLean.

| ANU-1642 | IN-3 | 3280 ± 80 a B.P. |
| ANU-1410 | IN-6B | 3230 ± 80 a B.P. |

Two calcareous sand samples from Ingram Island. Both bulk samples were moderately well sorted coarse sand collected from 2–5 cm beneath the cay surface. ANU-1642 was located at 50 m from the beachline in the centre of a low cuspate promontory on the western side of the island. ANU-1410 came from the highest part of the cay, 120 m northwest of its southeastern end. Coll. R. McLean.

Watson Island (14° 29′ 5:144° 56′ E)

ANU-1389	WN-103	510 ± 70 a B.P.
ANU-1390	WN-105	810 ± 70 a B.P.
ANU-1391	WN-109	1550 ± 70 a B.P.
ANU-1392	WN-110	1480 ± 70 a B.P.

A series of samples taken from the outer rampart to innermost fossil ridge across the southeastern end of Watson, a coral shingle island. All samples were loose surface shingle. ANU-1389 was a *Tridacna* shell from the highest part of the contemporary outer rampart. ANU-1390, a coral favid from the top of lower platform. ANU-1391 and ANU-1392 were tridacnids from the penultimate and innermost ridges respectively on the northern side of the island proper abutting a mangrove swamp. Coll. R. McLean.

Howick Cay (14° 30′ S:144° 57′ E)

| ANU-1605 | HCK-D3 | 2420 ± 70 a B.P. |

Coral *Diploastrea heliopora* from series of massive coral heads cemented into high calcarenite platform on southwestern corner of Howick Cay. Coll. R. McLean.

Houghton Reef (14° 32′ S:144° 58′ E)

| ANU-1287 | HON-D7A | 5850 ± 170 a B.P. |

Faviid coral head from area of fossil microatolls in mangrove swamp north-central part of reef, 220 m south of reef edge and 400 m east of sand cay. Coll. R. McLean.

ANU-1595	HON-D6	3330 ± 80 a B.P.
ANU-1596	HON-D5	2670 ± 70 a B.P.
ANU-1413	HON-D3	3550 ± 80 a B.P.

Vertical sequence from high calcarenite platform at western end of Houghton sand cay. ANU-1595, faviid coral head beneath calcarenite at juncture of inner edge of reef flat and rocky scarp. Coll. R. McLean. ANU-1596, calcarenite bulk sample containing calcareous sand and cement. Coll. R. McLean. ANU-1413, *Tridacna* shell from 20 cm thick cemented stick coral layer overlying calcarenite. Coll. H. Polach.

Coquet Island (14° 32′ S:145° E)

| ANU-1411 | CQT-D5 | 1070 ± 60 a B.P. |

Tridacna shell from surface of highest part of broad coral shingle ridge 80 m southeast of navigation light on main cay. Coll. R. McLean.

Leggatt Reef (14° 33′ S:144° 51′ E)

| ANU-1556 | LEG-105 | 2330 ± 70 a B.P. |

Moderately sorted coarse calcareous sand. Bulk sample from surface of high flat in centre of sand cay. Coll. R. McLean.

ANU-1286   LEG-D4A   5800 ± 130 a B.P.

Tridacna shell in growth position among high microatoll field exposed in area of fallen mangrove immediately east of sand cay and south of sand spit. Coll. R. McLean.

Hampton Reef (14° 33' S : 144° 52' E)

ANU-1207   HAM-D15   4870 ± 70 a B.P.

Faviid coral in growth position from area of high microatolls covered by *Rhizophora* mangroves and mangrove mud. Coll. B. G. Thom.

East Pethebridge Island (14° 45' S : 145° 05' E)

ANU-1384   EPE-D1A   2370 ± 70 a B.P.

Faviid coral in growth position from area of high microatolls in gap between cemented shingle platforms at southwestern end of island. Coll. R. McLean.

Turtle I Island (14° 44' S : 145° 13' E)

ANU-1597   TON-103   2480 ± 70 a B.P.
ANU-1598   TON-107   2760 ± 80 a B.P.

Calcareous sandy-gravel from soil pit 1 on lower terrace northwestern side of island 25 m south of beach line. Bulk samples. ANU-1597 at depth of 25–35 cm and ANU-1598 at depth of 85–100 cm. Coll. R. McLean.

ANU-1388   TON-111   3320 ± 80 a B.P.

Coral *Acropora* fragments from soil pit 2 on high shingle ridge 25 m south of pit 1. Sample from depth of 40–55 cm. Coll. R. McLean.

ANU-1477   TURT-028   1430 ± 70 a B.P.
ANU-1478   TURT-024   4420 ± 90 a B.P.

Tridacna shells from separate outcrops of cemented rampart rock between mangrove swamp and moat on southwestern part of reef. ANU-1477 from outer platform 10 cm below upper level of cementation. ANU-1478 from inner platform 24 cm below surface. Coll. D. Hopley.

ANU-1480A   TURT-030   1100 ± 80 a B.P.
ANU-1480B   TURT-030   2210 ± 170 a B.P.
ANU-1479   TURT-027   4910 ± 90 a B.P.

Samples from shallow borehole in small enclosed depression between two old coral shingle ridges at southeastern end of main island. Depression is occupied by living mangroves and floored by a 0.6 m thick deposit of black fibrous mud which overlies at least 0.3 m of coral shingle. ANU-1480, fibrous mud, was separated into two fractions: 1480A, coarse fibres, rootlets and bark retained of sieve meshes 10 and 25; and 1480B, fine grey clayey sediment which passed through 44 mesh and contained black organics but not fibres. ANU-1479, coral *Cyphastrea* from shingle containing corals and shells beneath mangrove mud horizon. Coll. D. Hopley and A. Bloom.

Nymph Island (14° 38' S : 145° 15' E)

ANU-1285   NY-D4A   3700 ± 90 a B.P.

Faviid coral in growth position from high microatoll field exposed in drainage outlet to large pool in southwest portion of island. Coll. R. McLean.

ANU-1592   NYM-033   3420 ± 75 a B.P.
ANU-1383   NY-D6   3540 ± 80 a B.P.
ANU-1602   NY-D5   2350 ± 70 a B.P.

Samples from high platform of cemented rampart rock located 100 m southeast of pond outlet in southwest part of reef. ANU-1592, *Tridacna* shell from cemented stick shingle layer at top of high platform. Coll. D. Hopley. ANU-1383, *Tridacna* shell firmly embedded in rock, 0.5 m from base of scarp. ANU-1602, calcitic cement from 30 cm beneath platform surface. Coll. R. McLean.

| ANU-1476 | NYM-035 | 520 ± 70 a B.P. |

Tridacna shell from cemented lower platform at southeastern end of small shingle island on southwestern part of reef. Sample embedded in rock 15 cm below upper surface of platform. Coll. D. Hopley.

Eagle Reef (14° 25′ S:145° 23′ E)

| ANU-1560 | EAG-101 | 2960 ± 70 a B.P. |

Well sorted coarse calcareous sand. Bulk sample from main sand ridge in centre of northern side of cay, 50 m from beach. Coll. R. McLean.

Two Isles (15° 03′ S:145° 27′ E)

| ANU-1558 | TWO-198 | 3900 ± 80 a B.P. |

Well sorted coarse calcareous sand. Bulk sample from depth of 45–60 cm in soil pit located beneath forest on backslope of high sand ridge 60 m from beach in north central part of main cay. Coll. R. McLean.

Low Wooded Island (15° 06′ S:145° 25′ E)

| ANU-1594 | LWI-D7 | 800 ± 60 a B.P. |

Coral *Porites* cf. *lobata* microatoll in growth position at junction of inner edge of contemporary shingle rampart and moat 190 m southeast of aeroplane wreckage at western end of island. Coll. R. McLean.

| ANU-1603 | LWI-010 | 6080 ± 90 a B.P. |

Coral *Platygyra lamellina* microatoll passing beneath lower cemented shingle platform at junction with moat in south central part of island. Coll. D. Hopley.

| ANU-1604 | LWI-D2 | 3320 ± 70 a B.P. |

Tridacna shell from surface of high cemented platform outcropping at southwestern end of enclosed pool 80 m west of beach on eastern side of island. Coll. R. McLean.

Three Isles (15° 06′ S:145° 27′ E)

| ANU-1641 | THR-105 | 2190 ± 70 a B.P. |

Moderately well sorted coarse calcareous sand. Bulk sample from depth of 20 cm in soil pit on lower terrace, 75 m west of beacon at western end of sand cay. Coll. R. McLean.

| ANU-1554 | THR-112 | 3640 ± 70 a B.P. |

Well sorted coarse calcareous sand. Bulk sample from depth of 60 cm in soil pit in shallow basin in high ridge 40 m east of beacon at western end of cay. Coll. R. McLean.

| ANU-1553 | THR-119 | 3350 ± 80 a B.P. |
| ANU-1414 | THR-122 | 3220 ± 80 a B.P. |

Well sorted coarse calcareous sand from two horizons exposed in 2.5 m high cliff at eastern end of cay. Bulk samples at 43–68 cm (ANU-1553) and 100–160 cm (ANU-1414) below top of cliff. Coll. R. McLean.

| ANU-1475 | THR-017 | 1460 ± 70 a B.P. |

Tridacna shell cemented 20 cm beneath surface of lower platform at exposed edge above moat on southeastern side of Third Island. Coll. D. Hopley.

ANU-1380	THR-D31	3750 ± 110 a B.P.
ANU-1381	THR-D32	2260 ± 80 a B.P.
ANU-1382	THR-D1	3050 ± 70 a B.P.

Sequence from high rampart rock platform outcropping in central eastern part of mangrove–shingle cay. ANU-1380, coral *Pavona* firmly cemented in basal facies of platform. ANU-1381, calcitic matrix surrounding ANU-1380. ANU-1382, *Tridacna* shell from loosely cemented 20 cm thick coral shingle veneer on upper surface of high platform. Coll. R. McLean.

East Hope Island (15° 45′ S : 145° 28′ E)

| ANU-1412 | EHO-110 | 3020 ± 70 a B.P. |
| ANU-1643 | EHO-116 | 2990 ± 80 a B.P. |

Moderately well sorted coarse calcareous sand. Bulk samples from 79–86 cm horizon in soil pit 1 located on highest ridge in centre of cay (ANU-1412) and 30 cm depth in pit 3 located on low sand terrace, 15 m from beach line at eastern end of cay (ANU-1643). Coll. R. McLean.

West Hope Island (15° 45′ S : 145° 27′ E)

| ANU-1599 | WHO-105 | 1210 ± 70 a B.P. |

Coral *Acropora* fragments from 40 cm beneath surface of easternmost ridge in shingle ridge sequence found on northeastern side of island. This ridge is being cut back exposing a 0.5 m high scarp immediately behind the present beach. Coll. R. McLean.

| ANU-1600 | WHO-108 | (?) > 850 ± 70 a B.P. |

Coral *Acropora* fragments from surface of highest shingle ridge in sequence 75 m from eastern beach and 25 m from mangrove swamp to west. Sample possessed large quantity of post-death contaminants. Coll. R. McLean.

Pickersgill Cay (15° 52′ S : 145° 33′ E)

| ANU-1606 | PIK-101 | 2330 ± 70 a B.P. |

Well sorted coarse calcareous sand. Bulk sample from top of unvegetated sand bank. Coll. R. McLean.

Low Isles (16° 24′ S : 145° 33′ E)

ANU-1607A	LOW-D18	740 ± 70 a B.P.
ANU-1607B	LOW-D18	560 ± 110 a B.P.
ANU-1601	LOW-D18C	380 ± 80 a B.P.

Samples from cemented coral shingle deposit forming bassett edges on inner edge of reef flat 250 m northwest of Green Ant Island on eastern side of reef. ANU-1607A, coral *Acropora*. ANU-1607B and ANU-1601 were high magnesium calcite matrix. Coll. R. McLean.

| ANU-1593 | LOW-108 | 800 ± 70 a B.P. |

Tridacna shell from surface of inner shingle rampart at southern end of mangrove–shingle island on eastern side of reef. Coll. R. McLean.

| ANU-1557 | LOW-106 | 2550 ± 70 a B.P. |

Moderately sorted very coarse calcareous sand. Bulk sample from sand cay surface 75 m west of lighthouse. Coll. R. McLean.

We wish to thank J. Head and J. Gower from the A.N.U. Radiocarbon Dating Laboratory for their technical assistance throughout this project; Dr J. Veron, Australian Institute of Marine Science, Townsville; and Dr D. R. Stoddart, University of Cambridge, for his leadership throughout the Expedition.

References (Polach *et al.*)

Callow, W. J., Baker, M. J. & Hassall, G. I. 1965 *Radiocarbon* **7**, 156–161.
Chappell, J. & Polach, H. A. 1972 *Quat. Res.* **2**, 244–252.
Chave, K. E. 1952 *J. Geol.* **60**, 190–192.
Craig, H. 1953 *Geochim. cosmochim. Acta* **3**, 53–92.
Craig, H. 1954 *J. Geol.* **62**, 115–149.
Craig, H. 1957 *Geochim. cosmochim. Acta* **12**, 133–140.
Damon, P. E., Long, A. & Wallick, E. I. 1972 In *Proceedings of 8th International Conference on Radiocarbon Dating*, Lower Hutt, New Zealand, pp. 44–59. Wellington: Royal Society of New Zealand.
Gillespie, R. & Polach, H. A. 1976 In *Proceedings, 5th International Conference on Radiocarbon Dating*, University of California, Los Angeles and San Diego, July 1976. (In the press.)
Lerman, J. C. 1972 In *Proceedings of 8th International Conference on Radiocarbon Dating*, Lower Hutt, New Zealand, pp. 612–624. Wellington: Royal Society of New Zealand.
McLean, R. F. & Stoddart, D. R. 1978 *Phil. Trans. R. Soc. Lond.* A **291**, 101–117 (this volume).
Mangerud, J. 1972 *Boreas* **1**, 263–273.
Olsson, I. U. 1970 In *Radiocarbon variations and absolute chronology.* p. 17. Stockholm: Almquist & Wiksell.
Olsson, I. U. & Blake, W. Jr 1962 *Saertr. norsk geog. Tidsskr. XVIII:* **12**, 1–18.
Olsson, I. U. & Osadebe, F. A. N. 1974 *Boreas* **3**, 139–146.
Polach, H. A. 1969 *Atomic Energy in Australia* **12**(3), 21–28.
Polach, H. A. 1972 In *Proceedings of 8th International Conference on Radiocarbon Dating*, Lower Hutt, New Zealand, pp. 688–717. Wellington: Royal Society of New Zealand.
Polach, H. A. 1974 In *Liquid scintillation counting: recent developments* (eds P. E. Stanley & B. A. Scoggins), pp. 153–171. New York: Academic Press.
Polach, H. A. 1976 In *Proceedings of Symposium on Scientific Methods of Research in the Study of Ancient Chinese Bronzes and Southeast Asian Metal and other Archaeological Methods* (ed. N. Barnard), pp. 255–298. Canberra: Department Far Eastern History, A.N.U.
Polach, H. A. & Stipp, J. J. 1967 *Int. J. appl. Radiat. Isotopes* **18**, 359–364.
Polach, H. A., Golson, J., Lovering, J. F. & Stipp, J. J. 1968 *Radiocarbon* **10**, 179–199.
Rafter, T. A. 1955 *N.Z. J. Sci. Technol.* B **37**, 20–38.
Rafter, T. A. 1968 *N.Z. J. Sci.* **11**, 551–589.
Rafter, T. A., Jansen, H. S., Lockerbie, L. & Trotter, M. M. 1972 In *Proceedings of 8th International Conference on Radiocarbon Dating*, Lower Hutt, New Zealand, pp. 625–675. Wellington: Royal Society of New Zealand.
Scoffin, T. & McLean, R. F. 1978 *Phil. Trans. R. Soc. Lond.* A **291**, 119–138 (this volume).
Shotton, F. W., Blundell, D. J. & Williams, R. E. G. 1970 *Radiocarbon* **12**, 385–399.
Thom, B. G. 1978 In *Coral reefs: research methods* (eds D. R. Stoddart & R. E. Johannes). Monographs on Oceanographic Methodology, Unesco.
Thom, B. G., Orme, G. R. & Polach, H. A. 1978 *Phil. Trans. R. Soc. Lond.* A **291**, 37–54 (this volume).

Phil. Trans. R. Soc. Lond. A. **291**, 159–166 (1978) [159]

Printed in Great Britain

Sea level change on the Great Barrier Reef: an introduction

By D. Hopley

Department of Geography, James Cook University of North Queensland, Townsville, Queensland, Australia 4810

Evidence for Holocene shorelines from the Queensland coast, off which the Great Barrier Reef lies, has epitomized the problems of eustatic fluctuations over the last 6000 years. While some areas of southern and central Queensland show evidence of no sea level higher than the present over this period, other areas, particularly within 150 km of Townsville on the mid-North coast, have provided radiometrically dated evidence for an emergence of up to 4.9 m. The area in which the 1973 Expedition worked has been described previously by several authors, and evidence for higher shorelines in the form of cemented platforms, raised reefs and related features suggesting higher sea levels, though without isotopic dating, has been noted. Research was aimed at confirming and accurately measuring and dating such evidence and relating it to the pattern described elsewhere. Any divergences must then be explained in terms of spatially and temporally varying oceanographic or geomorphic conditions and Earth movements of tectonic and/or isostatic origin.

The nature and magnitude of sea level changes over the Holocene period, and in particular the length of time sea level has been close to its present position, have direct implications for coral reef response. Unfortunately, it is over this period that the greatest divergence of views on sea levels exists, both on a world wide scale (e.g. Guilcher 1969; Mörner 1971*a*, *b*; Curray & Shepard 1972) and in Australia (Hails 1965; Thom, Hails & Martin 1969; Hopley 1971*a*; Gill & Hopley 1972; Thom, Hails, Martin & Phipps 1972; Thom & Chappell 1975) (figure 1).

Evidence from eastern Queensland

The evidence from the Queensland coast (figure 2) in many ways epitomizes the global problem (Hopley 1974*a*). To the south of the reef, at Coolum on the mainland, Thom *et al.* (1969) indicate that sea level has not been higher than present in the Holocene. However, 500 km to the north at Broad Sound, detailed mapping and dating of chenier sequences by Cook & Polach (1973) show a continuous progradation of the shoreline for 5500 years during which sea level has been continuously at or close to its present position. In contrast, at Stannage Bay on the peninsula to the east of Broad Sound, Jardine (1928) has described a raised beach rock rising to about 2 m, and offshore on Hunter Island along the same structural alignment, Steers (1937, 1938) has noted two higher cemented levels, the upper one rising to about 2.7 m. Although not dated, the similarity of these two sites to those described further north suggests that they record higher Holocene sea levels. No work pertinent to sea levels has been reported from the 90 km of coastline between Broad Sound and Mackay, but to the north the off shore islands of the Cumberland Group have been surveyed in detail (Hopley 1975). Within this island group, conclusive evidence of higher sea levels is lacking though at Cockermouth Island, a notch in a Pleistocene dune calcarenite at about 1 m, producing an enigmatic date of 6980 ± 130 a b.p., may represent a marginally higher mid Holocene sea level (see

Hopley (1975) for discussion). Elsewhere, the oldest Holocene deposits appear to be no more than 3000 years in age.

Immediately north of the Whitsunday Passage, from the northern extremity of the Cumberland Group, the offshore islands and mainland display an abundance of evidence related to higher sea levels in the form of raised reef and beach-rock on the islands, and mangrove peats and other depositional evidence from the mainland. Maximum levels exceeding 3 m are

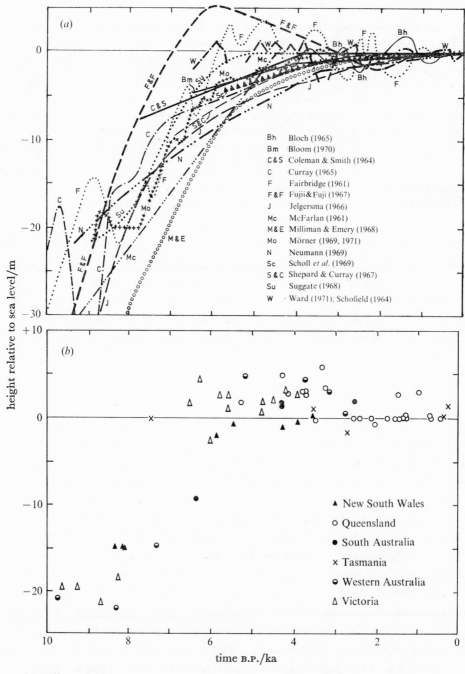

FIGURE 1. (a) Different interpretations of sea level history in the last 10 000 years. (b) Radiocarbon ages indicating sea level positions on Australian coastlines in the last 10 000 years.

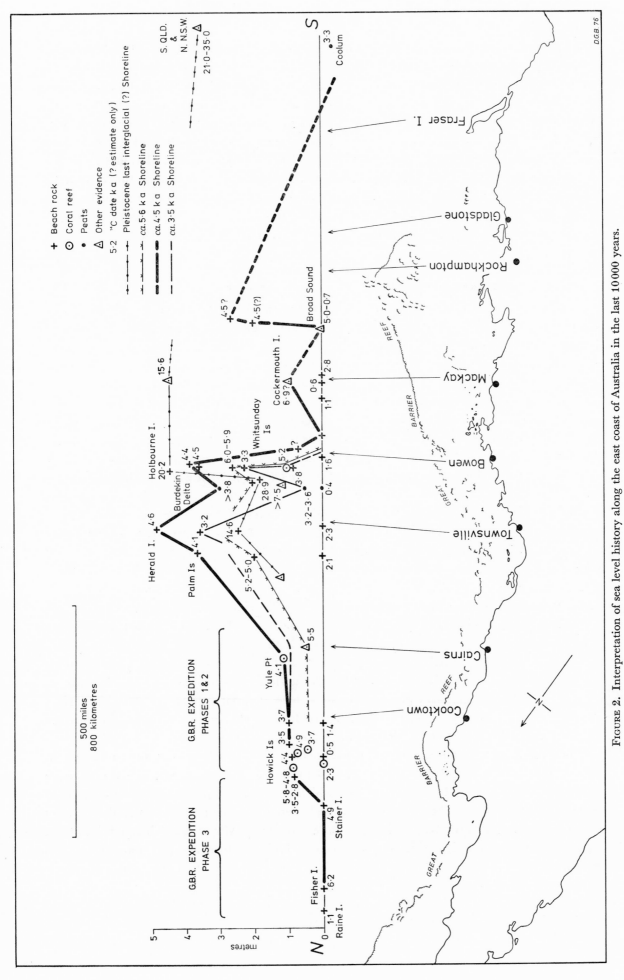

FIGURE 2. Interpretation of sea level history along the east coast of Australia in the last 10000 years.

maintained for 180 km at least as far north as Hinchinbrook Island. Significantly, shorelines of three separate ages may be recognized in this area: a largely terrigenous beach deposit of reworked corestones from the last glacial regolith marking the first time Holocene sea levels reached present level and dating between 6000 and 5000 a B.P.; a maximum level rising at its highest to 4.9 m and dating close to 4500 a B.P.; and a regression shoreline somewhat lower with dates in the order of 3500 a B.P. Islands in the Bowen area suggest higher sea levels 3.7–3.9 m above present (Hopley 1975), the Burdekin delta only 3.0 m (Hopley 1970), Herald Island near Townsville 4.9 m, and the Palm Islands 3.7 m (Hopley 1971 b). It is unfortunate that the coastline between the Palm Islands and Cairns has not been studied in detail, for it is clear from the work of Bird (1970, 1971 a, b) that at Cairns and at Yule Point just to the north, although a higher Holocene sea level is recorded, its maximum height does not exceed 1.2 m.

The area north of Cairns from pre-1973 literature

To the north of Cairns lies the area investigated by the 1973 Royal Society Expedition. In the light of the variation in levels displayed further south and the features described from the islands north of Cairns by earlier expeditions, evidence for higher sea levels might be expected. Research was thus aimed at confirming and accurately measuring and dating such evidence and relating it to the pattern of the whole Queensland coastline.

Most of the evidence described north of Cairns comes from the low wooded islands of the reef patches closest to the mainland. However, the cemented platforms which comprise much of this evidence have been related to reef flat level and not to the modern counterpart level of cementation. Indeed, the exact nature and origin of the platforms has not been determined. Thus Agassiz (1898) describes a 'coarse beach rock' elevated 2.4 m above h.w.m. on Three Isles and a similar feature on nearby Two Isles, while Steers (1929, 1937, 1938) and Spender (1930) describe two levels of cementation on these islands and others of the Turtle and Howick Groups further north, suggesting emergences in the order of 1.5 and 4.0 m. However, at Nymph Island at least, the difference in height between the two platforms is only 1 m (Steers 1938), and only on Nymph Island is there supplementary evidence in the form of in situ raised reef. Unfortunately, before the work of the 1973 Expedition, the only radiocarbon dates for this part of the Great Barrier Reef confused rather than clarified the sea level picture. Maxwell's (1969) dates of 250 ± 80 a B.P. for 'high beach rock' on Turtle Reef and 530 ± 80 a B.P. for reef-rock from Nymph appear anomalously young or cannot be related to higher sea levels.

To the north of the Howicks, Steers (1938) describes other low wooded islands, up to latitude 12° 53′ S (Chapman Island). However, although the 'low platform' related to the lower of his two emergences is ubiquitously present the higher level is apparently absent. Possibly parallel evidence from the Flinders Islands is mentioned by Lucas & de Keyser (1965), but they give no exact locations, heights or descriptions of their site.

Still further north the amount of data available becomes even sparser and more ancient in its source. Jukes (1847) and Rattray (1869) describe raised beach rock and reef on Raine Island and adjacent parts of the Great Barrier Reef which Fairbridge (1950) considers to indicate an emergence of 3.1–3.7 m. There are also reports of similar elevated features in the Torres Strait region (e.g. MacGillivray 1852; Haddon, Sollas & Cole 1894; Mayer 1918) though it is difficult to define precisely the height differences between the higher features and their modern counterparts.

Discussion of the evidence

The evidence presented suggests that not only are there discrepancies in the maximum height of sea level during the Holocene along the Queensland coast (variations between 0 and 4.9 m above present sea level) but also great variation exists in the time at which sea level reached its present position (variations between *ca.* 6000 and *ca.* 3000 years) (figure 2). Further, it would appear that sharp dislocations occur in the Holocene shorelines notably to the east of Broad Sound, and at the northern end of the Whitsunday Passage, though with the present spacing of sites studied in detail the dislocations may be spread over as much as 25 km.

Explanation of the variation should incorporate reasons for the sharpness of the dislocations. However, an even greater problem may be found by examining the height of Pleistocene last interglacial shorelines along the same coast (figure 2). With dislocations of almost 5 m in the Holocene shorelines, vertical variation of at least this amplitude might be expected in the older levels. This does not take place. As in the remainder of eastern Australia, a double barrier system representing beach ridge accumulation of Holocene and presumably last interglacial age is found intermittently along the entire Queensland coast. Surveyed heights of the Pleistocene shoreline do not exceed 5 m (figure 2). The vertical range is apparently less than that for the mid-Holocene shorelines. Moreover, the trend does not follow the pattern of the younger shorelines.

Three sets of reasons may be identified to explain these apparent anomalies in the Holocene shorelines.

1. *Error in measurement or interpretation*

The relatively narrow height range (5 m) within which the variations lie may be exceeded by short term oscillations of the sea surface by waves, tides, and occasional tsunami and storm surges. Exact height measurement should be made but this is not always possible, especially on a coast such as that of Queensland where bench marks and spot heights are rare. However, the relatively close network of tidal stations may allow heights to be related to tidal datum with confidence. Although some error may still be incorporated in the heights quoted, this is unlikely to be significant compared to the total height range of the data.

Error may also be associated with the radiocarbon dating technique. Variation may be produced by the non-uniform distribution of the amount of ^{14}C isotope in ocean waters and by geomagnetic and biomagnetic modulation of the production of radiocarbon in the Earth's atmosphere. Problems may be especially related to recrystallization of marine carbonates (see Thom 1973). In the majority of results quoted, however, the dated materials were highly aragonitic and stratigraphic cross checking was possible by obtaining a number of dates at each site.

A more serious error may be caused by misinterpretation of the evidence. Failure to recognize the height variation of shore platforms related to exposure conditions, compaction of peats or the high nature of storm beaches are obvious examples. In the region of the Great Barrier Reef shore platform evidence has not been utilized in spite of the widespread quotation in the literature of raised platforms (see, for example, Hedley 1925; Stanley 1928). Detailed examination of platforms in the Whitsunday area indicated that the majority of such features are high tide, near horizontal 'water layer weathering' type phenomena in which height variations of up to 1.5 m may exist along short stretches of coast. A major control is the height at which wet season spring lines emerge onto the coast, a factor related to the spacing of fracture zones and joints in the volcanic and granitic rocks. Problems may exist in the utilization of beach-rock which may be confused with other cemented materials (dune calcarenite, reef-rock, phosphatic cay sandstone, humic sandrock) but in general, beach bedding and a horizontal upper surface of cementation may be recognized in the sites utilized and the relation of the maximum elevation of the upper surface of cementation with m.h.w.s. appears firmly established. Corals also may be related to specific tidal levels though enclosure of reef flat waters behind shingle ramparts or algal rims may raise the levels of moated corals above those of their open water counterparts. A conservative estimate of the amount of exposure of raised corals has been made by comparing them with the highest level at which similar species are growing today, even where these may be moated.

2. *Real spatial variations in the level of ocean surface*

Even allowing for variations produced by waves and tides, it should be recognized that the ocean surface is non-horizontal; variations of up to 2 m occur with factors such as temperature and salinity (Fairbridge 1966), and significant slopes are associated with ocean currents as the result of centrifugal and Coriolis forces (see, for example, Hamon 1958). Tidal range may also change through time, an important factor when evidence utilized is related to tidal extremes (e.g. beach-rock and coral) and not to mean sea level. Changes have undoubtedly taken place in the last 6000 years in the configuration of the Great Barrier Reef, but no estimate can be made of the effect that this would have on tides along the Queensland coast. This may be an important factor on the south central coast where tidal ranges vary from about 3 m near Bowen to almost 10 m at Broad Sound. Also important along the coast are the meteorological influences on sea level. Storm surges up to 6 m have been recorded (Hopley 1974b) and many shingle ridges and other features on reef islands undoubtedly owe their origin to these short-lived events.

3. *Movements of the land*

Tectonic dislocation of the shorelines of eastern Queensland is an obvious cause of variation, particularly as the sharp dislocations coincide with major structural breaks and changes in the alignment of the coastline (e.g. the northern end of the Whitsunday Passage is a highly fractured zone of horst–graben structures (Hopley 1975)). Even more convincing is the apparent relation between the basement structure of eastern Queensland and the highs and lows in Holocene shoreline heights. The high shorelines near Townsville, for example, correspond with the major granitic axis of uplift, the lack of high evidence in the Cumberland Islands with the Tertiary Hillsborough Basin, the eastern side of Broad Sound with the northern limb of the south coast structural high; and the Coolum area of no higher shorelines with the Mesozoic Maryborough Basin (Hopley 1974a).

However, eastern Queensland is not recognized as a zone of high seismicity and the rates of tectonic deformation required to warp the Holocene shorelines are more compatible with a plate edge rather than mid-plate situation. Further, the heights of the Pleistocene shorelines appear to contradict any form of tectonic dislocation though it is possible that the inner barrier shorelines are interstadial levels uplifted to their present height (see Hopley 1974a). However, the coincidence in height of shorelines of differing ages and complete lack of higher and older shorelines in the areas of apparent uplift appears to militate against this explanation.

Isostatic warping may be more acceptable. Movements of hundreds of metres have long been recognized to be associated with the major Pleistocene ice sheets of the northern hemisphere.

More recently it has been calculated that the load applied to continental shelves by the Holocene transgression is sufficient to produce a response (Bloom 1967; Walcott 1972; Chappell 1974). As the nature of response depends on the morphology of the ocean basins, the shape of the continental shelf and the strength of crustal materials, geographical variation in the amount, pattern and time of isostatic warping may be expected. Such an explanation contributes much towards explaining the discrepancies along the Queensland coast. Notably, the evidence for higher shorelines comes from the high continental islands generally within 15 km of the mainland, the area in which Chappell (1974), at least, considers the most likely for upwarping. Further, the presence of three islands at varying distances up to 30 km off the coast of Bowen, all retaining evidence of three higher shorelines, has allowed the identification of greater slopes on shorelines normal to the coast than exist parallel to it, a situation which again might be expected from hydro-isostatic response models. As shelf morphology and structure are important determinants of hydro-isostatic response, it is not surprising that a correspondence exists between the pattern of shorelines and major structural regions. Apparently, however, much of the stress imposed on the shelf area has been released along pre-existing weaknesses. Sissons (1972) has described a similar but smaller dislocation in glacio-isostatically warped shorelines where they cross older fault lines in eastern Scotland. If a hydro-isostatic deformation is accepted, then Pleistocene shorelines may have undergone a similar amount of warping during the glacial low sea levels when the load was removed from the shelf. Only with the reapplication of the load are these shorelines resuming a near-horizontal disposition.

Local sediment loading may explain the lower elevations of Holocene shorelines in the Burdekin delta and the down-warping of the last interglacial surface beneath the present delta (Hopley 1970). Over 150 m of deltaic sediments are recorded in the delta, with up to 30 m of Holocene deposits along its eastern edge.

THE NORTHERN GREAT BARRIER REEF IN THE LIGHT OF THE DISCUSSION

The apparent explanation of discrepancies in Holocene shorelines along the Queensland coast by hydro-isostasy, possibly aided by temporal variations in oceanic factors such as tides, can only be confirmed by detailed study of the whole of the shelf area. In this respect the Great Barrier Reef offers a unique opportunity because, especially north of Cairns in the area of the 1973 Expedition, gauges in the form of coral reefs exist up to the very edge of the continental shelf. The reefs closest to the mainland may display both the longest record of Holocene sea levels at or close to their present level and any evidence related to higher sea levels of this period. Maximum subsidence on the outer edge of the shelf should produce the youngest reef surfaces here. From the literature, however, Raine Island, detached on the outer edge of the shelf and with recorded raised features, appears anomalous.

A multidisciplinary approach to this area of the Reef may provide answers to the sea level problem, which may in turn help in explaining the variable morphology of the Great Barrier Reefs, as for example, the decreasing degree of development of reef flats outwards across the shelf. Variable rates of reef growth, and the nature of the underlying karst surface over which the Holocene veneer has formed, should however, be combined with explanations of variable sea level curves in providing answers to Great Barrier Reef problems.

References (Hopley)

Agassiz, A. 1898 *Harvard Mus. comp. Zoology Bull.* **28**, 93–148.

Bird, E. C. F. 1970 *Aust. Geogr.* **11**, 327–335.

Bird, E. C. F. 1971 *a Search* **2** (1), 27–28.

Bird, E. C. F. 1971 *b Aust. geogr. Stud.* **9**, 107–115.

Bloch, M. R. 1965 *Palaeogeog. Palaeoclimatol. Palaeoecol.* **1**, 127–142.

Bloom, A. L. 1967 *Bull. geol. Soc. Am.* **78**, 1477–1494.

Bloom, A. L. 1970 *Bull. geol. Soc. Am.* **81**, 1895–1904.

Chappell, J. 1974 *Quat. Res.* **4**, 405–428.

Coleman, J. M. & Smith, W. G. 1964 *Bull. geol. Soc. Am.* **75**, 833–840.

Cook, P. J. & Polach, H. A. 1973 *Mar. Geol.* **14**, 253–268.

Curray, J. R. 1965 In *The Quaternary of the United States* (eds H. E. Wright, Jr & D. G. Frey), pp. 723–735. Princeton: Princeton University Press.

Curray, J. R. & Shepard, F. P. 1972 *Abstracts, American Quaternary Association Second National Conference*, pp. 16–18.

Fairbridge, R. W. 1950 *J. Geol.* **58**, 330–401.

Fairbridge, R. W. 1961 *Phys. Chem. Earth* **4**, 99–185.

Fairbridge, R. W. 1966 *Encyclopaedia of oceanography* (ed. R. W. Fairbridge), pp. 479–482.

Fujii, S. & Fuji, N. 1967 *J. Geosci. Osaka City Univ.* **10**, 43–51.

Gill, E. D. & Hopley, D. 1972 *Mar. Geol.* **12**, 223–233.

Guilcher, A. 1969 *Earth Sci. Rev.* **5**, 69–97.

Haddon, A. C., Sollas, W. J. & Cole, G. A. J. 1894 *R. Irish Acad. Trans.* **30**, 419–476.

Hails, J. R. 1965 *Aust. geogr. Stud.* **3**, 63–78.

Hamon, B. V. 1958 *Aust. Surveyor* (Sept.) 188–199.

Hedley, C. 1925 *Repts Great Barrier Reef Comm.* **1**, 61–62.

Hopley, D. 1970 *The geomorphology of the Burdekin Delta, north Queensland.* Department of Geography, James Cook University, Monograph Series, No. 1, 66 pages.

Hopley, D. 1971 *a Quaternaria* **14**, 265–276.

Hopley, D. 1971 *b Z. Geomorph.* N.F. **15**, 371–389.

Hopley, D. 1974 *a Proc. Second Int. Coral Reef Symp.* vol. 2, pp. 551–562.

Hopley, D. 1974 *b Aust. Geogr.* **12**, 462–468.

Hopley, D. 1975 In *Geographical essays in honour of Gilbert J. Butland* (eds I. Douglas, J. E. Hobbs & J. J. Pigram), pp. 51–84. Armidale, N.S.W.: Department of Geography, University of New England.

Jardine, F. 1928 *Repts Great Barrier Reef Comm.* **2**, 88–92.

Jelgersma, S. 1967 In *World climate from 8000 to 0* B.C., pp. 54–71. London: Royal Meteorological Society.

Jukes, J. B. 1847 *Narrative of the surveying voyage of H.M.S. Fly*, 2 volumes. London.

Lucas, K. G. & de Keyser, F. 1965 1:250 000 Geological series: Cape Melville, Queensland. *Bureau of Mineral Resources Explanatory Notes.* Canberra: Department of Natural Development.

McFarlan, E. Jr 1961 *Bull. geol. Soc. Am.* **72**, 129–158.

MacGillivray, W. 1852 *Narrative of a voyage of H.M.S. 'Rattlesnake' 1846–50*, 2 volumes. London.

Maxwell, W. G. H. 1969 *Sediment. Geol.* **3**, 331–333.

Mayer, A. G. 1918 *Carnegie Instn Wash. Dept. of Marine Biol. Papers*, **19**, 51–72.

Milliman, J. D. & Emery, K. O. 1968 *Science, N.Y.* **162**, 1121–1123.

Mörner, N.-A. 1969 *Sveriges Geol. Undersokn.* C, **640**, 1–487.

Mörner, N.-A. 1971 *a Palaeogeog. Palaeoclimatol. Palaeoecol.* **9**, 153–181.

Mörner, N.-A. 1971 *b Geol. Mijnb* **50**, 699–702.

Neumann, A. C. 1969 *Abstracts, VIII Inqua Congress*, Paris 1969, pp. 228–229.

Rattray, A. 1869 *Q. J. geol. Soc. Lond.* **25**, 297–305.

Schofield, J. C. 1964 *N.Z. J. Geol. Geophys.* **7**, 359–370.

Scholl, D. W., Craighead, F. C. Sr & Stuiver, M. 1969 *Science, N.Y.* **163**, 562–564.

Shepard, F. P. & Curray, J. R. 1967 *Progr. Oceanog.* **4**, 283–291.

Sissons, J. B. 1972 *Trans. Inst. Br. Geog.* **55**, 194–214.

Spender, M. A. 1930 *Geogrl J.* **76**, 194–214.

Stanley, G. A. V. 1928 *Repts Great Barrier Reef Comm.* **2**, 1–51.

Steers, J. A. 1929 *Geogr. J.* **74**, 232–257 and 341–370.

Steers, J. A. 1937 *Geogr. J.* **89**, 1–28 and 119–146.

Steers, J. A. 1938 *Repts Great Barrier Reef Comm.* **4**, 3, 51–96.

Suggate, R. P. 1968 *Geol. Mijnb.* **47**, 291–297.

Thom, B. G. 1973 *Prog. Geog.* **5**, 170–246.

Thom, B. G. & Chappell, J. 1975 *Search* **6** (3), 90–93.

Thom, B. G., Hails, J. R. & Martin, A. R. H. 1969 *Mar. Geol.* **7**, 161–168.

Thom, B. G., Hails, J. R., Martin, A. R. H. & Phipps, C. V. G. 1972 *Mar. Geol.* **12**, 233–242.

Walcott, R. I. 1972 *Quat. Res.* **2**, 1–14.

Ward, W. T. 1971 *Geol. Mijnb.* **50**, 703–718.

Phil. Trans. R. Soc. Lond. A. **291**, 167–186 (1978)   [167]

Printed in Great Britain

Sea level change in the Holocene on the northern Great Barrier Reef

By R. F. McLean,† D. R. Stoddart,‡ D. Hopley§ and H. Polach‖

†*Department of Biogeography and Geomorphology, Australian National University, Canberra,*
Australia 2600

‡*Department of Geography, University of Cambridge, Downing Place, Cambridge, U.K.*

§*Department of Geography, James Cook University of North Queensland, P.O. Box* 999, *Townsville,*
Queensland, Australia 4810

‖*Radiocarbon Dating Laboratory, Australian National University, Canberra, Australia* 2600

[Plates 1 and 2]

Detailed studies, utilizing a range of both well controlled sea level criteria and dates, are required if Holocene time–sea level curves are to be established with any degree of confidence. This paper is restricted to an interpretation of Expedition results from the northern Great Barrier Reef, excluding those from the drill core. Extensive colonies of emergent fossil corals in growth position indicate that present sea level was first reached about 6000 a B.P. Elevations of cay surfaces, cemented rubble platforms, microatolls, coral shingle ridges, reef flats and mangrove swamps, referenced to present sea level show an array of heights. However, levels of particular features are accordant on many reefs: it is believed that these can be related to particular sea levels. Radiometric dating provides the time framework. Ages of samples from similar deposits on different reefs are surprisingly consistent. Oscillations in sea level since 6000 a B.P., relative to present sea level, are identified with varying degrees of confidence. This history of relative sea level does not separate eustatic from non-eustatic components.

1. Introduction

While a massive rise in the level of the sea (> 100 m) since the maximum of the last glaciation is universally accepted, there is little agreement as to when this transgressing sea first reached its present position in the Holocene. Nor is there agreement as to the directions and magnitudes of sea level change since that time. Thus there is even conflict as to whether the most recent significant change has been a fall or rise in sea level. Detailed local studies, with the use of a range of both well controlled sea level criteria and dates, are required if Holocene time–sea level curves are to be accepted with any degree of confidence. Because reef areas possess an array of present and palaeo sea level markers together with abundant datable materials they are particularly appropriate sites for the investigation of recent sea level changes. Nevertheless, serious problems in identifying changes based on reef data exist, although they are not always readily acknowledged. During the Royal Society – Universities of Queensland Expedition to the northern Great Barrier Reef in 1973 we became particularly conscious of both the utility of reef data and problems associated with its interpretation. In this paper results from that Expedition are discussed. Our intention is to present the evidence for a recent sea level history by utilizing data solely obtained from the Expedition without recourse to earlier commentaries on this or adjacent areas of the Great Barrier Reef (for a review of these see Hopley 1978, this volume) nor to sea level histories from other regions. In addition we document the problems found and assumptions used in interpreting the evidence.

2. Problems associated with reef data

2.1. *Establishment of common tidal datum*

The most optimistic changes in sea level since the main Holocene transgression are of the order of ± 4 m relative to present sea level. Thus, except in areas of high tectonic activity, any evidence for recent sea level history is within a few metres of present mean sea level. Most deposits, surfaces and features which may be related to former sea levels fall within the present intertidal range or within the reach of contemporary storm wave levels. On the northern Great Barrier Reef the maximum altitude of any low island is less than 10 m above extreme low water level. Given the concentration of features within such a limited vertical range levelling accuracy and reference to a survey datum become very important. The problem of obtaining absolute levels between particular features on the same reef, and of comparing altitudes of equivalent deposits on surfaces between reefs is acute. Ideally, a precise datum is required on each reef visited. These do not exist in the northern Great Barrier Reef. Therefore, the establishment of a realistic common baseline as a substitute is necessary. For our work all levelled profiles were reduced to low water datum, which is the mean height of lower low waters at spring tides. Individual levels have been related to low water datum by reference to predicted tides at Cairns: for most of the area investigated the predicted tides do not differ significantly in amplitude or timing from those at Cairns, and corrections to the Cairns curves only need to be made north of Cape Melville.

It is important to realize that there are several sources of error in these reductions and we cannot accurately evaluate their magnitude. First, there may be differences in tidal curves between Cairns, the secondary stations for which predictions are available (these include several of the islands studied, notably Green Island, Hope Islands, Howick Island, Low Isles, Low Wooded Island), and the islands on which profiles were measured. Secondly, local tidal levels may have been distorted by meteorological effects: the Trades blew strongly through most of the Expedition, and it would be surprising if on some days at least local tidal levels were not distorted by up to 0.3 m. Third, and perhaps most important, all reductions to datum were made by observing a still water level at a known time on a particular profile and relating this to a predicted tide curve. Determination of a still water level is often difficult, either because of rough conditions, especially at middle and high waters, or because of ponding of water on reef flats at low tides. Relation of the still water level to the tidal curve is also more difficult on irregular low-amplitude neap tides than on springs.

In the absence of tidal records at each of the sites surveyed the height data represent the best estimate of elevations related to the mean level of lower low water springs as datum, and enable different sections to be directly compared. All heights are given in metres. For comparison, at Cairns, mean high water springs are at 2.3 m, mean high water neaps at 1.6 m, mean low water neaps at 1.2 m, and mean low water springs at 0.5 m; these figures probably apply to all the islands north to Cape Melville, from which the bulk of the evidence reported here comes.

2.2. *Relation of features to sea level datum*

If inferences are to be drawn regarding late Holocene sea level change it is essential that the deposits and levels of interest can be referred to a specific level of the sea such as mean low water neaps, mean sea level, extreme high water springs, etc., at the time of their formation.

This is by no means simple and often can only be done with coral microatolls in growth position. Sedimentary deposits for instance may show a variety of levels even on the one reef as a result of variations in exposure and cannot easily be related to a specific tidal level.

The problem of equating features with a specific level of the sea is brought sharply into focus if we consider the range of levels of features related to present sea level. For instance the heights of living corals vary considerably depending on whether they are in free draining situations such as on the reef flat or reef edge, or in areas of impeded drainage such as moats. Levels for the tops of open reef flat corals range through 0 to 1 m; and for moated corals range through 0.4 to 1.7 m. Likewise the heights to which island beaches and sedimentary deposits are built under present conditions are highly variable and depend on such factors as relative exposure, distance from reef edge, surface roughness between reef edge and deposit, size and shape of sediment, and storm frequency and intensity. For instance measured elevations for sand cay beaches range through 2.5 to 4.0 m and for outer shingle ramparts from 1.1 to 3.2 m. Furthermore, if we are to utilize the surfaces of cemented deposits such as beach-rock and rampart-rock as palaeo sea level markers, what do these surfaces represent? Are they erosional levels, depositional levels or levels of cementation? Can we identify an upper limit of marine cementation related to present tidal datum? Marine cements are known to precipitate up to the level of extreme high water spring tide and more commonly to mean high water spring tide. In the northern Great Barrier Reef these levels are 2.9 and 2.3 m respectively but we cannot be certain that such levels are either reached or not exceeded.

If we cannot identify with certainty levels of particular features with specific present sea levels – and it is clear there is a range of values within and between reefs – then we cannot expect the situation to have been greatly different in the past. Indeed, it is possible that variation is *less* now, as a result of a few thousand years of 'stable' sea level, than it was immediately following the Holocene transgression when great changes occurred on the reefs, and between-reef differences may have been more marked. Nevertheless, it is important and necessary to make an intelligent assessment of the relation between present sea level and contemporary features to serve as a baseline for comparison with older features. Such an assessment is made when the specific evidence is discussed below.

2.3. *Identification and interpretation of features*

Problems also arise in field identification and interpretation of features; some may be given a sea level connotation when it is not justified. For instance, raised beach-rock may be identified when the material is in fact either aeolianite or cay sandstone in which the cement is partly phosphatic, and which cannot be regarded as sea level markers. The question of corals in growth position versus transported corals is another example. Some transported corals can be deposited the right side up and could be interpreted as *in situ* when indeed they are not. Further difficulties arise when corals are covered by later deposits and observed only in section. Thus, the distinction between deposited corals and those in growth position as exposed in the basal unit of some rampart-rocks in the area was not always clear. Fortunately most of our dated *in situ* fossil corals were from areally extensive fields of microatolls where there was no doubt that they were in growth position. However, in such cases the problem relates to the height of microatolls. Rarely with the fossil microatolls could we determine for certain whether or not their growth environment was a free-draining or moated situation. A third problem focuses on distinguishing storm-wave from normal-wave shingle deposits and wind-deposited from wave-deposited sands. In the first case reef blocks and boulder tracts result from catastrophic storms, while shingle ramparts and ridges likely result from both rough and quiet weather conditions. In the second case lenses of drift pumice in sandy deposits proved useful indicators of the reach of wave-wash, but the presence of such material does not negate the possibility that wind may have had a rôle in cay building. Granulometric parameters of the sands provide useful clues, however, to environments of sedimentation.

2.4. *Time scale and radiometric dating*

There are also problems associated with establishing a time scale. When focusing on the last few thousand years it is necessary to establish a chronology which is accurate within tens, and certainly within one or two hundred years. The time scale used here is the radiometric one, and all dates are conventional ^{14}C ages B.P. All ages reported in this paper were determined by the Radiocarbon Laboratory, Australian National University, and are considered valid ages. Details of the multidisciplinary approach to the problems of obtaining valid radiocarbon dates as well as detailed sample descriptions and locations are given by Polach, McLean Caldwell & Thom (1978, this volume).

In the present context problems arise in the interpretation of ages, not in their determination. The most obvious question is: does the dated material yield a realistic age for the deposit, feature or surface from which it was obtained? Excluding bulk samples of calcarenite which consist of both constituent grains and cement and coral clasts known to contain secondary void infills, ages refer to the time of life of the organisms. With transported material such as shingle ramparts there may be a considerable discrepancy between that time and the time when the deposit accumulated. In addition, in bulk samples of cay sand the dated sediment comprises a suite of bioclasts and not just a single organism. Thus for the sands it is assumed that dates refer to the average age of all constituents which individually may have a wide age range. For both single and bulk samples from sedimentary deposits the reported dates must be regarded as maximum ages for the time of their formation. Multiple samples from equivalent units were dated and the consistency of these results together with geographic, geomorphic and stratigraphic evidence enable an assessment of the lag between the time-of-life of the organism and the time-of-formation and subsequent history of the deposit. This problem does not arise with dated *in situ* fossil organisms.

2.5. *Other problems*

The foregoing discussion is not exhaustive. There are additional problems relevant to the construction of a late Holocene sea level history for the area. These include the possibility that tidal range and/or climate have varied during the last few millennia and that tectonic movements and/or hydro-isostatic warping of the land and shelf may have taken place. Such possibilities on the northern Great Barrier Reef are discussed elsewhere by Hopley (1978, this volume) and Thom & Chappell (1978, this volume) and need not be developed further here.

3. SUMMARY OF PROBLEMS, ERROR TERMS AND ASSUMPTIONS

The establishment of a common tidal datum, difficulties in relating contemporary features to specific levels of the sea to serve as present baselines, problems in identifying and distinguishing between those fossil features which possess a sea level connotation from those that do not, problems in relating radiometrically dated samples to features relevant for a sea level history, as well as the possibilities of temporal variations in oceanographic factors and land movements, are common to all workers interested in documenting late Holocene sea level history but are rarely mentioned or acknowledged.

At most we are dealing with potential sea level variations of ± 4 m and a time period of a few thousand years. Our evidence for sea level change, if any, is thus to be found within or close to the present tidal range or within the range of contemporary storm surge levels. We therefore must acknowledge our error terms. Relative levels between features on the same traverse and between traverses on the same reef can be regarded as accurate to within a few centimetres. But between reefs there may be discrepancies of up to 0.5 m. Absolute levels in relation to Cairns datum may be in error by an equivalent amount. The statistical error term

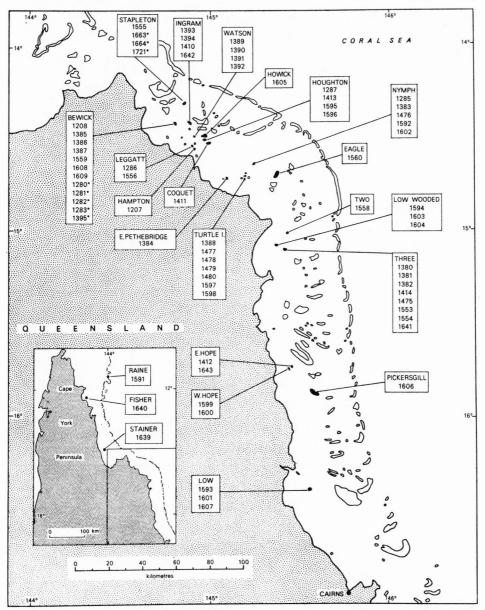

FIGURE 1. Location of radiometrically dated samples on the northern Great Barrier Reef identified with A.N.U. Radiocarbon Laboratory code numbers.

associated with the radiocarbon results ranges from ± 70 to 170 years. Because of this and because our time scale extends over some thousands of years it may well be easier to identify the dates of palaeo sea levels more accurately than their specific levels. Moreover, because the lower limit of most of our observations was low water level there is a bias toward recognizing features that would relate to higher rather than lower sea levels relative to present.

We must also acknowledge our assumptions, the major ones being: (1) features equivalent in form and composition, but at higher or lower levels than their contemporary counterparts, developed with a sea level that was higher or lower than present, the magnitude of sea level change being the difference between the two levels; (2) features inferred to be related to present sea level are indeed so related; (3) ranges in elevation between equivalent contemporary features for all reefs surveyed are similar to the ranges of relict features. We also assume in the first analysis that there have been no significant changes in tidal range or wave climate within the region, nor any land or shelf movement in the last few thousand years. That is, we assume the only land/sea changes have been eustatic and levels are relative to present sea level.

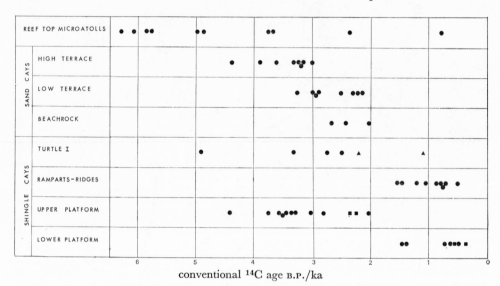

conventional ^{14}C age B.P./ka

FIGURE 2. Summary of radiocarbon dates: ●, skeletal carbonate; ▲, fibrous mud; ■ cement.

4. SUMMARY OF RADIOCARBON DATES

The locations of samples on the northern Great Barrier Reef radiometrically dated for the Expedition are given in figure 1. Figure 2 summarizes all dates from above low water level grouped in terms of features identified in the field. It is clear from these results that there is considerable clustering of data; the consistency in age of comparable features on different reefs gives confidence in interpreting the late Holocene history of reefs and reef islands in the area.

5. SUMMARY OF LEVELS

Figure 3 summarizes the surveyed levels in relation to datum of a variety of features from the reefs between Low Isles in the south and Stapleton in the north (figure 1). The range of any one feature is not less than 1 m. Nevertheless, it is clear that levels of equivalent features are generally accordant among the reefs such that particular forms extend over a definable vertical range. While overlaps between fossil and contemporary forms do occur, some are considerably higher than their modern counterparts and probably reflect changes in the level of the sea relative to present.

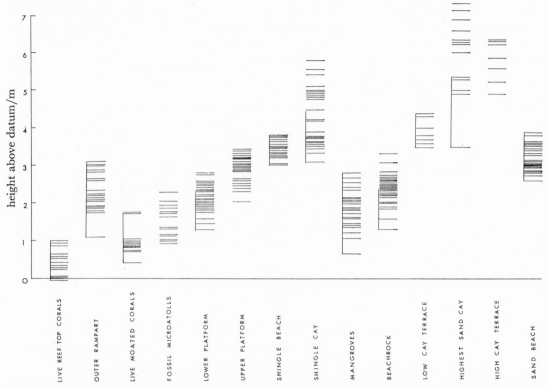

FIGURE 3. Levels of a variety of reef features on the northern Great Barrier Reef referred to tidal datum arranged from left to right in order of their occurrence across a reef top from the windward to leeward side. Vertical lines represent range of levels of features known or believed to be related to present sea level.

6. EVIDENCE FROM FOSSIL MICROATOLLS

Reef-top corals in growth position, particularly those of microatoll form, provide the most certain evidence of the position of local water levels. Scoffin & Stoddart (1978, this volume) give details of the utility of microatolls as water level recorders in the area. They conclude that mean low water springs (0.5 m) represents an effective upper limit to coral growth in free-draining reef flat locations, while for living moated microatolls the approximate upper limit is mean low water neaps (1.2 m) though *potential* altitudes of modern moated corals are higher.

Fossil corals and tridacnids from microatoll fields were dated from a number of reefs, and elevations from some of these sites in relation to living open reef flat corals and/or living moated microatolls are available (table 1). In most instances the dated corals and clams are from extensive fields of microatolls and are not just isolated occurrences. Fossil microatoll fields typically occur in central portions of reef tops within mangrove swamps where flat-topped colonies are emergent above the level of mangrove mud (figures 4 and 5, plate 1). Alternatively, they are exposed in the basal facies of rampart-rocks with individual corals sometimes emerging seaward of the rock scarps (figures 6 and 7, plate 1). In both cases, corals and clams are exceedingly well preserved with little evidence of post-death contamination, borings, encrustations, etc. Field observations and X-radiographs of corals from the Low Wooded Island site showed that the tops of some of the formerly rounded coral heads (e.g. *Platygyra*, ANU-1604) had been planed down as a result of later erosion, though this was not evident at the other dated microatoll sites. We envisage that modern counterparts of fossil microatoll environments are found in the central portion of Turtle IV and Low Isles reefs and in moats such as at Houghton and Watson reefs.

TABLE 1. AGES AND RELATIVE LEVELS OF MICROATOLLS

A.N.U. code	reef	age/a	comment
1640	Fisher	6310 ± 90	*Tridacna* resting on microatoll beneath shingle ridge. Level of fossil microatolls equivalent to living moated corals
1604	Low Wooded	6080 ± 90	*Platygyra* in growth position in microatoll field passing beneath rampart rock. Fossil corals are 0.6 m above present living reef edge corals
1287	Houghton	5850 ± 170	Faviid in growth position from microatoll field. Fossil microatolls are 1.1 m above highest living reef flat corals
1286	Leggatt	5800 ± 130	*Tridacna* in growth position from microatoll field. Fossil microatolls are 1.0 m above highest living reef flat corals and 0.35 m above highest living moated corals
1639	Stainer	4980 ± 80	*Favites* in growth position from microatoll field. Fossil microatolls are at a similar level to living moated corals
1639R		4960 ± 80	
1207	Hampton	4870 ± 70	Faviid in growth position from microatoll field. No height data, but microatolls definitely emergent above present living corals
1380	Three Isles	3750 ± 110	*Pavona* in growth position exposed at base of upper platform. *In situ* fossil corals are 0.8 m above moated corals
1285	Nymph	3700 ± 90	Faviid in growth position from microatoll field. Fossil microatolls are 1.3 m above living reef flat corals and 0.9 m above moated corals
1384	E. Pethebridge	2370 ± 70	Faviid in growth position from microatoll field. Fossil microatolls are 0.6 m above highest living reef flat corals
1594	Low Wooded	800 ± 60	*Porites* microatoll in moat. No height data but coral emergent by 20 cm above moat low water level

Radiometric ages and measured or estimated elevations of concordant coral colonies provide indisputable evidence that sea level in the northern region of the Great Barrier Reef first reached its approximate present position about 6000 a B.P. Dated materials of this age from four different reefs which cover some 3° of latitude indicate the spatial extent of the evidence and the likelihood that similar aged materials could be preserved on other reefs in the region. The levels of the extensive Houghton and Leggatt microatoll fields relative to living reef flat and moated corals suggest that by 5800 a B.P. the sea had passed above its present level. The possibility of moating at these two central reef sites at such an early stage in the development of reef-top features is considerably less than would be likely later on. The Stainer and Hampton corals also indicate that sea level was above present level a thousand years later (4900 a B.P.) while those from Three Isles and Nymph suggest that the highest level for which we have evidence, in excess of 1 m, was attained around 3700 a B.P. However, dates of transported clasts from cemented shingle ramparts of equivalent ages at these two sites illustrate that the fossil *in situ* corals could have been in moated situations. The level of the surface of the fossil microatoll field at East Pethebridge is 0.6 m above measured living reef flat corals, but this could also have been a moated situation. Nevertheless, the age of 2370 ± 70 a B.P. (ANU-1384) indicates that sea level was then close to or marginally above its present level at the time.

One obvious feature of the results in table 1 is the grouping of ages with gaps of about

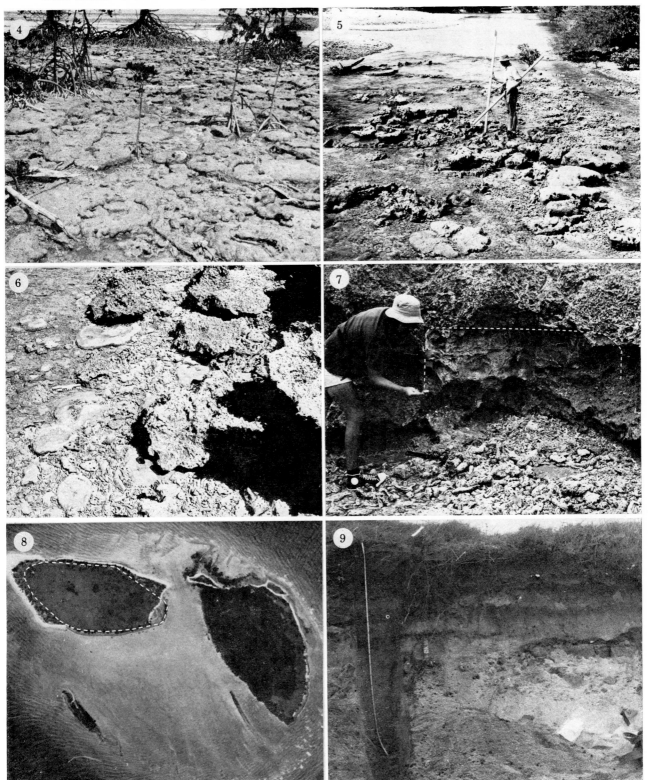

FIGURE 4. Field of emergent fossil microatolls at Leggatt Reef. A clam from this site dated 5800 a B.P.

FIGURE 5. Emergent fossil microatolls exposed in pond outlet at Nymph Reef. A coral from this site dated 3700 a B.P.

FIGURE 6. Fossil microatolls exposed beneath rampart-rock at Low Wooded Island. A coral from this site dated 6080 a B.P.

FIGURE 7. *Porites* microatoll exposed in scarp of upper platform rampart-rock at Three Isles. A coral from this site dated 3750 a B.P.

FIGURE 8. Aerial view of Three Isles showing sand cay upper left and shingle islands on right and lower left. Outline on sand cay represents 3000 a B.P. shoreline.

FIGURE 9. Sand cliff cut in high terrace at eastern end of Three Isles cay showing extent of surface soil development, buried soils and pumice layers. Two bulk sediment samples from this section were dated 3220 and 3350 a B.P.

FIGURE 10. Contemporary beach-rock at Waterwitch cay.

FIGURE 11. Relic beach-rock at Ingram cay.

FIGURE 12. Upper and lower platforms of rampart-rock at Watson Island.

FIGURE 13. Upper platform rampart-rock forming shore of shingle island at Three Isles. At this site a basal coral dated 3750 a B.P., a surface clam 3050 a B.P. and cement 2260 a B.P.

FIGURE 14. Bedded calcarenite overlying basal corals and topped by cemented coral shingle at western end of Houghton cay. At this site a basal coral dated 3330 a B.P., bulk calcarenite 2670 a B.P. and surface clam 3550 a B.P.

FIGURE 15. Residual block of upper platform rampart-rock on reef flat at Sinclair–Morris Reef.

1000 years between groups. These gaps may be an artefact of sampling or alternatively may reflect falls in water level. However, this evidence alone is insufficient to infer an oscillating sea level or the precise magnitude of sea level change. What is important is that we have definite evidence that sea level reached about its present position some 6000 years ago, that it subsequently rose above that level, and probably reached a maximum height around 3700 a B.P. The subsequent course of sea level and the time of its return to present level is not well documented although by 2300 a B.P. it was close to or still marginally above present level. Also of importance is the realization that reefs and superficial reef-top deposits at least on some reefs in the area have had up to six millennia in which to evolve and adjust to a sea level around its present position. Even if some of the corals were moated and hence reflect locally perched water levels these fundamental conclusions are not affected.

TABLE 2

cay	beach limit/m	low terrace/m	high terrace/m	difference/m
Combe	3.1	3.5	6.8	3.3
East Hope	3.5	3.6	6.3	2.7
Two Isles	3.2	3.5	6.0	2.5
Ingram	3.0	3.2	5.4	2.2
Three Isles	3.4	3.5	5.4	1.9
Howick	2.5	3.4	4.7	1.3
Leggatt	3.1	3.5	4.6	1.1
Houghton	3.3	3.4	4.5	1.1
Three Isles	3.6	4.5	5.6	1.1

7. EVIDENCE FROM SAND CAYS

Sand cays are present on many of the high reefs of the inner shelf of the northern Great Barrier Reef. Sediments making up these cays typically consist of moderately well sorted coarse sand composed of coral, Foraminifera, *Halimeda* and molluscan skeletal fragments which have been transported by wave action from surrounding reef top source areas to build the cays. Details of the nature and origin of cay sediments and history of cay sedimentation in the area are given by McLean & Stoddart (1978, this volume), while geomorphological features are considered by Stoddart, McLean & Hopley (1978, part B of this Discussion).

On a number of cays two distinct levels exist, here called the high terrace and low terrace. The altitude of these terraces where surveyed on continuous traverses together with the altitude of the beach/vegetation line are given in table 2.

On these traverses the present beach/vegetation limit reaches 3.6 m (mean high water springs 2.3 m) and is commonly within a few centimetres of the level of the lower terrace. Differences in elevation between the low and high terraces range from 1.1 to 3.3 m and are believed to reflect a relative change in the position of sea level within this range.

7.1. *Age of high terrace*

The high terrace is expressed in a variety of morphological types: as a relatively planar horizontal surface (e.g. Ingram), as an encircling continuous ridge (e.g. Three Isles), or as a single linear ridge (e.g. Stapleton). Bulk sand samples taken from the surface, pits or exposures in the high terrace were dated as given in table 3.

The high terrace is at a level well above the limit of contemporary cay sedimentation. The presence of thick or degraded soils, the extent and morphological homogeneity of the surface and the age range of samples suggest that reef flats were highly productive 3000–4000 years ago and that skeletal material produced at that time was swept from reef flats to greatly extend incipient cays and develop new ones. Because dated samples are generally from near-surface and island-edge locations the ages likely reflect the later rather than earlier phase of major cay development. Greater ages could be expected from the basal interior portions of cays, but no sediment was collected from such sites on simple cays. Thus by 3000 a B.P. the basic outlines of the larger cays in the area were well established (figure 8, plate 1). A sea level relatively higher than present is the simplest mechanism to account for the height of the high terrace. The presence of weathered drift pumice in surface soils and lenses of unweathered pumice at shallow depths argues against a wholly dunal origin for the high terrace and clearly indicates that wave-wash played a rôle in its formation (figure 9, plate 1).

TABLE 3

A.N.U. code	cay	age/a	elevation/m
1555	Stapleton	3240 ± 70	7.3
1558	Two Isles	3900 ± 80	6.3
1412	East Hope	3020 ± 70	6.3
1559	Bewick	4380 ± 80	5.9
1554	Three Isles	3640 ± 70	5.6
1410	Ingram	3230 ± 80	5.4
1553	Three Isles	3350 ± 80	5.0
1414	Three Isles	3220 ± 80	5.0

TABLE 4

A.N.U. code	cay	age/a	altitude/m
1557	Low Isles	2550 ± 70	4.5
1387	Bewick	2950 ± 80	4.4
1643	East Hope	2990 ± 80	4.2
1560	Eagle	2960 ± 70	4.2
1642	Ingram	3280 ± 80	3.7
1641	Three Isles	2190 ± 70	3.5
1606	Pickersgill	2230 ± 70	3.5
1556	Leggatt	2330 ± 70	3.4

7.2. *Age of low terrace*

On the larger cays the high terrace is frequently fronted by a lower surface, while on the smaller cays the main surface is at a comparable level to this lower terrace. The lower terrace and smaller cay surfaces are commonly developed within an altitudinal range little different from that reached by contemporary wave action and berm building but if anything above rather than below that level. Radiocarbon ages of bulk sand samples together with surface levels are listed in table 4.

As a group the ages for the lower terrace and smaller cays are younger than the higher surface, but there is overlap between the two groups. Thus for East Hope and Ingram the ages of samples from the two levels are not statistically significantly different. Moreover, the range of values for the low terrace is less than would be expected for deposits accumulating in sympathy with present sea level. In this case we would anticipate a greater range of dates with perhaps some samples reaching modern ages and certainly some being younger than 2000 a B.P.

The lack of such young samples, together with geomorphological evidence, strongly suggests that much of the sediment which has gone into building the lower berm terrace and smaller cays has been derived from reworking of the older cay sands with only a small addition of recent skeletal material. Thus it is impossible on this evidence alone to indicate precisely when the lower terrace of the larger cays and main surface of the smaller cays were formed except that they post-date 3000 a B.P. and have developed in harmony with a sea level comparable to or marginally above present level. In the cases of Two Isles and Three Isles there is comparative survey evidence to show substantial lower terrace construction in this century.

7.3. *Sea level change: interpretation*

Differences in elevation of 1 to 3 m between the high and low terraces of sand cays on the northern Great Barrier Reef suggest that sea level was relatively higher than present level by more than 1 m during the formation of the high terrace some 3500 years ago (average age of high terrace samples) but was similar to the present level or marginally above during formation of the low terrace which post-dates 3000 a B.P. It could be argued that the differences in level are not the result of a change in sea level but rather reflect a reduction in exposure and wave energy on the leeside of reefs as a result of the development of shingle ramparts and ridges on the windward side at a time between the accumulation of the high and low terraces. Obviously, such features have influenced leeside refraction patterns and caused local variations in the spatial distribution of at-cay wave energy convergence and divergence zones through time; erosion and accretion patterns so created are easily recognizable in cay plan geometry and morphology. However, we believe that the magnitude of this effect has not been sufficient to account for the required difference in wave run-up heights in all cases, over 2 m in the case of Ingram, East Hope, Two Isles and Three Isles where both the high and low terraces are most clearly and extensively developed. All these reefs contain broad reef flats and possess negligible or relatively minor windward shingle accumulations in relation to the extent of open reef flat. In the case of Ingram and East Hope, bare open reef flat accounts for 98 % of the reef top and for Two Isles and Three Isles the corresponding figures are 74 and 75 % respectively. Furthermore, the ages of the older shingle islands are similar to the sand cays (see below) indicating synchronous formation of windward and leeward structures. They do not fall between the time of development of the high and low terraces as a hypothesis invoking local variations in wave energy necessitates.

On this evidence, sea level was more than 1 m above present around 3500 a B.P., but we lack data from the sand cays to indicate when this level was first achieved. The oldest cay sand dated, from the surface of the highest ridge on Bewick cay (ANU-1559, 4380 ± 80 a B.P.), is nearly 1000 years older. While suggesting that the sea was at a comparable level to that pertaining about 3500 a B.P., neither the age nor elevation are alone sufficient to establish the palaeo sea level at the time with any certainty. Likewise, sand cay data relating to the subsequent pattern of sea level change is not strong, except for the obvious fact that it has fallen below its earlier level and subsequently reached its present position. The presence of sediments dated 3020 ± 70 a B.P. (ANU-1412) from the high surface at East Hope suggests that sea level was still relatively higher at least 3000 years ago. Moreover, the ages greater than 2000 a B.P. for the lower terrace sediments could indicate that sea level was at or above present in the 2000–3000 a B.P. interval. This may well be the period when the major fall from the level at 3500 a B.P. took place.

The fact that none of the dated cay sands were younger than 2100 a B.P. bears close scrutiny. Although this may be a result of sample selection the age of clean sand taken from the small unvegetated cay, Pickersgill (ANU-1606, 2230 ± 70 a B.P.), does suggest that the production and recruitment of post-2000 a B.P. sediment has been minor. One obvious explanation for the deficit of sand cay building sediment at least on some reefs is that reef-top sand supply areas have been reduced considerably. The development of shingle ramparts and ridges some of which encircle the reef and enclose a pond, and particularly the expansion of mangroves across the reef flat in sheltered environments has resulted in an incremental diminution of sand supplies. At present on Houghton, Coquet and Newton reefs, for example, little sand is being produced, while on Bewick, Nymph and Low Wooded Island reefs virtually none is being produced. In these cases the extant sand cays are essentially fossil forms.

An alternative view which would account for the deficiency in post-2000 a B.P. sand supply and cay nourishment, is that many reef tops were emergent for much of the last 2000 years and have only recently been reoccupied by the sea as a result of a small rise in sea level. Much of the geomorphological evidence is consistent with such a view. First, the larger sand cays, notably those possessing the high terrace, retain much of their fundamental pre-3000 a B.P. outlines. Preservation and retention of such outlines through to the present would be greatly enhanced by a fall in sea level. Secondly, a contemporary rise in sea level would result in the rectification of cay outlines, particularly on those cays on relatively open reef flats. Thus shore cliffing, exposure of relict beach-rock, accumulation and extension of fresh berms and down-drift spits on such cays as Three Isles, Two Isles, Ingram, Sinclair–Morris and Combe may be quite recent in origin. Such a suggestion is also consistent with only a small dilution of pre-3000 a B.P. sediments by contemporary materials, thus bringing the radiocarbon ages for these deposits into the 2000–3000 a B.P. bracket.

8. Evidence from beach-rock

Cemented sand in the form of beach-rock is exposed around many of the sand cays in the northern Great Barrier Reef. The nature and formation of beach-rock in this area is discussed by Scoffin & McLean (1978, this volume).

In the field, two types of beach-rock were recognized on morphological grounds: (1) high or horizontal beach-rock and (2) inclined beach-rock, the latter being similar to that usually described in the literature. The horizontal type is believed to represent cementation within the berm or backshore of a beach immediately landward of the seaward dipping foreshore bands. Contemporary examples of single beach-rock outcrops passing abruptly from inclined to horizontal surfaces were seen on Waterwitch and Combe cays (Figure 10, plate 2), while older examples, where both surfaces were preserved, occurred on Two Isles, Three Isles and Ingram cay (Figure 11, plate 2). At other sites on these three cays and also at Howick and Bewick the horizontal surface was truncated to seaward by an erosional scarp which dropped steeply to the reef flat. In such cases removal of the inclined portion is indicated, one destination for the eroded slabs being the sand ridge immediately islandward of the horizontal exposure.

Extreme high water spring tide is the likely upper limit for beach-rock cementation, though the more frequently obtained level of mean high water springs may be a more realistic limit. At present these levels are 2.9 and 2.3 m respectively on the northern Great Barrier Reef.

TABLE 5

cay	top/m	base/m	type	comment
Leggatt	3.1	1.2	inclined	relict
Bewick	3.1	1.4	horizontal	relict
Three Isles	2.8	n.a.	horizontal	relict
Three Isles	2.6	1.6	horizontal	relict
Three Isles	1.8	0.8	inclined	contemporary
Two Isles	2.7	n.a.	horizontal	relict
Two Isles	1.6	0.5	inclined	contemporary
Ingram	2.7	1.5	horizontal	relict
Ingram	2.5	1.3	horizontal	relict
Ingram	2.3	0.9	inclined	contemporary
Eagle	2.3	0.8	inclined	contemporary
Combe	2.2	0.9	inclined	contemporary
East Hope	1.8	0.4	inclined	contemporary

TABLE 6

A.N.U. code	cay	age/a	comment
1605	Howick	2420 ± 70	coral head
1386	Bewick	2030 ± 70	*Tridacna*

Elevations of tops and bases of beach-rock exposures together with a field classification of outcrops are given in table 5.

Clearly, these data do not provide positive evidence of sea level change. On the one hand the maximum surface levels fall generally within the range of present extreme spring tides, and on the other considerable variation in outcrop morphology has been reduced to two types, inclined and horizontal. Moreover, the relict/contemporary classification allows accusations of circular argument, though it is based on a suite of criteria including the presence or absence of terrestrial or marine vegetation, surface weathering and solution features, occupied or unoccupied boreholes, surface coloration, etc. Despite these problems, both the upper and lower levels of horizontal beach-rock are invariably higher than inclined equivalents. Where both are present in close proximity on the one cay as at Two Isles and Three Isles, a difference of approximately 1 m is found. We believe that outcrops designated as contemporary reflect formation during a sea level similar to present, while those designated as relict could have formed when sea level was higher than present. In the absence of a suite of dates on bulk samples of beach-rock, evidence from the location of outcrops with respect to loose cay sands immediately behind them suggests that beach-rock formed soon after the older cays developed, thus stabilizing their early outlines. In other words, at least some of the relict beach-rock represents part of the shores of cays when they were at their high terrace stage some 3500 years ago, whereas the inclined beach-rock equates with the lower terrace. This is well illustrated at Three Isles and Ingram (Figure 11), while at Bewick relict beach-rock which forms the western shore of the cay is surmounted on its inner edge by a high sand ridge whose constituents were dated at 4380 ± 80 B.P. (ANU-1559). At the western end of Houghton cay (figure 14, plate 2), beach-rock is exposed overlying reef flat corals and is in turn topped by a thin cemented coral stick veneer yielding ages of 3300 ± 80 and 3550 ± 80 a B.P. (ANU-1595, ANU-1413). A bulk sample of calcarenite from this beach-rock was aged 2670 ± 70 a B.P., which because of the presence of cement gives a minimum age for its constituents. Individual bioclasts from the surface of two other beach-rock outcrops were dated as shown in table 6.

These smaller ages may illustrate a later phase of relict beach-rock formation, although the large size of the dated materials, which contrast markedly with the calcarenite below, indicate deposition during stormy conditions and need not represent the period of sand accumulation and cementation.

9. Evidence from shingle cays, ridges and ramparts

9.1. *Shingle cays*

On the reefs of the Turtle Group cays built of a mixture of both sand- and shingle-sized bioclastics have developed (McLean & Stoddart 1978, this volume). On the larger cays of this Group two surfaces are apparent. Levels of the present beach/vegetation limit, low and high terraces and differences between the terrace elevations on continuous traverses are given in table 7. Samples from the high and low terraces on Turtle I were dated as given in table 8.

Table 7

cay	beach limit/m	low terrace/m	high terrace/m	difference
Turtle I	3.4	4.4	5.6	1.2
Turtle I	3.5	4.1	5.6	1.5
Turtle II	3.5	3.5	4.7	1.2
Turtle V	3.3	4.0	5.4	1.4
Turtle VI	3.6	3.8	5.1	1.3

Table 8

A.N.U. code	age/a	altitude/m	comment
1388	3320 ± 80	5.5	high terrace
1598	2760 ± 80	4.2	low terrace
1597	2480 ± 70	4.2	low terrace

Table 9

A.N.U. code	age/a	altitude/m	comment
1480A	1100 ± 80	2.1	fibres, rootlets
1480B	2210 ± 170	2.1	organic mud
1479	4910 ± 90	1.5	coral shingle

Both ages and levels accord with those from the sand cays and it is believed that they reflect a comparable sea level and geomorphic history with that described above for the sand cays. Samples from a shallow borehole in a small enclosed depression between two coral shingle ridges near the centre of the island indicate the antiquity of cay development and the presence of mangroves. The depression is occupied by living mangroves and is floored by a deposit of mixed sand and coral shingle overlain by a black fibrous mud. Levels and ages of these materials are shown in table 9. These ages and levels indicate that superficial sands and shingle were accumulating on this reef top nearly 5000 years ago and incipient cay development likely dates from around this time, while mangroves were present on the reef top before at least 2000 a B.P.

9.2. *Shingle ridges and ramparts*

In contrast to the cays of the Turtle Group which possess broad, near-planar core surfaces, some of the shingle islands are built up of a series of closely spaced shingle ridges separated by narrow swales. Ridges are generally at accordant levels and suites of 5–15 ridges are common. Watson and West Hope provide the best examples of islands made up of such ridge sequences, though comparable sequences are found on some of the more complex islands such as Nymph and Low Wooded Island. On other shingle cays, particularly those on the windward side of low wooded island reefs, more discrete asymmetrical ramparts occur, not greatly different in height and dimensions from ramparts presently accumulating. Sediments making up these features are principally fragments of branching corals, broken from reef flat and reef edge colonies during storms and built into ridges and ramparts during rough and quiet weather conditions. Ages and surface elevations of dated samples are given in table 10.

TABLE 10

A.N.U. code	island	age/a	altitude/m	comment
1391	Watson	1550 ± 70	3.6	inner ridge
1392	Watson	1480 ± 70	4.4	inner ridge
1390	Watson	810 ± 70	2.1	loose platform shingle
1389	Watson	510 ± 70	3.1	outer rampart
1600	West Hope	850 ± 70†	3.2	inner ridge
1559	West Hope	1210 ± 70	3.3	outer ridge
1411	Coquet	1070 ± 60	n.a.	main ridge
1593	Low	800 ± 70	2.0	inner rampart
1608	Bewick	760 ± 65	3.5	inner rampart

† Sample contaminated (see Polach *et al.* 1978, this volume).

TABLE 11

reef	lower platform/m	upper platform/m	difference/m
Houghton	2.0	3.2	1.2
Watson	2.1	3.3	1.2
Three Isles	2.0	3.1	1.1
Nymph	2.1	3.2	1.1
Watson	2.1	3.2	1.1
Low Wooded	2.5	3.5	1.0
Turtle I	2.5	3.5	1.0
Bewick	1.6	2.6	1.0

Radiometric ages of components and levels of these ridges and ramparts do not provide clear evidence of a sea level significantly different from present. What the ages do indicate is that coral shingle accumulated and islands were formed or extended on the windward side of reefs during the last 1500 years or so, the period when leeside primary sand production and accumulation was at a minimum. Whether or not reef flats were emergent during part of this period, which was one possibility earlier discussed with reference to sand cays, cannot either be substantiated or negated on this evidence because of the rôle of storms in the origin of shingle ramparts and ridges.

10. EVIDENCE FROM RAMPART-ROCKS

One rather unique feature of the low wooded islands of the northern Great Barrier Reef is the presence of cemented shingle ramparts, here termed rampart-rocks. These exposed limestones occur typically but not exclusively on the windward side of reefs. In the field, two surface levels were recognized and called the upper platform and lower platform (figure 12, plate 2). On some reefs the distinction between the two surfaces was not always clear and on other reefs only one, the lower platform, was present. Considerable variation in the absolute elevation of platforms between reefs was noted. However, where continuous traverses were run across both platforms, results were as given in table 11.

In the above instances a distinctive scarp separated the upper from lower platform. A difference in level of about 1 m between the two features is indicated. The possibility that these platforms are raised reef flats is discounted on grounds of stratigraphy and fabric. Questions relevant for a time/sea level interpretation of the platforms include (1) the age of the constituent rubble components, (2) the nature, age and origin of the cementing matrix, and (3) the significance of the levels of the cemented surfaces. These aspects are considered in detail by Scoffin & McLean (1978, this volume).

TABLE 12

A.N.U. code	reef	age/a	altitude/m
1383	Nymph	3540 ± 80	3.8
1592	Nymph	3415 ± 75	3.8
1604	Low Wooded	3320 ± 70	3.5
1478	Turtle I	4420 ± 90	3.5
1413	Houghton	3550 ± 80	3.1
1595	Houghton	3330 ± 70	3.1
1380	Three Isles	3750 ± 110	3.0
1382	Three Isles	3050 ± 70	3.0
1208	Bewick	2840 ± 70	2.6
1609	Bewick	2050 ± 70	2.6

10.1. *Age of upper platform*

The surface of the upper platform is either nearly horizontal or promenade-like such as at Three Isles and Houghton or consists of a highly irregular karstic topography such as at Watson Islands. Within the upper platform three and locally four vertical facies are recognized, the last being a veneer of stick coral deposited after the main mass. Ages of constituent components indicate that accumulation of the initial rampart was both a rapid and a discrete event. Where present and dated, the ages of the basal corals in growth position (ANU-1285, 3700 ± 90 a B.P.; ANU-1380, 3750 ± 110 a B.P.) have been dealt with earlier. Ages of bioclastic components other than these together with the altitude of the surface at the sample location are given in table 12.

Excluding the Bewick results, where the altitude of the field-identified upper platform and age of constituent components are less than the others, the results suggest a large accumulation of rampart rubble before 3000 a B.P. This would be a minimum age for the major influx of coral shingle, because in most instances samples were taken from the outer scarp or outer surface of the platform and not from inland exposures, except at Turtle I where the material is 800–1000 years older than the other sites. Thus in the period 4500–3000 a B.P. the windward flats and crests of many of the reefs in the area between Three Isles and Bewick possessed large thickets of living branching corals which were broken up and accumulated as loose shingle ramparts. At least for Nymph, Houghton, Low Wooded Island and Three Isles, this last event took place around 3300–3600 a B.P.

Once the ramparts were formed and stabilized, cementation began. Details of this process are given elsewhere (Scoffin & McLean 1978, this volume). The cementing matrix suggests that lithification took place in an intertidal environment adjacent to mangrove swamps. Ages of two cements indicate that cementation was completed within about 1000 years of rampart formation (table 13). These ages are from the cementing matrix towards the base of the platforms at their outer exposed seaward edges and probably reflect the last phase of upper platform cementation. Older cements are likely to occur in outcrops further inland.

TABLE 13

A.N.U. code	reef	age/a	level/m
1602	Nymph	2350 ± 70	3.8
1381	Three Isles	2260 ± 80	3.0

TABLE 14

A.N.U. code	reef	age/a	altitude/m
1475	Three Isles	1460 ± 70	2.0
1477	Turtle I	1430 ± 70	2.4
1385	Bewick	640 ± 70	1.6
1476	Nymph	520 ± 70	2.1
1607A	Low Isles	740 ± 70	1.7
1607B	Low Isles (cement)	560 ± 110	1.7
1601	Low Isles (cement)	380 ± 80	1.7

10.2. *Age of lower platform*

The lower platform whose surface elevation is generally between 1.5 and 2.1 m, but which may reach 2.8 m, has an extremely variable morphology. In places it abuts the outer scarp of the upper platform while in others it is separated from the latter by a moat. Elsewhere it occurs as a promenade fringing an island or an isolated unit on the reef flat which in some cases has a 'bassett edge' morphology. Ages of components from four sites classed as lower platform and of components and cementing matrices from one site classed as bassett edges are given in table 14. Ages of bioclasts accord with those from the loose uncemented shingle ridges and ramparts discussed earlier, while the ages of cementing matrices indicate the rapidity of rampart lithification.

10.3. *Sea level change: interpretation*

It is not known what precisely controls the height of platform surfaces and therefore whether they can be used as definitive sea level markers. Scoffin & McLean (1978, this volume) reject an explanation invoking marine erosion and suggest an origin by deposition and/or lithification up to a level to account for the fairly uniform platform surfaces. Marine cements are known to precipitate today at heights up to extreme high water springs (2.9 m), although mean high water springs is a more common upper limit. In this region of the Great Barrier Reef the latter is 2.3 m above datum. However, platform surfaces need not necessarily represent the upper limits of lithification. Instead they may relate more closely to the height of the original ramparts which could be below the potential cementation ceiling. The assumption that platform surfaces do reflect the upper level of cementation therefore tends to be conservative, but is adopted here.

Thus we believe that the surface of the lower platform that occurs within present mean spring tide range illustrates the level of rampart-rock formation associated with present sea level. On this evidence no change in sea level relative to present is thought to have taken place in the last 1500 years. On the other hand the upper platform is developed at levels in excess of 1 m above adjacent lower platforms and its surface at 3 m above datum is above the range of contemporary cementation. Our evidence suggests that the shingle ramparts which accumulated in the period 4500–3000 a B.P. and were initially bonded at that time to form the upper platform did so with a sea level about 1 m higher than present. The cluster of ages between 3300 and 3700 a B.P. (mean age of eight samples whose surfaces are all above 3 m is 3546 a B.P.) may represent the time when sea level was at its highest.

At present, upper platform limestones are not in equilibrium with the conditions and environments in which they formed. Instead, the processes are essentially of a destructive nature. Where exposed as the inner edge of windward reef flats, notches occur at the base of scarps and outer upper surfaces display a karstic topography (figure 13, plate 2). On some reefs (e.g. Sinclair–Morris, Leggatt), small high isolated blocks of upper platform rock occur as residuals of what were formerly far more extensive features (figure 15, plate 2). These observations suggest that destructive processes have been operating for a considerable time and are consistent with a fall in sea level since the formation of the upper platform some 3000 years ago. The younger ages and lower levels of the dated Bewick site classed as upper platform in the field may indicate the period when sea level was falling from its earlier higher level and certainly indicate that rampart-rock formation continued after the main phase of development. It is also likely that some of the features classified as lower platform in the field, particularly the higher, more extensive, promenades, also date from this time. Thus in the period 3000–2000 a B.P. the distinction between lower and upper platform is blurred. As we envisage platform formation as a relatively continuous process, the assignment of platforms to either the upper or the lower category is somewhat arbitrary in cases where there is an overlap in levels between the two. Nevertheless the end-members are quite discrete and the distinction in level between the older upper platforms and younger lower platforms is clear.

11. Discussion and conclusions

In the first part of this paper we outlined the problems, limitations and assumptions on which our analyses have been based. These included problems in establishing a contemporary tidal datum, recognition of the range in levels to which features are developed with modern sea level, and discussion of problems associated with the interpretation of field evidence and the use of radiocarbon dates. Before any results purporting to show sea level changes can be accepted as valid, a necessary first step is the recognition of the limitations of the data. In addition we have stated a number of assumptions, one being that there has been no movement of the land and shelf in the last few thousand years and that departures from present sea level are of a eustatic nature. Nor have we attempted to test any hypothesis relating to sea level change, being aware that any such stance would prejudice our interpretation. Indeed, if we did have a framework it was to see how far back in time it was possible to go before it became necessary to invoke any sea level departure from present. Thus, we have tended to be conservative and minimize rather than magnify the resulting palaeo sea levels. While we fully accept that our results do have limitations, fewer ambiguities have arisen than initially expected, considering the large number of surveys and radiocarbon dates which could obscure rather than clarify any pattern. Results in level and age of any one feature on different reefs are surprisingly consistent, and perhaps even more surprising is that different types of features produce complementary rather than contradictory sea level histories:

1. Levels and ages of microatolls at four sites in the northern Great Barrier Reef provide firm evidence that the present level of the sea was attained in the Holocene some 6000 years ago and that shortly after (by 5800 a B.P.) it was marginally higher than at present.

2. We have no evidence to indicate the pattern immediately following, but certainly less than 1000 years later emergent microatolls at two other sites aged 4900 a B.P. show sea level was still higher at that time. It is likely that incipient sedimentary features began accumulating on at least some reef tops in the area about this time.

3. In the following two millennia incipient sand and shingle cays were greatly extended so that by about 3000 a B.P. the basic outlines of many of the larger leeside sand cays and windward shingle ramparts were established. Reef-top productivity reached its zenith during this interval. Many of the fossil reef-top features on a large number of reefs have ages clustering between 3200 and 3800 a B.P. There is considerable evidence to indicate that in this period the highest sea level in the Holocene was reached. Levels of two extensive emergent fossil microatoll fields dated 3700 a B.P., the high terrace of the larger cays (average age of 8 samples 3500 a B.P.) as well as the cemented shingle ramparts of the upper platform (average age of 8 samples 3546 a B.P.), indicate that sea level was at least 1 m above present.

4. It remained at that elevation at least until 3000 a B.P. and continued higher than at present for the following 1000 years, though falling gradually from its maximum during this period. Slightly emergent fossil microatolls at East Pethebridge, lower levels and younger ages of clasts in the surface of the upper platform at Bewick and of beach-rock at Howick and Bewick, bulk sands from the lower terrace on a number of sand cays and upper platform cements at Nymph and Three Isles, all of which have ages in the 2000–3000 a B.P. bracket, are believed to represent the falling sea level.

5. Coming closer to the present, in the last 2000 years the pattern of sea level change becomes less clear. Perhaps the most startling feature is the deficiency of material dated younger than 2000 a B.P., excluding clasts from loose and cemented windward shingle ramparts and ridges which owe at least part of their origin to storm wave activity. Moreover, in our records there is a complete absence of material dated in the 2000–1500 a B.P. range. Notwithstanding the build-up of shingle ramparts and ridges on some reefs and the extension of mangrove swamps on others, sometime around 2000 a B.P. reef tops shifted from depositional to erosional environments. Such a shift could result from a slight fall in sea level to present from a marginally higher level at 2000 a B.P., or alternatively could result from a fall to below present level with a late contemporary rise. While we tend to favour the last hypothesis the evidence is not at all conclusive.

6. It is thus the most recent phase of sea level change, the direction and magnitude of change that has given our present level, where the major ambiguities have arisen. Moreover, it is obvious that while we have been able to document certain past times and levels higher than present, we have not been equally able to document levels lower than present. Nevertheless, it would appear that for a substantial proportion of the last 6000 years sea level has been above its present level for longer than it has been below.

7. The foregoing study provides the most extensive regional investigation of late Holocene sea level change based essentially on reef-top features.

REFERENCES (McLean *et al.*)

Hopley, D. 1978 *Phil. Trans. R. Soc. Lond.* A **291**, 159–166 (this volume).
McLean, R. F. & Stoddart, D. R. 1978 *Phil. Trans. R. Soc. Lond.* A **291**, 101–117 (this volume).
Polach, H., McLean, R. F., Caldwell, J. R. & Thom, B. G. 1978 *Phil. Trans. R. Soc. Lond.* A **291**, 139–158 (this volume).
Scoffin, T. P. & McLean, R. F. 1978 *Phil. Trans. R. Soc. Lond.* A **291**, 119–138 (this volume).
Scoffin, T. P. & Stoddart, D. R. 1978 *Phil. Trans. R. Soc. Lond.* B **284**, 99–122 (part B of this Discussion).
Stoddart, D. R., McLean, R. F. & Hopley, D. 1978 *Phil. Trans. R. Soc. Lond.* B **284**, 39–61 (part B of this Discussion).
Thom, B. G. & Chappell, J. 1978 *Phil. Trans. R. Soc. Lond.* A **291**, 187–194 (this volume).

Phil. Trans. R. Soc. Lond. A. **291**, 187–194 (1978)   [187]

Printed in Great Britain

Holocene sea level change: an interpretation

By B. G. Thom† and J. Chappell‡

†*Department of Geography, Faculty of Military Studies, University of New South Wales,*
Duntroon, Australia 2600

‡*Department of Geography, School of General Studies, Australian National University,*
Canberra, Australia 2600

Interpretation of factors responsible for land–sea level change in areas such as the Great Barrier Reef involve an appreciation of not only the field evidence purporting to show change, but also the theoretical models which attempt to explain depth variations in shorelines of a given age. Relative movements in sea level in Holocene time may result from a number of factors operating either external to the study area (e.g. glacio-eustatic, and broad-scale hydro-isostatic deformation of the globe resulting from the last deglaciation and sea level rise), or those whose effects are essentially local (e.g. changes in circulation and tidal levels within partially enclosed water bodies induced by sedimentation or biogenic reef growth, meteorological changes affecting the magnitude and frequency of storminess, regional flexures and/or faulting, and hydro-isostatic deformation of shelves and adjacent coasts accompanying the Postglacial Transgression). In this paper, data from the northern Great Barrier Reef Province are evaluated in relation to various causes of sea level change. Emphasis is placed on explaining variations in relative sea level position by hydro-isostatic theory. Deflexion in the ocean margin 'hinge zone' varies with continental shelf geometry and rigidity of the underlying lithosphere. The fact that the oceanic crust meets the continental crust quite abruptly east of the study areas, dictates that moderately strong flexures occur, and that variations in Holocene hydro-isostatic flexure in the Great Barrier Reef Province are partly explainable in these terms.

Introduction – identifying datum in reefal materials

Interpretation of factors responsible for land–sea level change in areas such as the Great Barrier Reef involves an appreciation of not only the evidence purporting to show change, but also the theoretical models which attempt to explain depth variations in shorelines of a given age. Relative movements in sea level in Holocene time may result from a number of factors operating either external to the study area, or whose effects are essentially local.

It is quite clear that many time–depth curves which have been used to show changes in sea level in Holocene times vary markedly from place to place, and even between areas remote from glacio-isostatic effects there are pronounced differences. Before seeking to explain these, it is necessary (i) to reach agreement about the most useful datum from which measurements of sea level changes should be made, and (ii) critically evaluate the geomorphic and stratigraphic elements upon which interpretations of palaeo sea levels are based. Once a particular facies unit is well dated and its height/depth relations to the chosen modern datum are established, estimates of its associated palaeo-datum can be made, as long as certain factors are borne in mind. For coral reef environments, these include the following.

The geometry of reefs and reef islands during Holocene times has changed, so that relations of certain zones to datum has altered. For example, upward growth of a barrier crest may have lagged behind sea level rise during the postglacial transgression, with the crest becoming stable at some stage during the last 6000 years when relative sea level changes have been minor. In such cases, the changing relation between the particular reef zone and datum should be recognized from close examination of composition and structure (see, for example, Chappell & Polach 1976). In other situations, such as on reef flats, false interpretation might stem from past temporary super-elevation, above m.l.w.s., of shallow coral meadows or microatolls as a result of moating. Only careful examination of the plan-view reef flat structure will reveal this. Finally, high-tide beach and supratidal storm deposits are also susceptible to changing reef geometry, among other factors. The height and horizontal depth to which these are built is affected by the degree of retardation of storm waves across the developing reef flat, as well as by variations in the productivities of different reef faunas. Especially in a very broad region of reefs such as the Great Barrier Reef Province, the sweep of storm waves is likely to have declined as reef crests, intertidal flats and supratidal islands have achieved their present forms during later Holocene times. Thus Neumann (1972) has suggested the existence of a Holocene 'high energy interval', during which 'deposits dating back to this early high energy window still rest where they were emplaced – a few feet higher than those of the more protected present coast – not because they were deposited during a higher stand of the sea, but during an interval of higher energy when the new coast was less protected' (Neumann 1972, p. 42).

The second factor which would influence environment of reef growth and sedimentation is more problematic: the possibility of changing storminess in the last 6000 years or so. Significant climatic changes in higher latitudes have been interpreted from glacial and other evidence elsewhere in the world (Denton & Karlen 1973; National Academy of Sciences 1975). Such changes may also have had an effect on patterns of storminess affecting Australian coasts, as Thom (in preparation) infers from dated pulses of transgressive dune formation on the New South Wales coast. Consequently, there is a possibility that storminess within the Great Barrier Reef Province may have varied in Holocene times, with concomitant variations in patterns of supratidal storm ridge and rampart formation. However, storm rampart deposits on the low wooded islands are not adequate evidence themselves of relatively stormy epochs, as their construction depends in part on reef geometry, as mentioned above, and on productivity of shallow-water branching corals. In any case, such ramparts may reflect the least frequent, large, events within a magnitude–frequency distribution of storms which might have remained constant in Holocene times. Storminess history might be investigated by mapping and dating a large number of the large blocks of forereef and buttress-zone material, ripped out during major storms, which lie scattered on patch reef flats.

The foregoing remarks indicate that many reefal facies may have varied in their relation to datum in Holocene times. Bearing these points in mind, and before proceeding to identify the course of relative sea level change for a particular site, the datum itself must be chosen. In coral reef environments it is tidal extremes rather than mean sea level which can most easily be related to particular facies, especially spring tide low water level which delimits the top of the reef buttress zone and which regulates the upper growth limit of many corals of the outer reef flat. The high tide level may, in certain circumstances, be recognized from sedimentary structures preserved from ancient intertidal beaches. Care must be taken, however, when comparing time variations of ancient low water level datum from different places, because the tidal range may have varied. While this is probably negligible for outer shelf and deep ocean positions, in Holocene times, it can be significant for major bays and archipelagic regions where the coastal and bottom geometries may have changed during and since the Postglacial Transgression. It is possible that tidal range may have altered within some inner regions of the Great Barrier Reef, a possibility which can be tested only by very careful examination of the shallow three-dimensional structure of the largest reef islands.

CAUSES OF DIFFERENCE BETWEEN CURVES OF HOLOCENE DATUM CHANGES

Given agreement about the interpretation of the evidence for palaeo-datum positions, it is possible to examine the proposition that these palaeo sea levels represent any or all of: (i) tectonic movements; (ii) isostatic movements; (iii) true eustatic changes resulting from ice volume changes. Before anything positive can be said about oscillations or trends in sea level due to changes in ice volume over the last 6000 years, it is important to examine the first two factors which can be responsible for regional variation of apparent sea level histories. Such variations on the Queensland shelf and coast may be explained by hydro-isostatic effects (Chappell 1974), or by tectonic movements (Hopley 1974), or by a combination of both. In this paper, an attempt will be made to apply hydro-isostatic theory developed by Walcott (1972) and Chappell (1974) to the northern Great Barrier Reef province, in order to see if the different elevations of palaeo sea levels can be attributed to isostatic movements.

In most studies of sea level change, the hydro-isostatic factor is generally neglected. Hydro-isostatic depression of ocean basins and flexure of their margins is a consequence of the post-glacial increase of ocean volume. There are two mechanisms of load compensation: (i) instantaneous elastic adjustment of the broadest-scale Earth figure; and (ii) progressive slow creep of the figures. For the last 600 years, glacial volume changes are minor and the elastic effect is insignificant for deformation at ocean margins. However, continuing isostatic creep causes significant flexure near ocean margins in the last 6000 years (Walcott 1972; Chappell 1974). The exact pattern of flexure may be calculated for a given rheological model of the upper Earth, and a given history of ocean volume change. Theory and methods of calculation for simpler cases have recently been presented by Chappell (1974).

The creeping flexure of the ocean floors and margins involves deep flow in the upper mantle (asthenosphere) in compensation for the surface redistribution of the load. Physical constants pertaining to this flow have been estimated by many geophysicists for deformations ranging in size from glacial-lake Bonneville, through postglacial rebound of Pleistocene glaciated regions, to the Earth's equatorial bulge (see review by Walcott 1973). During and following deglaciation, the ocean basins subside gradually while the continents rise. Somewhere near the margins between the two is a relatively stable zone, the hinge zone. Strong flexure near this hinge affects apparent or relative sea level changes in coastal and continental shelf regions.

The mean course of ocean basin subsidence by the creep mode of compensation, as well as mean updoming of continents, can be calculated for a given history of glacial ice melt, and a reasonably realistic model of the Earth's rheology. Assuming a simple deglaciation of curve, these calculations show ocean basins subsiding towards the centre of the Earth *ca.* 8 m in the last 6000 years, while the mean continental surfaces rise *ca.* 16 m. The latter result is consistent with modern geodesic relevelling rates determined by Hicks (1972) for continental U.S.A.

The effect of ocean floor subsidence on sea level change recorded near ocean margins must be compounded with the hydro-isostatic flexural movement of the margins themselves. These can similarly be calculated, given appropriate rheological models for the marginal zones. In figure 1, three cases of isobase deformation are depicted for different upper Earth models, continental lithosphere, ocean (less rigid) lithosphere (from Chappell 1974), and an oceanic–continental transition case. The results for the transitional case are preliminary and *must* be treated with caution.

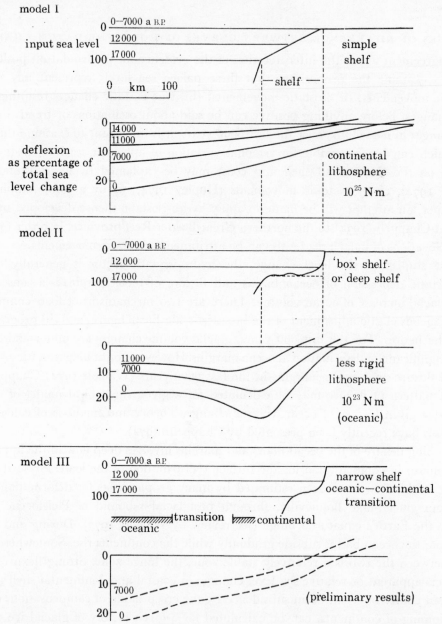

FIGURE 1. Hydro-isostatic deformation of continental margins possessing different rheological properties. Isobase divergence is strongest in the case of a deep shelf with an 'oceanic' (or less rigid) lithosphere.

Two main points arise from figure 1. First, divergence of isobases from the ocean margin to the coast increases with time. Divergence is strongest in the extreme case of ocean lithosphere and box shelf geometry, and weaker in those cases of rigid lithosphere and narrow planar shelf. It also appears that divergence is weak in the transitional model case, indicating limited deformation of outer and inner shelf perhaps because the rate of continental lithosphere bending is dependent on movement in the oceanic crust. The second point is that the hinge or crossover from subsidence to uplift migrates inland with time. This means that all points near the coast are effectively subsiding. If the shelf and near-coastal subsidence equals the mean subsidence of ocean basins, then there is no apparent sea level change for the last 6000 years

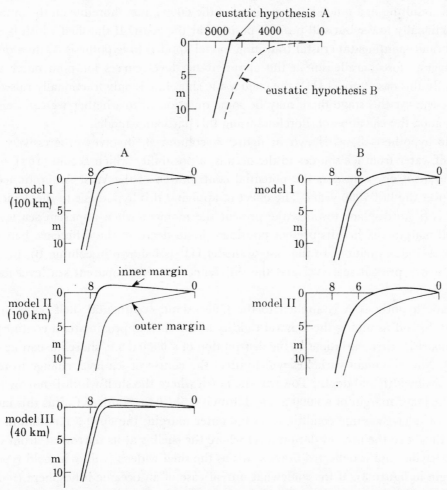

FIGURE 2. Relative sea level curves for inner and outer margins of continental shelves, for two ice-melt (eustatic change) hypotheses. Three rheological models are depicted: I, continental lithosphere; II, oceanic lithosphere; III, continental–oceanic transition lithosphere. Curves are derived for continental shelf widths of 100 km in the case of models I and II, and 40 km in the case of model III.

(assuming zero ice volume change). If near coastal subsidence is less than that of the mean ocean basin, net shoreline emergence should be apparent over the last 6000 years. However, if the shelf is subsiding at a rate which exceeds the mean ocean floor (e.g. the case of ocean lithosphere and wide shelf), then shelf submergence will have occurred over this period. For a given ocean volume history, shelf geometry, and Earth structure, apparent sea level curves can be constructed for both outer and inner margins of the continental shelf, by methods given in Chappell (1974, fig. 7).

The exact course of glacio-eustatism has been extensively debated, principally because of difference between Holocene sea level curves from different places. Let us examine two eustatic hypotheses on hydro-isostatic terms. Eustatic hypothesis A views the ice-melt contribution to sea level change as effectively ceasing about 7000 years ago (cf. Bloom 1971). The apparent courses of sea level, given a broad, sloping shelf and a continental lithosphere, are shown in figure 2, column A, as model I. At the coast (inner shelf margin) sea level falls from a high 6000 years ago to the present time, but at the outer margin it is continually rising. In model II an oceanic lithosphere in invoked. The degree of downward deflexion of the shelf edge increases in this case resulting in a much lower position of the 6000 a B.P. shoreline on the outer margin, and a fractionally lower but still positive position at the coast. If the shelf width is narrowed, and an oceanic–continental crustal transition model applied to hypothesis A, then preliminary results suggest close parallelism in the apparent sea level curves for both outer and inner shelf sites. In this case (model III), the 6000 a B.P. shoreline is only fractionally raised (to 1 m) above present. At this stage there may be some question as to whether we can achieve such a resolution on the elevation of shorelines using this particular model.

Eustatic hypothesis B, as shown in figure 2, column B, involves progressively declining addition of water from ice sources to the oceans, a possibility which Bloom (1971) regards as being relevant if one considers the potential contribution of the West Antarctic ice sheet to sea level over the last 7000 years. The effect of applying this hypothesis to rheological models I and II is (i) to displace towards the present the moment when apparent sea level on the inner shelf margin reached its present position; (ii) to decrease the difference between inner and outer shelf deformation. In the case of model III (not shown in column B), the degree of elevation above present sea level and the difference between apparent sea level for the two sites is negligible.

If eustatic hypothesis A is applied to the Queensland continental shelf between latitudes 14 and 20° S, and assuming the flexural rigidity of the shelf approximates a continental lithosphere (model I), then variations in the distribution of a 6000 a B.P. shoreline can be estimated (figure 3). Now the main variables which affect the course of sea level change from place to place are shelf width and depth. Towards the north where the shelf width is narrow, the elevation of the inner margin of a 6000 a B.P. datum is 0.5–1.0 m. South of Cairns this increases to 1.0–1.5 m. Under the same conditions on the outer margin, the 6000 a B.P. datum would be at or a little above the present datum level where the shelf is at its narrowest, from Cooktown to Cape Melville, but is deflected downwards as the shelf widens (values in bold type at outer shelf margin in figure 3). If the somewhat unreal case of an oceanic lithosphere (model II) is assumed beneath the north Queensland shelf, the pattern changes. At the outer margin the 6000 a B.P. datum would lie at a greater depth (values in parentheses in figure 3). The elevation of inner margin shorelines would be lowered only slightly.

The effect of employing eustatic hypothesis B on the Queensland shelf would be to lower even further the elevation of 6000 a B.P. shoreline on the shelf outer margin. It would occur below present sea level on the inner margin or coast sites. It is of interest to note that the shoreline which would achieve maximum elevation (ca. 1.0 m) at the inner margin under this eustatic hypothesis would date between 3000 and 4000 years ago.

In summary, various hydro-isostatic models as applied to the northern Great Barrier Reef Province suggest that the 6000 a B.P. shoreline should be elevated the highest relative to present sea level by the order of 1–2 m, under eustatic hypothesis A. But if hypothesis B is applied, then this shoreline should lie below present sea level, and the 3000–4000 a B.P. shoreline should be the highest (ca. 1 m). When these conclusions are confronted with the field evidence (McLean, Stoddart, Hopley & Polach 1978, this volume) certain discrepancies emerge:

(i) where it has been recognized, the 6000 a B.P. low water datum lies 0–0.5 m above present;

(ii) the highest raised or elevated deposits which possibly could be attributed to higher relative sea levels by McLean were deposited 3000–4000 years ago (+1 to 1.2 m).

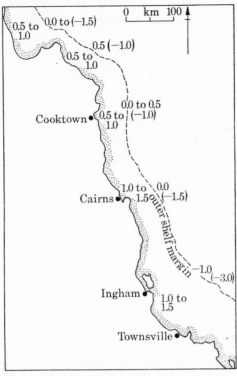

FIGURE 3. Preliminary estimates of elevations of a hypothetical 6000 a B.P. shoreline datum on the north Queensland coast applying eustatic hypothesis A (figure 2), and assuming no tectonic movements. The portion of this shoreline as shown here would be only dependent on glacio-eustatic and hydro-isostatic factors. At the outer margin figures not in parentheses represent a continental lithosphere shelf (model I); parenthetical figures represent the elevation of the 6000 a B.P. shoreline assuming an oceanic lithosphere shelf.

Thus, it appears that some compromise between eustatic hypothesis is called for. More exact resolution will be possible when apparent palaeo-datum curves for several Great Barrier Reef sites are agreed upon, and when hydro-isostatic calculations are completed for the oceanic–continental transitional model.

There are two significant consequences to the understanding of hydro-isostasy in relation to areas such as the Great Barrier Reef. First, for students of late Quaternary sea level changes, differences between sites are to be expected and are at least partly explainable in these terms. Secondly, hydro-isostatic flexures cause the relative rates of sea level change over the last 6000 years to differ substantially between inner and outer zones of shelves, and this surely is an ecological factor of some importance with a strong potential influence on the nature of reef development.

In conclusion, it is now possible to view an apparent or relative sea level curve for any particular stretch of coast as being the product of several factors or mechanisms. The contribution of each factor changes with time. For instance, between 6000 and 9000 years ago, glacio-eustasy accounted for up to 80 % of observed sea level change; over the last 3000 years this factor was probably responsible for no more than 10 % or 10 cm of observed change, while hydro-isostasy becomes relatively more important. This principle is summarized in figure 4. The exact proportions for the case of the northern Great Barrier Reef cannot yet be stated, and the course of sea level shown here is no more than a first approximation derived from somewhat ambiguous evidence at and above present sea level (McLean *et al.* 1978, this volume), and below present sea level (Thom, Orme & Polach 1978, this volume).

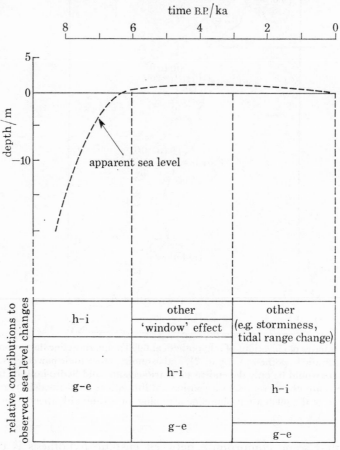

FIGURE 4. The approximate position of apparent or relative sea level over the past 8000 years based on evidence from 'low wooded' islands of the northern Great Barrier Reef Province. The relative contribution of a variety of factors or mechanisms responsible for change is indicated as changing with time: h–i, hydro-isostasy; g–e, glacio-eustasy. The exact degree of explanation by each factor is unknown.

REFERENCES (Thom & Chappell)

Bloom, A. L. 1971 In *Late Cenozoic glacial ages* (ed. K. L. Turekian), pp. 355–380. New Haven: Yale University Press

Chappell, J. 1974 *Quat. Res.* **4**, 429–440.

Chappell, J. & Polach, H. 1976 *Bull. geol. Soc. Am.* **87**, 235–240.

Denton, G. H. & Karlen, W. 1973 *Quat. Res.* **3**, 155–205.

Hicks, S. O. 1972 *J. geophys. Res.* **77**, 5930–5934.

Hopley, D. 1974 In *Proc. 2nd Int. Symp. Coral Reefs, 1973*, vol. 2, pp. 551–562.

McLean, R. F., Stoddart, D. R., Hopley, D. & Polach, H. 1978 *Phil. Trans. R. Soc. Lond.* A **291**, 167–186 (this volume).

National Academy of Sciences 1975 *Understanding Climatic Change. A Program for Action.* U.S. Committee for the Global Atmospheric Research Program, National Research Council, Washington, D.C., 239 pages.

Neumann, A. C. 1972 In *Abstracts American Quat. Assoc. 2nd National Conference*, Miami, Florida, pp. 41–44.

Thom, B. G., Orme, G. R. & Polach, H. 1978 *Phil. Trans. R. Soc. Lond.* A **291**, 37–54 (this volume).

Walcott, R. I. 1972 *Quat. Res.* **2**, 1–14.

Walcott, R. I. 1973 *A. Rev. Earth planet. Sci.* **1**, 15–37.

THE NORTHERN
GREAT BARRIER REEF

PART B

Phil. Trans. R. Soc. Lond. B. **284**, 3–21 (1978)   [**3**]
Printed in Great Britain

Ribbon reefs of the Northern Region

By J. E. N. Veron† and R. C. L. Hudson‡

*Department of Marine Biology, James Cook University of North Queensland, P.O. Box 999,
Townsville, Queensland, Australia 4810*

[Plates 1–4]

Ribbon reef is a general term applied to most of the shelf-edge reefs of the Northern
Region. The principal morphological characters of these reefs are described from aerial
photographs. Their position in relation to the Queensland Trench is described from
direct observation with the use of Scuba and from soundings. The depth and position
of major reef zones and their substrates and dominant biota are described from
transects of Tijou and Great Detached reefs.

Introduction

Of the many well defined reef systems within the Great Barrier Reef (G.B.R.) province, the
northern shelf edge reefs are by far the most inaccessible. They are remote from any port or
township and for most of the year are pounded by heavy seas driven by the SE trade winds.
Thus they have remained unstudied and are almost completely undescribed. However, they
form one of the major physiographic units of the G.B.R. province and are potentially one of
the most important to the understanding of the G.B.R.'s complex morphology, structure and
evolution.

Maxwell (1968) divided the G.B.R. into three regions on the basis of bathymetry and geomorphic changes and correlated these with differences in reef density, morphology and development. The Northern Region, north of 16 °S, is characterized by shallow water, the presence
of an outer reef system and prolific reef growth across the shelf, induced by the proximity of
a steep continental slope. This paper, the first of a series of three, summarizes aspects of the
physical environment of the Northern Region that are relevant to the development of the shelf
edge system, gives a general description of that system up to the northern limit of the 'ribbon
reefs' (Figure 1) and gives the results of a study of two ribbon reefs, Tijou Reef and Great
Detached Reef. The second paper (Veron 1978a, this volume) describes the shelf edge reefs
north of the ribbon reefs, the 'deltaic' reefs and the 'dissected' reefs (Figure 1) while the third
(Veron 1978b, this volume) gives an account of the evolutionary history of the latter, far
northern reefs.

Methods

Field work with the use of the following methods was carried out from R.V. *James Kirby*
(a modified 17 m steel trawler) during Phase III of the 1973 expedition and during a second
northern voyage in November 1974.

† Present address: Australian Institute of Marine Science, P.O. Box 1104, Townsville, Australia 4810.
‡ Present address: Department of Zoology, University of Melbourne, Parkville, Victoria, Australia 3052.

(a) Aerial photography

Physiographic features of the outer edge ribbon reefs were assessed from Commonwealth aerial photographs covering the outer reefs from their southern limit north to 12° 35′ S. A further series of three runs cover part of the northern deltaic reefs. R.A.A.F. aerial photographs of a greatly reduced quality are available for most far northern reefs.

Specific areas of study in the vicinity of Tijou Reef were photographed from a chartered aircraft by using 35 mm and Polaroid cameras.

Zonation patterns are clearly discernible in aerial photographs, as for example illustrated in figure 2, plate 1.

(b) Bathymetry

Detailed data on reef profiles were obtained by Scuba divers using depth gauges and underwater measuring lines laid perpendicular to reef faces.

Bathymetric data in the vicinity of reefs were obtained from R.V. *James Kirby* using a Furuno F850 mark III sounder. Positions of the vessel relative to reefs were measured directly by a measuring line for distances of 200 m or less and by radar from a reflector mounted in a dinghy for greater distances.

Bathymetric data at greater distance from reefs were obtained from the Hydrographic Office of the Royal Australian Navy and from R.A.N. and British Admiralty charts.

Off-reef bathymetric profiles were positioned so as to be continuous with the reef transects (see below).

(c) Reef transects

Rope transect lines laid perpendicular to the reef front were used by Scuba divers to measure the width of major reef zones. The depth, nature of the substrate, dominant biota and percentage cover of dead and living coral were estimated every few metres. These transects were continued across the reef flat and down the outer slope (eastern side) and reef back slope (western side).

(d) Sampling

Detailed collections of corals were made at the position of each quantitative transect (see below) and at all well defined biotopes studied. The results of these collections are included in Veron & Pichon (1976) and subsequent monographs in the series *Scleractinia of eastern Australia*.

Dominant species of all other biota were collected along reef transects.

Samples of the substrate along reef transects were collected by divers. At greater depths, substrate and coral samples were collected by using custom-made anchor dredges.

(e) Quantitative transects

Quantitative measurements were made at Tijou and Great Detached Reefs of the percentage area cover of dominant biota and substrate types, as well as of the percentage area cover of dominant coral species, by use of a 30 m measuring line laid parallel to the reef and transverse to the transects. Measurements were recorded as percentages of the 30 m line.

(f) Supplementary surveys

Each locality studied was selected on the basis that it was representative of a large area of reef. Selection was made according to data obtained from aerial photographs, aerial surveys, and surveys by Scuba divers towed behind small boats.

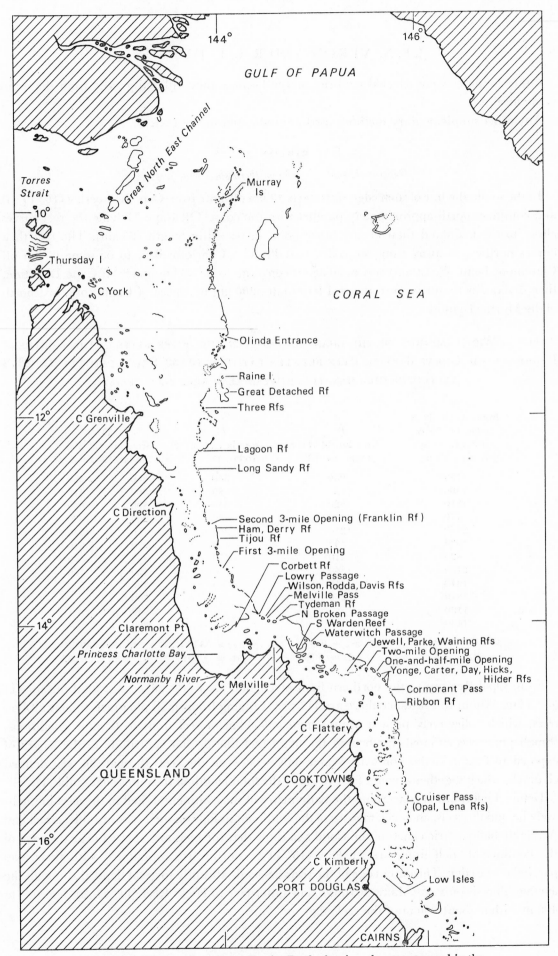

FIGURE 1. The northern Great Barrier Reefs, showing place names used in the text and the extent of the barrier reef system.

Other localities were selected for brief survey because they differed greatly from the main study localities.

Any other supplementary methods used are indicated in the text.

The ribbon reefs

General characters of the northern shelf edge reefs

In the south, the line of shelf edge reefs starts 43 km offshore from Cape Kimberley (16° 17′ S) and continues north approximately parallel with the coast. Off Cape Melville the reef curves closer to the mainland than at any other point in its entire length (25 km). The reef then curves northwards away from the coast, past Princess Charlotte Bay, to be some 65 km off Claremont Point. Coast and reef once again converge to be 28 km apart off Cape Direction, then diverge increasingly until the reef terminates 205 km northeast of Cape York, just north of the Murray Islands.

TABLE 1. Width (metres) of the channels between the outer reefs in the Northern Region of the Great Barrier Reef between latitude 15° 39′ S and latitude 12° 23′ S (excluding the channels between rear reef plug reefs).

A from lat. 15° 39′ S to lat. 14° 55′ S (from Lena to Ribbon Reef)	B from lat. 14° 41′ S to lat. 14° 13′ S	C from lat. 13° 27′ S to lat. 13° 15′ S	D from lat. 13° 2′ S to lat. 12° 23′ S	
1180	950	340	850	380
1500	820	85	300	410
1810	820	340	890	680
770	1300	340	260	850
2150	1210	210	70	340
1630	430	470	270	640
820	610	470	470	510
2150	1340	810	420	760
1940	1340	380	380	550
1810	260		300	
1120			740	
1430			260	
av. = 1526 s.d. = ± 475	av. = 908 s.d. = ± 391	av. = 383 s.d. = ± 201	av. = 491 s.d. = ± 232	

From Opal Reef up to the northern limit of Lagoon Reef (12° 23′ S) is a total distance of 555.5 km. Within this distance there are 117 reefs ranging from small plug reefs to long ribbon reefs, which collectively present approximately 536 km of reef front to oceanic conditions. (Small plug reefs situated just behind the main reef line do not contribute to the length of exposed reef front in this context.) Water moves between the reefs through 86 well defined channels, which together account for 116 km of the total length of the outer edge (table 1).

Detailed information concerning the bathymetry of the continental shelf in the area immediately behind the reefs, and even more so of the continental slope on their seaward side, is lacking. Available bathymetric charts of the Coral Sea show that the reefs are situated near the edge of the continental shelf from Cruiser Pass (15° 40′ S) northwards, orientated with their axes parallel to the edge. The outer edge of the reefs descends abruptly to great depths, although the average slope is only 4° (Maxwell 1968). The outer face of Hicks Reef (14° 27′ S) descends to 400 m within about 300 m of the reef edge.

In the back reef area, the depth of the water is variable and, while it does not generally exceed 36 m, it is more commonly in the order of 28 m.

Southwards from Cruiser Pass, between Lena Reef and Opal Reef, the reefs occur increasingly close to shore and further from the edge of the continental shelf. They are not orientated with respect to the continental margin, but in a generally southeast direction. There appears to be a plateau at about 55 m between the outer limit of these reefs and the continental slope, probably a remnant of the 58 m (32 fathoms) strand line (Maxwell 1968). The sea floor shows a distinct declination eastwards to the outer reefs and, in the back-reef area, is deeper than that found further north. The depth of water is now greater than 36 m and it is possible to distinguish the 'Marginal Shelf' in Maxwell's terminology, as defined by the 36 and 92 m (20 and 50 fathom) contours.

The depth of water in the channels between the reefs is largely unknown. Where data are available, the depth would appear to be similar to, or greater than, that of the back reef area. Marked exceptions occur where strong currents flow and where former rivers have scoured deep channels, as for example in Lowry Passage between Wilson and Rodda Reefs (13° 57' S). The ends of the reefs normally descend very abruptly.

North of Olinda Entrance (11° 14' S), the appearance of the outer reefs changes greatly, from 'ribbon' reefs into 'deltaic' reefs, and then into 'dissected' reefs. The term 'ribbon' reef is used in this paper in its most generalized sense to include all the elongate shelf edge and detached reefs of the Northern Region south of the deltaic reefs.

Morphological characteristics of ribbon reefs

The enormous variability in shape, size and development of reefs within the G.B.R. province has been classified by Maxwell (1968) into a scheme which associates the present surface appearance of reefs with developmental stages. These stages appear to be well established, at least in principle, and the scheme thus appears to have a general, functional application.

The stages of reef development indicated by Maxwell represent the results of an interplay of all the factors, past and present, which influence the reef. In localities where these influences are approximately uniform all round, a symmetrical platform reef is formed. Continued expansion ultimately leads to the formation of a shallow lagoon, as conditions in the centre of the reef fall below those necessary to maintain coral and algal growth. In other localities where, for example, the bathymetry is limiting, reef expansion can occur only in certain directions. If this occurs along a drop-off contour, the asymmetrical elongate reefs variously known as 'wall', 'linear' or 'ribbon' reefs are formed. These reefs change, and are changed by, patterns of water movement. Strong currents flowing around the ends cause those ends to curve back, and may lead to prong and buttress formation in the back reef slope. Eventually the arms of the reef may join to form a lagoon. A number of other growth forms are described by Maxwell. As a result of continued growth, hydrological and biological conditions can change to such an extent that an adjacent reef or reefs may 'resorb' or degenerate by one process or another.

The shelf edge reefs are mostly elongate. They occur in an almost continuous series, separated by channels of varying widths as described below. The considerable variation in size and surface shape found amongst these reefs is indicated by the ten distinguishable types (elaborating on Maxwell's scheme) illustrated in figure 3. They range from simple ribbon reefs (types A and B) to closed ring reefs (type G), a variety of plug reefs (types H, J and K), and finally to a few reefs that show 'resorption' (type L). Such a classification does have limitations, however, as reef forms with intermediate features common to two or more types are frequently found. With the exception of the plug reefs (type J only) the different reef morphologies all appear to be derived initially from a basic building unit: the short wall reef (type A). Plug reefs develop in the openings between reefs or are remnants of earlier existing reefs. Coral growth occurs parallel with the current and so leads to the formation of a triangular reef with the apex directed upstream.

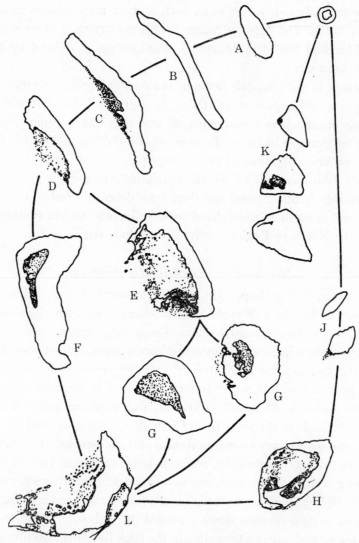

FIGURE 3. Morphological variation in outer edge reefs, elaborating Maxwell's (1968) classification (see text).

The following descriptions of reefs are best read with relevant navigational charts. However, most of the information given here has been obtained from aerial photographs and is not usually obtainable from the charts.

The reefs lying between and including Opal Reef at the southern limit of the region and Cruiser Pass (15° 41′ S) represent the transition from the scattered platform reef development, so characteristic of the Central Region, to the linear reefs that typify the Northern Region.

Within this section the reefs are generally crescentic with, in some cases, considerable secondary coral growth in the back reef area. They appear, however, to be undergoing 'resorption' and are thus of type L. Their orientation would appear to be determined more by the influence of the SE trade winds than the direction of the continental shelf edge.

North of Cruiser Pass the outer reefs are orientated along the very edge of the continental shelf. Consequently they face in any direction between north and southeast; they are not aligned according to dominant wind direction.

Several natural divisions occur in the line of reefs that mark recognizable changes in the pattern of reef development. These divisions correspond roughly with the five major water exchange sites referred to above. The line of reefs extending north from Cruiser Pass to Ribbon Reef consist of a series of elongate reefs, mainly type D, showing thickening of the ends and some infilling of the back reef area with secondary coral growth. The average length of these reefs is 7930 m (range: 14190–2750 m). The width at their narrowest central part ranges from 950 to 470 m (mean: 720 m). Ribbon Reef itself is a type C reef. It is extremely long (35030 m) but only 1300–610 m wide. One, two or three plug reefs (type J) are located in almost every channel between the linear reefs, some distance behind the reef front. A group of three plug reefs is located radially behind the channel at the south end of Ribbon Reef. The occurrence of these reefs is indicative of strong currents, especially to the north and south of Ribbon Reef which inhibits E–W water flow by its great length. Between Lena Reef and the south end of Ribbon Reef, the ratio between the length of exposed reef front and the width of the channels between the outer reefs is 4.8 : 1.

The groups of reefs, which includes Yonge, Carter, Day, Hicks and Hilder Reefs, between Ribbon Reef and Two-mile Opening are type E, with two exceptions. Hilder Reef, situated between Two-mile and One-and-half-mile Openings, is a large advanced plug reef, type H. The small reef in Cormorant Pass is a small plug reef, type J. Type E and H morphologies are relatively unusual in the reef line, being found nowhere as well differentiated as in this section. The marked curving of the ends of these reefs and the extensive back reef margin are indicative of good water circulation. The 'Open Ring' formation, in Maxwell's terminology, develops progressively as one proceeds northwards in this section, suggesting that this region is one of the major sites for water exchange between the inner reef and oceanic waters and that the major currents pass through Two-mile and One-and-half-mile Openings. The possible influence of the southeast trade winds is seen in the greater coral growth in the southern recurved arms. The reef flats are 780–600 m wide and reefs appear similar to the simple wall type. Their average length is 6630 m (range: 3460–8650 m). The ratio between length of reef front and channel width from Yonge Reef to Hilder Reef is 5.4 : 1.

North of Two-mile Opening the reef continues as a series of narrow ribbon reefs, types C and D, to Waterwitch Passage (14° 12′ S). The strong currents at the southern end of this section and the long fetch over unobstructed water in the inner reef region for the prevailing southeast trade winds, appear to have led to the inward extension of Jewell Reef and its continuation in a line traversing the inner shelf by Parke Reef and Waining Reef. These appear very similar to the linear reefs, type A and type B respectively, of the outer edge. That strong currents flow in a northwesterly direction between these inner reefs is indicated by the presence of sand levees. A kite-shaped plug reef (type J) occurs as an outer reef off the northern end of Jewell Reef. It is followed by four elongate reefs, 3780–13580 m (mean: 8050 m) in length with an average breadth of 900 m (range: 690–1150 m).

A single large irregular type C reef, 49 050 m in length with an extremely variable width of 1340–3370 m, separates Waterwitch Passage from the group of Passes from North-Broken Passage to Lowry Passage (collectively referred to as Melville Pass). This is the longest and widest reef in the Northern Region. Two reefs similar to ribbon reefs, of which the innermost is S. Warden Reef, have developed across the shelf off the southern end of this reef. Sand levees have formed behind and to the sides of the channels between them. The development of the narrow inner shelf reefs, the presence of sand levees, the line of back-reef development, and the breadth of the outer reef all indicate that strong currents flow northwesterly on the inside of this reef towards Melville Pass.

The three reefs situated in Waterwitch Passage and the six reefs in Melville Pass are characteristic of areas subject to strong currents, where conditions favourable to growth are good but not symmetrical. Thus the central reef in Waterwitch Passage and Wilson Reef in Melville Pass are of the closed ring type where the ends of the reef have grown around to enclose a lagoon (type G). Apart from the obvious plug reefs present in both passes, Tydeman, Davie and Rodda Reefs in Melville Pass, and the northernmost reef in Waterwitch Pass, all appear to be derived from short linear reefs (type A), modified by considerable back reef development, either in the form of thickening along the entire axis as seen in Tydeman Reef, or by prolongation of the back reef to form a triangular reef with its base seawards (type K). All these reefs have sand banks or unvegetated cays at their northwesterly ends.

The position and curvature of Corbett Reef on the inner shelf and the backward northwesterly directed extension of the southern end of the first reef north of Rodda Reef indicate that there are still strong currents flowing northwards inside the reef. North of this point, however, these currents appear to dissipate. The contours of the channels between the reefs up to First Three-mile Opening and the extent of coral development in the back reef areas suggest that some portion of the northerly current exists through the reefs here. In addition, there is substantial water flow through First Three-mile Opening and another channel to its south, that has enhanced back reef growth of the reefs immediately to their south. As a result, these reefs have developed type F morphologies.

The line of reefs extending north from Rodda Reef to First Three-mile Opening consist of types A, D, E, F and K. The influence of currents flowing on the inside of these reefs is seen

---

DESCRIPTION OF PLATE 1

FIGURE 2. Wilson Reef (13° 15′ S), illustrating comparisons between zones visible in aerial photographs and those observed on the reef. These zones are: (A) Sandy ocean floor with large clumps of broken reef-rock. Coral cover approximately 10%, dominated by *Millepora tenera*. (B) Approximately 50% cover of ramose *Acropora*, dominated by *A. intermedia* and *A. formosa*. (C) Flat, consolidated limestone with much filamentous algae and approximately 1% coral cover of many species. (D) Flat, consolidated limestone with sand and rubble; no coral. (E) Mostly sand with approximately 5% coral cover, mostly ramose *Acroporas*. (F) The lagoon, extending to a depth of approximately 5 m. The substrate was clean, white, calcareous sand with occasional outcrops of ramose *Acropora* and *Porites*. (G) Clean, white calcareous sand. (H) A mixture of sand and *Acropora* debris with irregular patches of ramose *Acropora*. (I) Cemented *Acropora* rubble. The outer edge of this zone was bordered by a 5 m wide strip of actively growing *Acropora* sp., the same species which comprises most of the cemented rubble. (J) Flat, limestone covered by filamentous algae and some sand. This is the zone of maximum wave action. (K) Coral cover increased was approximately 30% and was dominated by *A. hyacinthus*, *A. humilis* and *A. palifera*. The latter species became very dominant at depths greater than 3 m. The whole zone was characterized by spur and groove formations, with the grooves up to 3 m deep. Vertical exaggeration in lower part of figure is 1 : 20.

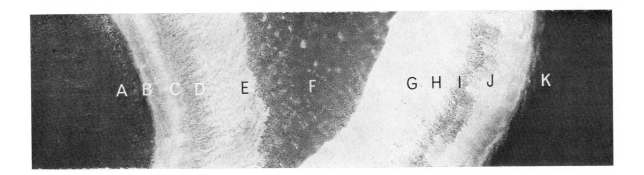

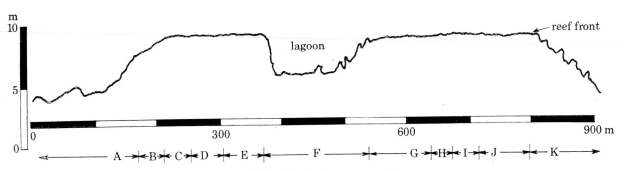

FIGURE 2. For description see opposite.

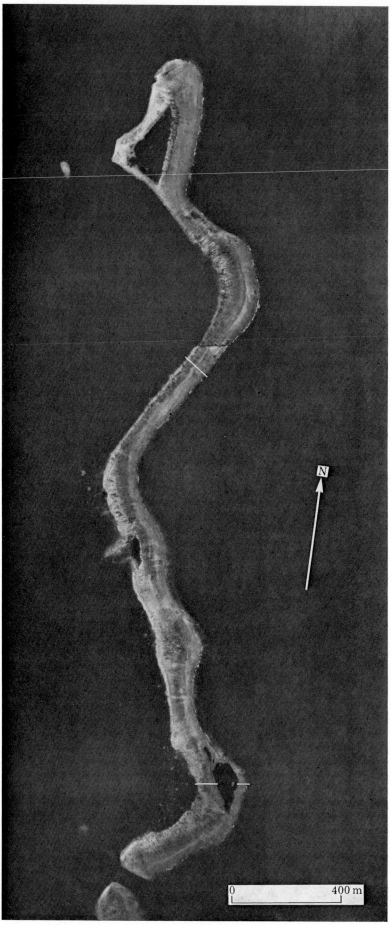

FIGURE 4. Tijou Reef. The positions of the reef transects (figures 6 and 7) are indicated.

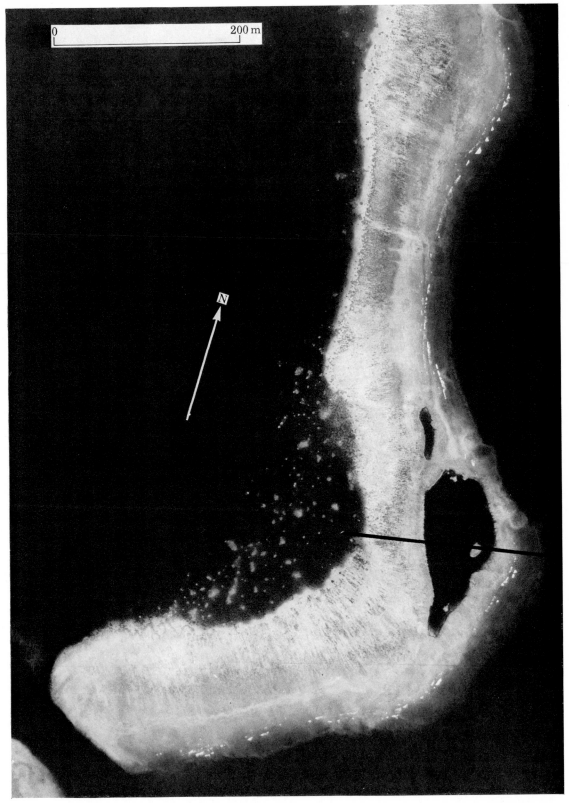

FIGURE 5. The south end of Tijou Reef. The position of the reef transect (figure 6) is indicated.

Phil. Trans. R. Soc. Lond. B, volume 284

Veron & Hudson, plate 4

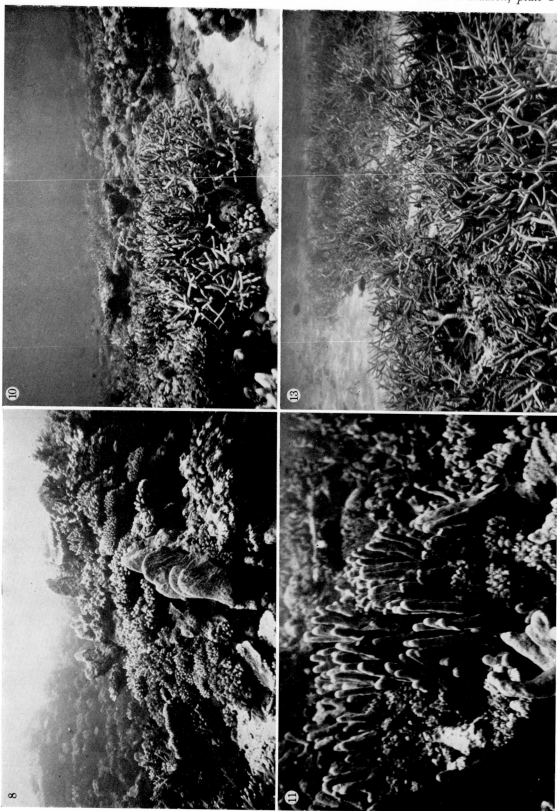

FIGURES 8, 10, 11 AND 13. For description see opposite.

in the development of broad reef flats, varying between 1200 m and 2660 m wide. The development of these reefs compares with that of the long reef between Waterwitch and Melville Passes. Their average length (excluding type K reefs) is 8990 m, varying from type A, 3440 m long, to type D, 15310 m long.

North of First Three-mile Opening, the character of the outer reef changes again to become a long irregular line of reefs with similar breadths following closely one after another. The reefs are mostly of types A–D, interspersed with types J and K. The linear reefs are generally shorter, except for Tijou Reef (13° 8′ S) which is 27800 m long, and narrower than the reefs to the south. Excluding the various plug reefs and Tijou Reef, the average length of the linear reefs is 6600 m (range: 11510–3160 m) and their average minimum width 830 m (range: 1150–590 m). In addition to the above features, the channels between the reefs are significantly narrower than those of the southern sections. The reef back margin is markedly reduced.

Strong tidal currents occur in Second Three-mile Opening which interrupts the outer barrier to the north of Tijou Reef and is the only site of major exchange between reef and oceanic waters in this section. Strong currents also appear to occur in the narrow channels between the reefs. Two small type K reefs – one unnamed, the other Franklin Reef – are situated in Second Three-mile Opening. These reefs, in common with the two larger type K reefs (Ham and Derry) that continue the line north, have sand banks on their west or north-west sides. A number of small plug reefs (type J) have become established close to the rear margins of the reefs in the strong currents running through several narrow channels. These currents have in places led to marked but narrow rearward growth of the ends of some reefs, while in others the ends of the reefs have thickened. In some cases, e.g. Long Sandy Reef (12° 31′ S), the ends of the linear reefs remain unmodified. Presumably the currents around these reefs are slight. Several short linear reefs (type A) show overall thickening to create short, broad, rectangular reefs, similar in appearance to those described above in Melville Pass and Waterwitch Passage, although smaller. These reefs suggest that greater water movements and exchange occur in their vicinity.

In this section of the reef there appears to be a reduction in the total width of the channels penetrating the reef line. The ratio between the length of exposed reef front and the width of the channels is 9.1 : 1. Lagoon Reef at the northern limit of this section is unusual. It is approximately triangular, has reef front along two sides, and has a deep lagoon towards its apex which is open to the sea. In some respects it resembles a large, well developed plug reef, type H. However, it is at about this point that the long linear reefs give out and the outer reefs continue as a series of short reefs which cease to show strong affinities with the elongate linear reefs to the south.

DESCRIPTION OF PLATE 4

FIGURE 8. The outer slope of Great Detached Reef at 15 m depth.

FIGURE 10. The inner reef flat of Great Detached Reef approximately 300 m from the front, along the line of the transect.

FIGURE 11. The back reef slope of Great Detached Reef, showing the characteristic growth form of *Acropora palifera*.

FIGURE 13. The reef flat of the far northern end of Great Detached Reef, showing the dominance of ramose *Acropora* species in a sandy substrate approximately 300 m from the reef front along the line of the transect.

Tijou reef

Tijou Reef (figure 4, plate 2) is a long narrow reef with an exposed reef front 27 800 m in length. The width of the solid reef mass varies from 1550 m at its maximum to 640 m at the narrowest part near the north end. The general orientation of the reef is north–south, but as the reef is very irregular in shape, different sections face in directions ranging from southeast to northeast.

In terms of the reef types illustrated in figure 3, Tijou Reef combines features of types B, C and F. At the northern end, the strong currents in Second Three-mile Opening appear to have induced the formation of a lagoon studded with large coral heads (type F). From this lagoon south to the approximate midpoint of the reef, it is basically type B, being almost uniformly narrow (about 680 m) and having little secondary coral growth in the back reef area. The southern half of the reef is type C. It is more variable in width and secondary coral growth has occurred along the rear margin, particularly in the protected angle where the reef curves through to the southwest, and in the region of the obtuse curve of the reef at about the midpoint. The southern end is not thickened in any significant way.

Tijou Reef, south end

The south end of Tijou Reef is characterized by an enclosed deep lagoon (figure 5, plate 3). The principal surface features of the reef and lagoon are illustrated in figure 6.

The outer slope

The outer slope was not examined in detail by divers. The seaward margin is dominated by deep spur and groove formations that penetrate the reef front at an oblique angle. At their seaward end, the grooves are up to 7 m deep and 10 m wide.

The reef flat and lagoon

The principal features of the reef flat and lagoon are indicated in figure 6. The eastern side was dominated by deep spur and groove formations running obliquely to the reef front. At their seaward end the grooves were up to 7 m deep and 10 m wide. The reef margin was penetrated for distances of up to 80 m by grooves that gradually diminished in size. The dominant corals were *Acropora humilis* (Dana) and *A. palifera* (Lamarck). A wide diversity of Alcyonaria was present in similar abundance. Green filamentous algae and hydroids dominated the wave-battered reef flat between the spur and groove zone and the edge of the lagoon.

The reef flat up to 100 m east of the lagoon was mostly composed of partly cemented rubble and sand. A narrow band of rubble sloped steeply into the lagoon and was replaced by loose calcareous sand grading into mud in the lagoon itself.

The small reef patch in the lagoon shown in figures 5 and 6 had irregular, vertical or undercut walls of limestone to approximately 5 m depth which became buried in steeply sloping, soft, muddy sand. The western edge of the lagoon also had an irregular vertical wall to about 15 m, where it merged into a soft sandy floor. The extraordinarily great depth of the lagoon, 43 m, was measured by lead-line at four positions along the line of the transect.

The reef flat was dominated by Alcyonaria and coralline algae up to 70 m west of the lagoon, and thereafter by very irregular patches of ramose *Acropora* species, interspersed by patches of sand, dead coral and rubble.

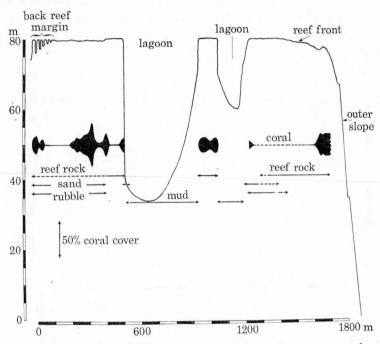

FIGURE 6. Profile of Tijou Reef, south end, showing the distribution of substrate types and estimated percentage cover of coral. The position of this profile is indicated in figures 4 and 5. Vertical exaggeration is 1 : 20.

The back reef margin

As indicated in figure 6, the back reef margin consisted of an extensive area of secondary coral development. A brief examination showed these to be similar to the back reef margin of the northern end of Tijou Reef, described below.

TIJOU REEF, NORTH END

The principal surface features of the north end of Tijou Reef at the position indicated in figure 4 are summarized in figure 7.

The outer slope

The outer slope below the spur and groove zone declined at an average angle of approximately 45°, reaching a depth of 70 m at a distance of 80 m from the front. Beyond that the reef descended more steeply, reaching a depth of 127 m at 100 m from the front.

Coral cover was between 50 and 100 % over most of the outer slope, to a depth of at least 45 m. At approximately 5 m depth, *Acropora humilis* (Dana), *A. palifera*, *Pocillopora verrucosa* (Ellis and Solander) and an unidentified *Acropora* were dominant. *Porites lichen* (Dana) also became dominant at 10 m. At 30 m, *P. lichen*, *A. palifera* and *Tubipora musica* (Linnaeus) shared approximately equal dominance. At greater depths, species diversity greatly increased and no species showed marked dominance.

The reef flat

Coral cover in the spur and groove zone (indicated in figure 7) averaged approximately 30 %, being greatest on the spurs and least in the grooves which are approximately 1–1½ m deep. At approximately 20 m west of the well defined front, the coral cover decreased to approximately 5 %, then to less than 1 % at 40 m. *Acropora humilis* and *A. pyramidalis* Klunzinger are markedly dominant.

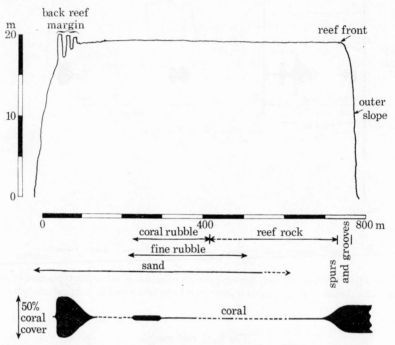

FIGURE 7. Profile of Tijou Reef, north end, showing the distribution of substrate type and estimated percentage cover of coral. The position of the profile is indicated in figure 4. Vertical exaggeration is 1 : 20.

TABLE 2. PERCENTAGE OF TOTAL AREA COVERED BY THE BIOTA AND SUBSTRATES (INDICATED) OF REEF SURFACE, AT THE POSITIONS INDICATED, AS DETERMINED BY 30 m TRANSECTS LAID PARALLEL TO THE REEF FRONT, NORTH END OF TIJOU REEF.

Position of 30 m transect	dead coral (consolidated and loose)	sand	Scleractinia	Alcyonaria	hydroids	sponges	filamentous algae	coralline algae
5 m depth, reef front	66.4	—	32.8	0.8	—	—	—	—
10 m depth, reef front	30.7	—	62.9	5.9	—	0.5	—	—
15 m depth, reef front	34.2	—	50.8	13.5	0.2	1.3	—	—
200 m behind crest	6.2	64.1	21.5	—	—	0.8	5.5	1.9
300 m behind crest	93.6	—	6.4	—	—	—	—	—
450 m behind crest	77.8	—	16.0	2.8	—	3.4	—	—
600 m behind crest	7.2	87.8	3.5	—	—	1.5	—	—

Very little coral occurred over the rest of the reef flat. Limestone, rubble and sand zones can be readily identified in aerial photographs. Hydroids are common to 140 m in from the front. Massive species of corals, mostly Faviidae, form an ill defined zone 560–630 m from the front. At 660 m from the front, *Acropora humilis* became markedly dominant again and remained the major reef-building species up to the reef back margin.

Table 2 summarizes the results of quantitative (30 m) transects, run parallel to the reef front, at three depths down the outer slope and at increasing distances west of the front, on the reef flat. There is a marked discrepancy between the results obtained by this method and those obtained from visual estimations by the authors (cf. figure 7). In view of the patchiness of coral distribution and the limited nature of such quantitative samples, it is concluded that 30 m transects of this type are inadequate for assessing the cover of biota and substrates of these reefs. Analysis of data from many hundreds of metres (each requiring perhaps a day's work) would appear necessary for each transect.

TABLE 3. PERCENTAGE SURFACE AREA COVER OF LIVE CORAL, DEAD CORAL AND SAND ON THE FRONT
OF GREAT DETACHED REEF AT THE THREE DEPTHS INDICATED (see text).

depth below reef front/m	*Acropora palifera*	*Acropora* sp.	*Stylophora pistillata*	*Acropora pyramidalis*	*Porites lichen*	*Pocillopora damicornis*	*Pocillopora verrucosa*	other Scler- actinia	total coral cover	dead coral	sand
5	10.5	1.8	1.3	2.3	—	—	1.7	10.1	27.7	64.3	7.6
10	16.2	11.0	3.8	—	—	2.7	—	12.2	45.9	50.3	0
15	1.8	0.7	3.5	—	4.0	—	—	14.6	24.6	62.0	10.2

The back reef margin and back reef slope

Echo soundings west of the back reef margin indicated an even sandy floor sloping gently from a depth of 5 m at a distance of 20 m, to 30 m depth at 100 m distance and then to 34 m depth at a distance of 200 m.

The reef back margin was composed of a well defined line of deeply and irregularly dissected limestone hillocks and ridges which were penetrated by sand to varying degrees and which supported a lush and varied coral growth, especially on their sides.

Species diversity was very great, and no species was conspicuously dominant. A single 25 m transect across the top and down the side of one such dissected hillock 3 m high, gave area covers (including overlapping) for dead coral of 47.0 %, live coral 43.3 %, coralline algae 40.0 %, Alcyonaria 8.3 %, and sand 1 %. Average coral cover on the side (alone) was 63 %, with reduced areas of dead coral (22.5 %) and coralline algae (17.5 %).

GREAT DETACHED REEF

The Great Detached Reef is one of the northernmost, and one of the longest, of the ribbon reefs. The reef front is approximately 30 000 m long and forms the eastern wall of a sandy plateau 175 km² in area that is detached from the continental shelf by a 6–7 km wide trench. The depth of this trench is unknown, but is at least 280 m. It is bounded on both sides by distinct lines of small reefs and sand banks which respectively compose the western edge of Great Detached Reef and the eastern limit of the continental shelf. The latter reefs and sand banks are a continuation of the more southerly ribbon reefs, including Three Reefs. The trench curves around the northern limit of Great Detached Reef to form the deep Raine Island Entrance.

The outer slope

At depths below approximately 5 m, the substrate consists of poorly cemented coral rubble interspersed with irregular pockets of *Halimeda* sand. The coral cover was approximately 50 % and (as is perhaps indicated in table 3) was irregular both in species composition and density.

In most areas *A. palifera* were dominant to depths of approximately 15 m (figure 8, plate 4), after which *Porites lichen* became markedly dominant to about 30 m depth. No species were clearly dominant below 30 m.

At depths of less than approximately 5 m, the substrate consisted of increasingly well cemented limestone carved into irregular spurs and grooves up to 2 m deep. This spur and groove zone was approximately 30 m wide, the grooves becoming gradually shallower with decreasing depth. *Acropora palifera*, *A. humilis* and *A.* cf. *pyramidalis* were dominant, all with the extremely flattened growth forms indicative of areas exposed to extreme wave action.

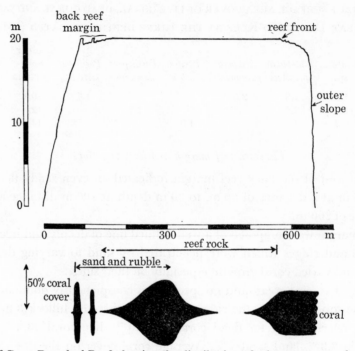

FIGURE 9. Profile of Great Detached Reef, showing the distribution of substrate type and estimated percentage cover of coral. Vertical exaggeration is 1 : 20.

The reef flat

The principal features of the reef flat indicated in figure 9 are probably generally representative of most northern ribbon reef flats orientated in a north–south direction, which are exposed to prevailing winds and which are uncomplicated by close proximity of channels, other reefs, etc.

Coral cover decreased rapidly behind the spur and groove zone and became completely absent 100 m behind the reef front. In this region the substrate consisted of flat, very hard limestone sparsely covered with green filamentous algae and hydroids.

Corals first reappeared approximately 180 m behind the reef front and continued to increase in density to form a zone of coral growth which is distinctly visible in aerial photographs of many northern ribbon reefs. In this area, *A. formosa* (Dana) was markedly dominant, although brief inspection of neighbouring areas indicated that several different species of ramose *Acropora* may dominate this zone.

Coral cover steadily decreased beyond about 300 m from the reef front and tabular and subtabular species and growth forms of species (especially *A. hyacinthus* (Dana) and *A. humilis*) became increasingly dominant (figure 10, plate 4). There is a narrow zone of massive species, composed primarily of Faviidae, at the very back of the coral zone.

Beyond 450 m behind the reef front, the now poorly consolidated limestone surface became increasingly replaced by meandering valleys with sandy floors which increased gradually in depth and width towards the back reef margin.

The back reef margin and back reef slope

The back reef margin (figure 11, plate 4) was composed of a well defined line of irregularly dissected limestone standing above a substrate of calcareous sand. West of the back reef margin, the sand sloped relatively steeply to approximately 27 m depth, then gently to reach a constant depth of 34 m, 180 m from the reef back.

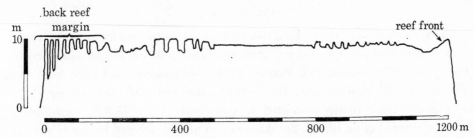

FIGURE 12. Profile of Great Detached Reef, far northern end. Vertical exaggeration is 1 : 20.

THE NORTH END OF GREAT DETACHED REEF

The north end of Great Detached Reef, shown in profile in figure 12, is well protected from the prevailing seas but is adjacent to the deep water of Raine Island Entrance. The reef front had a dense coral cover, mainly of *A. humilis* and *A. pyramidalis*. This decreased abruptly till, 20 m behind the front, the cover was about 1 %. The substrate was mostly loose rubble. At a distance of 80 m behind the front, *A. palifera* became dominant, with individual colonies growing to over 1 m in height. Sand was the main substrate from 100 m onwards behind the front, with *A. palifera* remaining dominant until at 210–250 m, it was replaced by ramose *Acropora* species (figure 13, plate 4). *Acropora palifera* again became dominant at 300 m and remained so over the rest of the reef, in most areas showing extremely lush growth, with individual colonies sometimes reaching 1 m in height.

THREE REEFS

The profile of Three Reefs differs greatly from all other ribbon reefs observed (figure 14). The reef surface consisted almost entirely of limestone with deep spur and groove formations on the eastern side. These grooves extended across to the western side of the reef flat, decreasing gradually in depth. Corals and calcareous algae were nowhere abundant.

These reefs appear to be in the path of strong tidal currents; they are also exposed to very strong wave action. Perhaps as a consequence, the depth of water over the reef was approximately 1 m greater than that over the main transect of Great Detached Reef, itself approximately 0.5 m deeper than corresponding areas of the protected north end of Great Detached Reef.

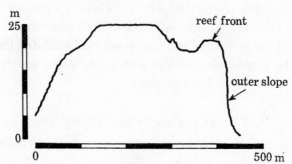

FIGURE 14. Three Reefs, shown in profile. Vertical exaggeration is 1 : 20.

DISCUSSION

The northern outer reefs of the G.B.R. have remained largely undescribed by reason of their remote and inaccessible position. Hence, their omission from the discussions of earlier writers on the theory of reef evolution (see Veron 1978*b*, this volume) and even from more recent accounts of the G.B.R. itself (notably Dakin 1963; Maxwell 1968; Bennett 1971).

Darwin himself (1842) mostly restricted his comments on the G.B.R. to the brief observation that '. . . if instead of an island, the shore of a continent fringed by a reef were to subside, a great barrier reef, like that of NE Australia would be the necessary result'. Darwin's concepts of reef development have indirectly influenced many subsequent attempts to define and categorize reef systems within the G.B.R. province. The most recent classification (Maxwell 1968) is generally applicable to the G.B.R. (and probably to other shelf reef systems, as distinct from Darwin's oceanic reefs). Maxwell divided the G.B.R. into the three regions already noted, and each region into six more-or-less bathymetrically defined, semi-meridional zones. Of these, the outermost is the 'marginal shelf' zone which is bordered by the 'shelf edge' reefs. These occur in the Southern Region as the outer reefs of the 'Pompey Complex' (discussed in Veron 1978*a*, this volume), in the Northern Region as the ribbon reefs, and in the far north as the deltaic and dissected reefs.

These reefs are, in a sense, 'barrier' reefs. Certainly they form a barrier to the interchange of water between the continental shelf and the open ocean. However, the term 'barrier reef', used in the Darwinian sense, essentially means a line of reefs separated from the mainland by a lagoon. Such barrier reef–lagoon combinations are well known and form a large proportion of the world's coral reefs. In most cases the lagoons are shallow and contain platform reefs of various types. In these respects the whole Northern Region could be considered as an extremely large lagoon, and the shelf-edge reefs as Darwinian barrier reefs. The difference is primarily one of size and complexity. As with other shelf reef systems, the G.B.R.'s origins are associated with tectonic subsidence of the continental margin. As with many other barrier reefs, the shelf edge reefs of the Northern Region are situated on, and are orientated to, the edge of the continental shelf and form an actively growing barrier to oceanic waters. As with many other large reef lagoons, the inter-reef areas of the Northern Region have substrates of soft mud or sand, with a very low density of sedentary fauna and flora.

The unproductive, lagoon-like qualities of the Northern Region are best developed behind the shelf edge reefs where there is a 10–25 km wide expanse of open, mostly reef-free water with an extremely sparse sedentary fauna. They are least developed in the far north, where major exchange of water occurs through Torres Strait and also at the principal sites of exchange of oceanic water (described above) where reefs tend to border or invade the deep ocean passes. Away from areas of oceanic input, major reefs develop only where the north–south setting tidal currents are strong or in areas of very shallow water around islands etc.

Thus, much of the reef development in the Northern Region is associated with currents, either those of the distant past which have led to the establishment of the present bathymetric elevations and associated reef developments, or those of the present, which transport oceanic waters of a suitable quality for reef growth into the region. As yet, little is known about those aspects of water quality which affect or govern reef growth. Inorganic nutrient enriched waters associated with upwelling at the edge of the continental slope, high levels of organic nutrient transport from the open ocean and current-induced sediment removal have each been suggested to play a part in stimulating coral growth.

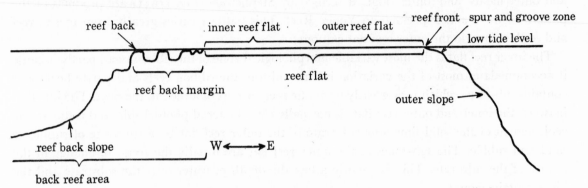

FIGURE 15. Diagrammatic profile of a ribbon reef, showing the principal morphological zones and indicating the terminology used in the text.

Ribbon reefs differ substantially from other reefs within the G.B.R. province. Their position and general shape is determined bathymetrically, but is modified by hydrodynamic influences which are mostly of their own making. Their surface features are further modified by climatic factors, dominated by the SE trade winds which generate a heavy oceanic swell for approximately 8 months of the year.

Figure 15 indicates the principal morphological zones of ribbon reefs and the terminology applied to them in this study. The outer slope and the reef back margin both support a dense coral growth usually dominated by *A. palifera*. This one extremely polymorphic species forms much of the consolidated substrate of the reef and probably much of its volume also. At the top of the outer slope there is usually a well defined zone of spur and groove formations, orientated to the direction of the prevailing winds. The grooves penetrate the reef flat, often for long distances, as 'drainage grooves'.

The wave-pounded reef front of most ribbon reefs is sparsely covered by a thin layer of corals and calcareous algae. In some areas the algae are dominant over the corals but nowhere do they form hard, smooth, pavement-like 'algal ridges', 'algal ramps' or 'algal platforms', such as have been described on many outer Indo-Pacific Reefs (e.g. Tracey, Ladd & Hoffmeister, 1948; Chevalier 1973; Stoddart 1973; Pichon 1979). The zone of maximum wave action is, however, situated well behind the reef front. Waves build up over the front and break on the outer reef flat, an area of hard, flat, denuded limestone which appears to be characteristic of all shelf edge reefs of the Northern Region which are exposed to the ocean swell. This zone on

Yonge Reef has been termed the 'reef crest' by Stephenson, Stephenson, Tandy & Spender (1931), and appropriately described by them as 'A pavement of solid coral rock, swept clear of coral debris, over 3 miles [*ca.* 5 km] in length and over 160 yards [*ca.* 145 m] in breadth'. This type of outer reef flat, or its close equivalent, has also been described from exposed reefs from other parts of the world where they are frequently termed algal ramps or algal platforms. A similar zone (from the exposed reefs of Tuléar, Madagascar) has been termed 'reef glacis' by Clausade *et al.* (1971). This is perhaps the least ambiguous of the established terms for it.

The seaward edge of the outer reef flat may show 'moats' or other irregularities in height which vary from reef to reef but which appear to be associated with the degree and type of exposure to wave action. Maximum wave action occurs closest to the reef front in reefs which face the prevailing winds and which have a steep outer slope. Where the outer slope is gentle, wave build-up is gradual and wave energy is dissipated over a greater area. Thus the 'inner and outer moats' and 'outer ridge' described by Stephenson *et al.* (1931) are products of the pattern of wave action peculiar to Yonge Reef. This pattern varies greatly from reef to reef and so therefore do the outer reef flat contours.

The inner reef flat is the most variable morphological zone of the ribbon reefs, partly because it accommodates most of the variation in the width of the ribbon reefs and partly because it contains sub-zones which vary greatly from one reef, or part of a reef, to the next. The junction between the inner and outer reef flats is normally clear in aerial photographs and is due to the replacement of the solid limestone substrate of the outer reef flat by a substrate of rubble or sand and rubble. The sub-zones of the inner reef flat are usually the result of change in the nature of the substrate. This frequently alters the depth of water over the substrate and the biota contained on it.

Yonge Reef, one of the southern ribbon reefs and the only one to have been previously described (Stephenson *et al.* 1931), has been re-studied by the authors and others and will be re-described in detail elsewhere. The second paper of this series (Veron 1978a) describes the shelf edge reefs north of the ribbon reefs, the deltaic and the dissected reefs.

The authors wish to thank Mr L. D. Zell for assistance in all aspects of this work, Mr L. Brady for preparing the figures, Miss A. Addison the text and Mr J. Amess for library assistance. Comments on the manuscript from Professor W. G. H. Maxwell and Professor M. Pichon are gratefully acknowledged. Field work was partly assisted by members of the expedition, especially Mr J. D. Collins and the crew of R.V. *James Kirby*.

This project was supported by the Royal Society, James Cook University, the Australian Research Grants Committee and the Australian Institute of Marine Science.

REFERENCES (Veron & Hudson)

Bennett, I. 1971 *The Great Barrier Reef.* (183 pages.) Melbourne: Lansdowne Press.
Brandon, D. E. 1973 In *Biology and geology of coral reefs* (eds O. A. Jones & R. Endean), vol. 1 (Geology I), pp. 187–232. New York. Academic Press.
Chevalier, J. P. 1973 In *Biology and geology of coral reefs* (eds O. A. Jones & R. Endean), vol. 1 (Geology I), pp. 113–141. New York: Academic Press.
Clausade, M., Gravier, N., Picard, J., Pichon, M., Roman, M., Thomassin, B., Vasseur, P., Vivien, M. & Weydert, P. 1971 Coral reef morphology in the vicinity of Tulear (Madagascar): contribution to a coral reef terminology. (74 pages.) *Téthys*, supplément 2.
Dakin, W. J. 1963 *The Great Barrier Reef and some mention of other Australian coral reefs*, 2nd edn. (176 pages.) London: Angus and Robertson.

Darwin, C. R. 1842 *The structure and distribution of coral reefs.* London: Smith, Elder & Co.

Highley, E. 1967 *Oceanic circulation patterns off the east coast of Australia.* Tech. Pap. Div. Fish. Oceanogr. C.S.I.R.O. Aust., 23.

Maxwell, W. G. H. 1968 *Atlas of the Great Barrier Reef.* (258 pages.) Amsterdam, London and New York: Elsevier.

Pichon, M. 1979 *Atoll Res. Bull.* (In the press.)

Stephenson, T. A., Stephenson, A., Tandy, G. & Spender, M. 1931 *Scient. Rep. Gt Barrier Reef Exped. 1928–29,* **3,** 17–112.

Stoddart, D. R. 1973 In *Biology and geology of coral reefs* (eds O. A. Jones & R. Endean), vol. 1 (Geology I) pp. 51–92. New York: Academic Press.

Tracey, J. I., Ladd, H. S. & Hoffmeister, J. E. 1948 *Bull. geol. Soc. Am.* **59,** 861–878.

Veron, J. E. N. 1978*a Phil. Trans. R. Soc. Lond.* B **284,** 23–37 (this volume).

Veron, J. E. N. 1978*b Phil. Trans. R. Soc. Lond.* B **284,** 123–127 (this volume).

Veron, J. E. N. & Pichon, M. 1976 *Scleractinia of eastern Australia,* Part I. (86 pages). Aust. Inst. Mar. Sci. Monogr. **1.**

Veron, J. E. N. & Pichon, M. *Scleractinia of eastern Australia,* Part III. (In the press.)

Veron, J. E. N., Pichon, M. & Wijsman-Best, M. 1977 *Scleractinia of eastern Australia,* Part II. (233 pages.) Aust. Inst. Mar. Sci. Monogr. **3.**

Phil. Trans. R. Soc. Lond. B. **284**, 23–37 (1978)   [23]

Printed in Great Britain

Deltaic and dissected reefs of the far Northern Region

By J. E. N. Veron†

Department of Marine Biology, James Cook University of North Queensland, P.O. Box 999, Townsville, Queensland, Australia 4810

[Plates 1 and 2]

The outer barrier reefs north of the ribbon reefs are composed of two distinctly different reef types, here called 'deltaic' reefs and far northern 'dissected' reefs. Two reefs of each of these types are described. The deltaic reef system is composed of 96 km of reef front characterized by the presence of regular, well defined channels containing very strong tidal currents, and also by the presence of a deltaic pattern at the reef back. The dissected reefs are the northernmost of the barrier reefs. They are composed of many small E–W elongate reefs interspersed by many wide channels. Both reef types are part of the one structure, their differing surface morphologies being attributed to bathymetric and hydrodynamic factors of the present and of the past.

Introduction

The previous paper (Veron & Hudson 1978) was concerned with the long series of more or less elongate reefs collectively known as 'ribbon' or 'wall' reefs. These reefs are characterized by a general similarity which allows them to be readily distinguished from other reef types including other types of barrier reefs. The ribbon reefs extend northward to about the level of Olinda Entrance (latitude 11° 14′ S) (figure 1), whereupon the barrier undergoes a gradual change from a ribbon system into a system of very different reefs that are here called 'deltaic' reefs, following Maxwell's (1968) terminology.

The deltaic reefs extend northward for a distance of approximately 96 km. They are composed of approximately 28 major reefs, 0.4–3.7 km in length, which are interspersed with about 33 major channels. Over this distance the deltaic appearance becomes increasingly more complex; major channels become less distinctive and are increasingly confused with an interlocking network of smaller channels. Thus at latitude 10° 10′ S the deltaic pattern consists of a thoroughly confused network or irregular elongate patches intermixed with a mass of channels, most of which are small and shallow. The general appearance is broadly similar to the outer edge of a mature river delta.

From this point north, the barrier line becomes increasingly simplified. The reefs and channels retain their irregular appearance but the channels become less interwoven. At about latitude 10° S the barrier consists of alternating elongate reefs separated by relatively well defined, straight channels. For the purpose of this account, these reefs are called 'dissected' reefs.

At its northern limit the barrier line consists of an irregular row of very small reefs which become increasingly difficult to distinguish in aerial photographs. Beyond the visible northern limit, Chart AUS 377 indicates a shallow area annotated 'strong ripplings'.

The only previous description of these reefs known to the author is that of Captain Blackwood

† Present address: Australian Institute of Marine Science, P.O. Box 1104, Townsville, Australia 4810.

(1844): 'From Pandora entrance the reef runs N.E.b.N., seven miles, to Olinda Entrance and is intersected by small but narrow openings unfit for shipping, and then gradually turns away N.b.W. running for a space of nearly ninety miles in an impenetrable line of reef, until Murray Island is approached'. Navigational charts (2354 and AUS 377) indicate the position of the reef line but no detail given can be recognized in aerial photographs. More recent reconnaissance maps (SC 54–16, 55–13, 54–12, 55–9, 55–5) from satellite photographs distinguish the approximate shape of major reefs, but again, detail is lacking. The present paper gives a simplified description of a major section of the Great Barrier Reef which has hitherto remained almost completely unknown.

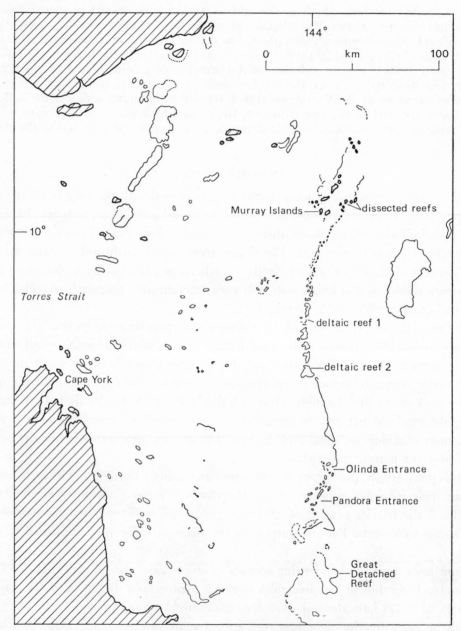

FIGURE 1. The northern shelf edge reef, showing place names referred to in the text and the positions of the deltaic and dissected reefs.

METHODS

This study was undertaken during the second far northern cruise of R.V. *James Kirby*. Methods used are as described in the previous paper except that no reconnaissance was made from chartered aircraft and no quantitative transects were attempted.

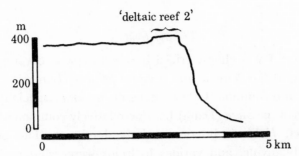

FIGURE 2. Profile of the deltaic reefs at the latitude of 'deltaic reef 2'. Drawn from soundings taken during the present study. Vertical exaggeration is 1 : 20.

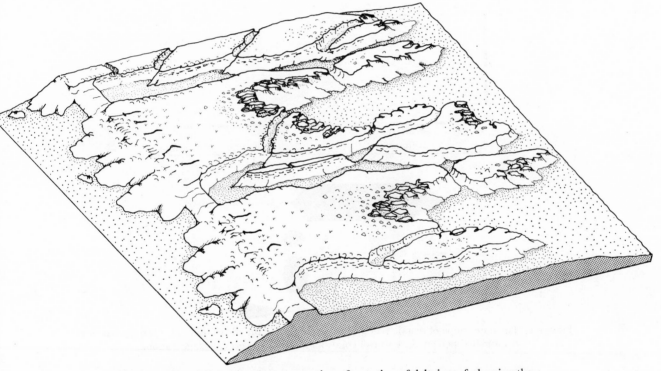

FIGURE 3. Three-dimensional reconstruction of a section of deltaic reef, showing the approximate spatial relation of features referred to in the text.

THE DELTAIC REEFS

The deltaic reefs are 140–150 km from the mainland, almost due east of Cape York. The continental shelf between the barrier reefs and the mainland is poorly chartered but seldom appears to be more than 30 m deep (figure 2) and contains large numbers of patch reefs. As noted above, the deltaic reefs at their northern extremity consists of a complex of interwoven narrow channels and reefs that, from aerial photographs, become confused. A few kilometres further south the deltaic pattern illustrated in figure 3 emerges.

Two reefs, here called 'deltaic reef 1' and 'deltaic reef 2' (figures 4 and 5, plates 1 and 2 respectively), were selected for study. They both lie in the northern half of the deltaic system, where the deltaic pattern is best developed, and have approximate latitudes of 10° 44' and 10° 50' S respectively.

The reef front

The front of deltaic reef 1 was characterized by a very sparse coverage of coral reaching a maximum of only 5 % at 3.5–5 m depth; *Acropora palifera* (Lamarck) was dominant. An unidentified sponge was also common. Other species of *Acropora*, especially *A. humilis* (Dana), *A. hyacinthus* (Dana), and *A. surculosa* (Dana) became relatively common at depths to approximately 11 m after which *Halimeda* became entirely dominant. The burrowing echinoid, *Echinostephus molaris* de Blainville, and various hydroids were relatively abundant on solid reef substrates. Calcareous algae were scarce.

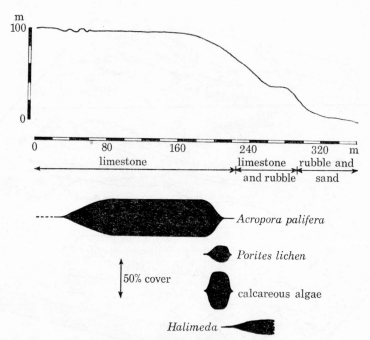

FIGURE 6. The outer slope of deltaic reef 2, showing in profile the distribution of substrate type and estimated percentage cover of corals and algae. No vertical exaggeration.

In marked contrast with reef 1, deltaic reef 2 (figure 6) had well defined zones of coral and algae. The substrate was similar, consisting of solid, wave-worn rock followed by a mixture of limestone and rubble, then rubble and sand. Coral zonation was distinct, *Acropora palifera* being dominant to a depth of approximately 15 m followed by a narrow zone of *Porites lichen* (Dana) which formed characteristic horizontal plates extending from the reef surface. Calcareous algae were very abundant in the *P. lichen* zone, the two forming an almost continuous cover. Below this zone the reef surface became increasingly composed of coral rubble and sand with *Halimeda* completely dominant.

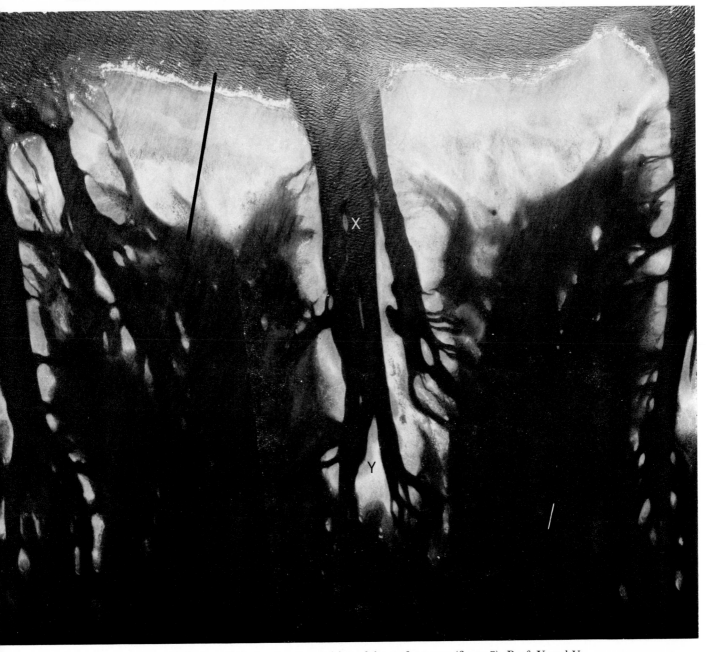

FIGURE 4. 'Deltaic reef 1', showing the position of the reef transect (figure 7). Reefs X and Y, referred to in the text and in figures 10 and 12, are indicated.

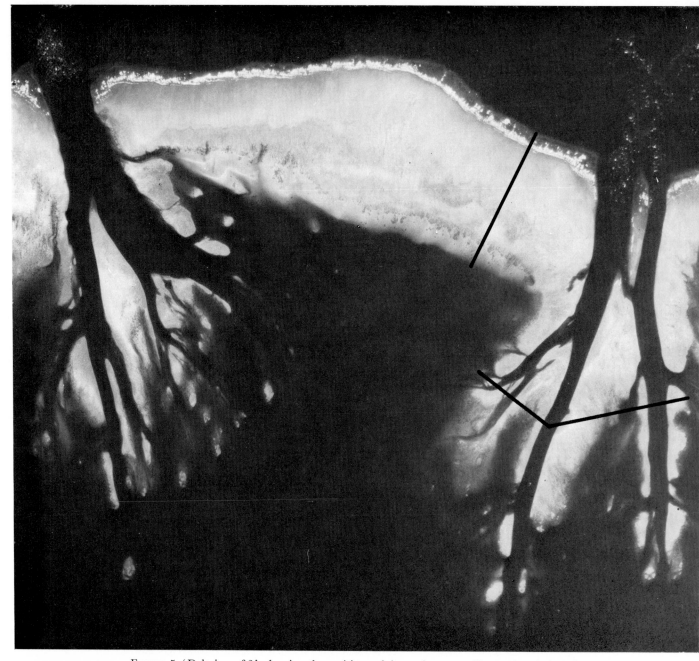

FIGURE 5. 'Deltaic reef 2', showing the positions of the reef transect (figures 6 and 8) and the transect across the deltaic system (figure 9).

The reef flat

As can be seen in figure 4, the reef flat of deltaic reef 1 is asymmetrical in shape and, like the majority of other deltaic reefs, is penetrated by irregular channels from the reef back. Figure 7 shows the correspondingly irregular profile, and indicates the nature of the reef surface and the percentage coral cover.

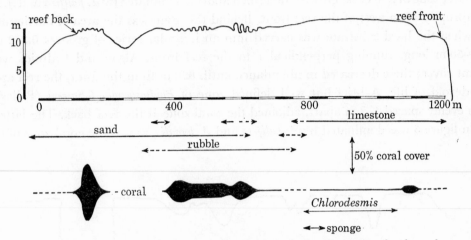

FIGURE 7. Deltaic reef 1, showing in profile the distribution of substrate type and estimated percentage cover of corals and other biota. Vertical exaggeration is 1 : 20.

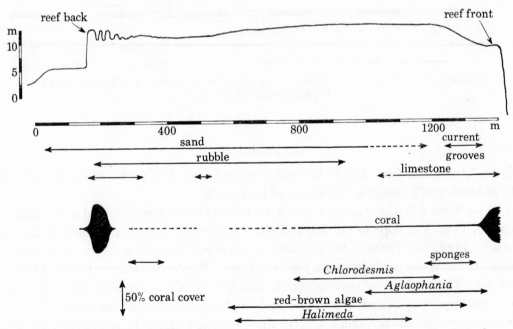

FIGURE 8. Deltaic reef 2, showing in profile the distribution of substrate type and estimated percentage cover of corals and other biota. Vertical exaggeration is 1 : 20.

The reef surface up to 120 m west of the reef front contained a series of current-worn grooves approximately 5 m apart and 15 cm deep. The surface of all the solid reef was mostly covered with close cropped filamentous algae, hydroids, occasional colonies of ramose *Millepora*, and *Acropora* spp., predominantly *A. humilis*. *Acropora palifera* was not present.

A zone of coral growth, mostly of ramose *Acropora*, occurred 570–820 m west of the reef front, *A. palifera* being dominant for the last 100 m. The substrate in this zone was mostly sand and *A. palifera* rubble. A second, narrow band of coral growth occurred at the reef back, consisting primarily of ramose *Acropora* species interspersed with large *Porites* colonies.

Deltaic reef 2 differed from reef 1 in being wider and more uniform in shape and appearance, but the surface features, as indicated in figure 8, were nevertheless essentially similar.

The outer zone of rich coral growth (described above), dominated by *A. palifera* and *A. humilis*, ended abruptly 50 m west of the reef front. Behind this zone was the area of maximum wave action when the hard substrate was carved into an irregular series of grooves 0.3–2 m deep and 30–50 m long, running perpendicular to the reef front. Algae and hydroids were the dominant cover; these decreased in abundance until, 820 m from the front, the reef appeared almost devoid of life. A brief but well defined zone of *Phyllospongia foliascens* (Pallas), with colonies evenly spaced 1–2 m apart, adjoined the coral zone at the reef back. The latter zone, visible in figure 5 was dominated by *A. palifera* and *A. humilis*, as was the coral zone of the reef front.

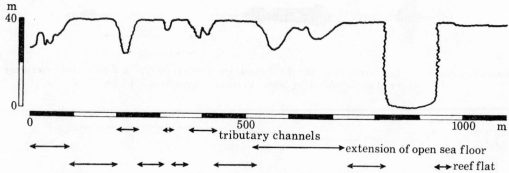

FIGURE 9. Profile of part of the deltaic system along the line indicated in figure 5 (see text). Vertical exaggeration is 1 : 5.

The deltaic system

Figure 9, compiled from direct observation using SCUBA, is a profile of the deltaic system south of deltaic reef 2 along the line indicated in figure 5.

At its northern end, the transect covered a terrain of soft calcareous sand and rubble from a depth of 18 m. The slope to the first reef flat was covered by extensive outcrops of *Acropora intermedia* (Brook) and *Porites andrewsi* (Vaughan) at depths below 3 m. Above 3 m, *A. palifera* became dominant and, with *Stylophora pistillata* (Esper), *Pocillopora verrucosa* (Ellis and Solander) and *A. humilis*, formed a coral cover of approximately 80 % of the area of the first 100 m of reef. Calcareous algae were abundant and the whole reef surface was well cemented.

Between 100 and 330 m (figure 9) were three tributary channels, respectively 50, 20 and 60 m wide. These channels had rugged, irregularly eroded surfaces of reef rock and coral debris with little or no faunal or algal cover. The upper edges of the channels were covered with corals of the species dominant on the reef flats plus thickets of *Millepora tenera* (Boschma). Their floors were covered with sand and rubble except for the small channel which had a cemented limestone floor.

Between 330 and 720 m there were two converging areas of reef flat which make up the southern and northern walls of the third tributary channel and the first main channel respectively (see figure 5). Coral cover on these flats was approximately 20 % and was dominated by *A. palifera*, *A. humilis* and *Porites lobata* (Dana), with some *Stylophora pistillata*. The reef surface was very hard and well cemented by calcareous algae. Toward the edge of the main channel the reef became corrugated by small regular grooves about 30 cm deep.

The deeper area between these reef flat areas, which represents an extension of the open sea and sea floor, was mostly flat and was covered with sand and rubble with large patches of blue-green algae and *Acropora* species.

The reef flats adjoining the main channels were similar to the first reef flat area described above. Coral cover decreased markedly with increasing distance from the channels. The transition zone between the reef flat and the deeper areas that are extensions of the open sea floor (i.e. those areas appearing dark in figure 5) was composed mainly of poorly cemented reef with approximately 5 % coral cover, dominated by *Porites* spp. and the hydroid *Agloaphania cupressina*. Below approximately 6 m the reef surface became covered with soft calcareous sand and an almost continuous, although sparse, cover of *A. formosa* (Dana).

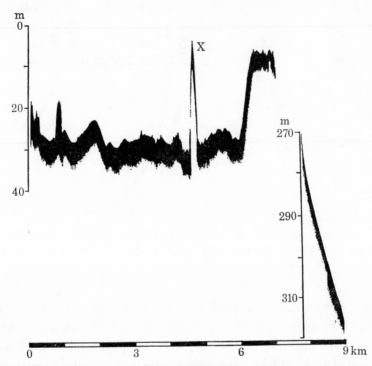

FIGURE 10. Echo sounding down the channel south of deltaic reef 1. Reef X is indicated in figures 4 and 12 (see text). Vertical exaggeration is 1 : 134.

The major channels

The major channels reaching the reef front are mostly uniform in size and general appearance. Those of the northern half of the deltaic system are up to 0.7 km across and 5.5 km long. Further south the channels become wider and shorter and the deltaic pattern less distinct.

Depth soundings taken along the centre of four channels (one channel south of deltaic reef 1, two south and one north of reef 2) indicated that their depths were mostly uniform throughout their length with an overall range of 18–35 m.

The sounding south of reef 1, reproduced here as figure 10, is characteristic. At the outer edge of each channel was a ridge connecting the two adjacent outer slopes. The ridge was saddle-shaped with the axis of the saddle lying along the reef front. Figure 3 gives an impression of the ridge as determined from soundings, from observation from small boats and from SCUBA diving on the ridge.

Tidal currents in the channel were very strong, especially within 2–3 h of low tide when little water movement occurs over the reef surface. One estimate of 3.8 m/s was made from the *James Kirby* while maintaining a constant position inside the outer ridge on an ebb tide. On the ridge itself the current forms standing surface waves up to approximately 2 m high. These waves are visible in figure 5 where they are seen to extend as finger-like projections from the channels for distances of up to 1.7 km.

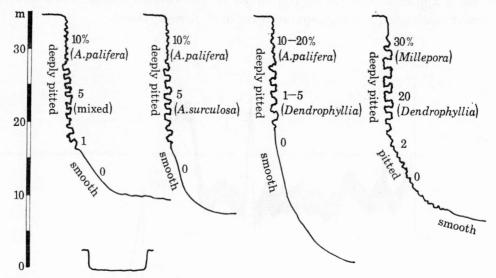

FIGURE 11. Surface features and shapes of the walls of different major channels. The dominant corals and percentage cover are indicated. Inset: the general shape of major channels shown without vertical exaggeration.

The surface of the outer ridge was hard and well cemented. To a depth of approximately 7.5 m about 60 % of the surface was almost exclusively covered with flat, encrusting *A. palifera* colonies which gave the appearance of a flat, irregular pavement. Mixed coral species then occurred to a depth of 12 m, at which point the coral cover was approximately 10 %. Between 12 and 21 m (on the front of the outer ridge) the coral cover decreased to < 1 %; soft corals, especially *Lobophytum*, were dominant. Only occasional corals, mainly faviids, were found below 21 m; the still hard substrate had a sparse cover of filamentous red–brown algae, filamentous green algae, *Halimeda* and fine hydroids.

Various profiles of major channels are compared in figure 11. In each case the walls descended vertically to a depth of 12–19 m, then curved in a regular fashion to form the channel floor at depths of approximately 25–33 m. The vertical walls were always deeply pitted; sometimes the pits were in the form of horizontal, current-worn grooves up to 2 m deep. The curved part of the channel walls were mostly slightly pitted or smooth. The flat channel floors were mostly smooth and hard; in some places regular current worn undulations were formed, in others, ridges of coarse sand and or rubble.

Small, elongate reef patches frequently occurred in the centre of major channels. Figures 3 and 10 show these reefs in lengthwise profile. Their bases were much more elongate than their tops, so much so that the bases of successive reefs appeared to connect to form a central ridge running along much of the centre of most major channels. Figure 12 is a N–S profile across the middle of one such patch, 925 m long at the surface, indicated in figure 4. The reef appeared to be completely symmetrical and to have walls essentially similar to the walls of the major channels.

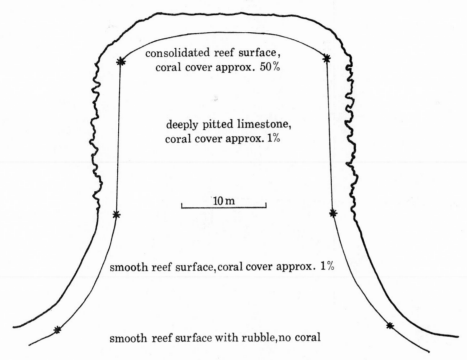

FIGURE 12. Profile across reef X, shown in figures 4 and 10.

A second type of reef frequently occurred at the backs of major channels where they branched into small channels leaving roughly triangular reefs with sharply defined apices pointing east, and irregularly shaped inner slopes.

The apex of reef Y (figure 4) is similar to the eastern point of reef X (figure 4). It gradually sloped to a depth of 31 m. To a depth of 2 m, the coral cover was approximately 10 % and consisted almost exclusively of *A. palifera*. The coral cover increased to approximately 60 % at 2–10 m depth with encrusting species of *Millepora* and *Porites* becoming dominant. At 10–18 m the consolidated reef gave way to rubble at which point soft corals, especially *Lobophytum*, became dominant. Few hard corals were observed below 26 m.

The western side of reef Y (figure 4) is well protected from wave action and currents and is essentially similar to the backs of the major reefs. The floor consisted of soft calcareous sand which sloped gently to a depth of 7.5 m, 170 m from the centre of the reef back. Very large massive *Porites* spp. and *P. andrewsi* colonies, visible in figure 4, occurred between approximately 50 and 170 m. Beyond 170 m from the reef back the depth increased and the substrate became a coral-free mixture of sand and rubble.

THE DISSECTED REEFS

The dissected reefs, as noted above, extend northward from the northern limit of the deltaic reefs for a distance of approximately 35 km. Over this range they become progressively simplified in appearance, so that toward their northern end (figures 13 and 14) they resemble a continuing series of small elongate plug reefs.

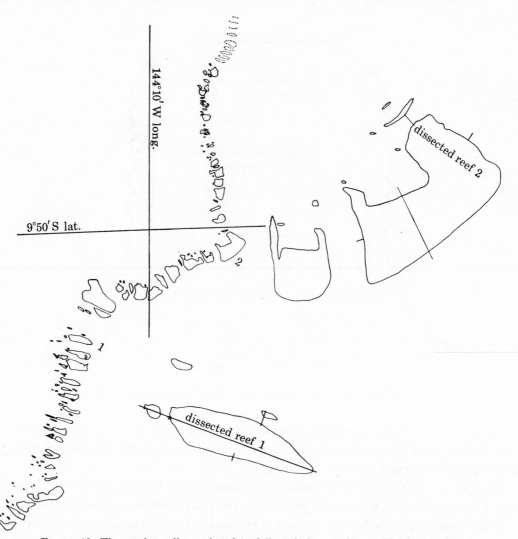

FIGURE 13. The northern dissected reefs and (insets) the two dissected reefs studied, showing positions of transects referred to in the text.

Two reefs indicated in figure 13, here called 'dissected reef 1' and 'dissected reef 2', were selected for study. They both lie eastward of the Murray Islands and have approximate latitudes of 9° 53′ S and 9° 50½′ S respectively. Reef 2 is the northernmost major reef of the barrier system; reef 1 is a small elongate reef which is more representative of other reefs in the area.

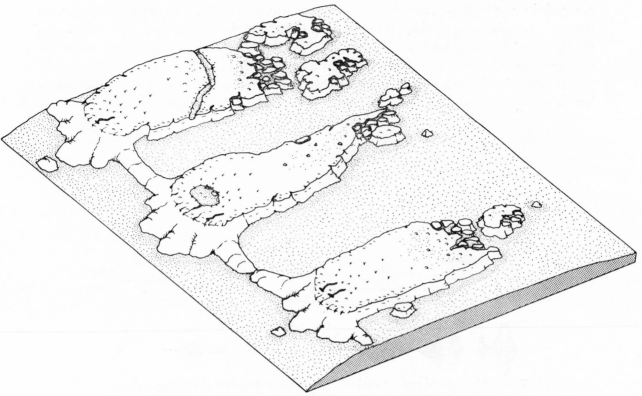

FIGURE 14. Three-dimensional reconstruction of a section of dissected reefs, showing the approximate spatial relation of features referred to in the text.

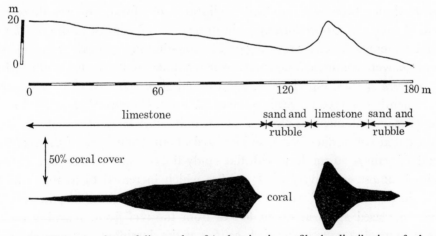

FIGURE 15. The outer slope of dissected reef 1, showing in profile the distribution of substrate type and estimated percentage cover of corals. No vertical exaggeration.

The reef front

The front of dissected reef 1 (figure 15) had a very gradual slope. At 6 m depth it was poorly cemented and had a coral cover of approximately 5%. At 9 m the face became horizontal and the coral cover increased to 30%. *Acropora palifera* was the dominant species and (unusual for an outer reef face) *Goniopora* was also abundant. At 12 m the coral cover reached about 50%

but ended abruptly in a small valley of sand and rubble at 15 m. At 100 m from the crest there was a slight elongate ridge rising to 14 m depth and densely covered with coral, again mostly *A. palifera*. At greater distance the coral gradually gave way to sand and rubble, the last corals, of very mixed genera, occurring at a depth of 21 m.

The front of reef 2 was only very briefly observed. Corals were abundant near the surface but were largely replaced by sand and rubble at depths greater than 10 m.

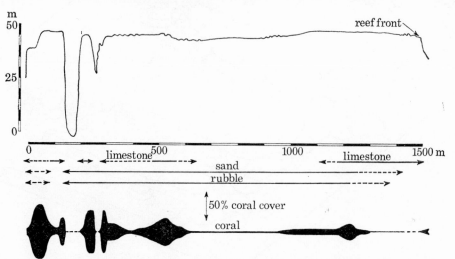

FIGURE 16. Dissected reef 1, showing in profile the distribution of substrate type and estimated percentage cover of corals. Vertical exaggeration is 1 : 20.

The reef flat

As figure 16 shows, there was no clearly delineated reef front, nor was there any marked development of spur and groove formation in dissected reef 1. Most of the reef face consisted of dead coral cemented by calcareous algae. Shallow drainage grooves with smooth surfaces about 1 m across were common. The coral cover remained at 1 % or less and consisted almost entirely of *Acropora* species, especially *A. palifera* and *A. humilis*. Dead colonies up to 11 cm high were frequently observed to within 80 m of the front. Clumps of *Xenia*, 2–30 cm diameter, and patches of sparse, filamentous algae, were the only other biota observed.

The consolidated reef surface began to be pitted about 220 m west of the reef front and at 270 m contained valleys 60 cm deep with flat sandy floors. Between 270 and 350 m the coral cover consisted almost exclusively of *A. palifera* which increased to as much as 30 % then decreased again as the area of eroded reef, filled with sand and rubble, increased. *Acropora palifera* cover increased again at about 950 m from the reef front, reaching 50 % cover at 1000 m. At 1050 m, *A. humilis* replaced *A. palifera* as the dominant species and, apparently as a result, the area of consolidated reef greatly decreased. The reef flat remained a varied mixture of *Acropora* species, especially *A. humilis*, and rubble, until the lush coral of the reef back margin was reached.

Dissected reef 2 had a more clearly defined front and a greater coral cover than had dissected reef 1 (figure 17). Again, there was little development of spur and groove formation. Behind the front the reef flat was well consolidated and had the drainage grooves noted above. The coral cover decreased rapidly until, 130 m west of the reef front, it reached zero. Small clumps of *Xenia* and small amounts of algae occurred to about 80 m. The reef flat between 130 and

250 m west of the reef front, which consisted only of flat, hard limestone, was almost devoid of visible life.

At 250 m the limestone gradually became covered by sand along with a varied biota dominated by small spherical coral colonies. At 310 m the reef flat was exclusively sand and rubble with a coral cover of 1–5 %, consisting mostly of *Acropora* series dominated by *A. humilis*. Small valleys with sandy floors started to occur at 430 m, getting wider and deeper as the depth of the reef surface started to increase. At 580 m the reef surface consisted of extensive areas of dead and living *A. palifera*, penetrated by vertical-sided valleys averaging 2 m deep and 10 m wide. Towards the reef back the surface was variable in depth and consisted of irregular patches of *A. humilis* or *A. palifera* dominated reef and valleys of sand and rubble. The reef back margin occurred abruptly at 650 m.

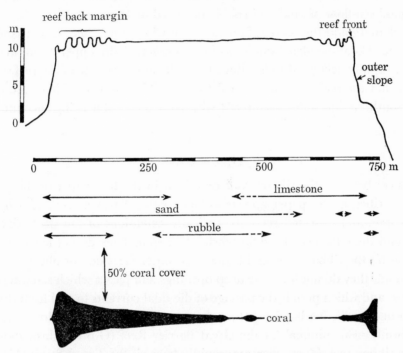

FIGURE 17. Dissected reef 2, showing in profile the distribution of substrate type and estimated percentage cover of corals. Vertical exaggeration is 1 : 20.

The reef back margin and sides

Change in coral cover with depths at the reef back margin of dissected reef 1 is illustrated in figure 16. At 1220 m west of the reef front *A. hyacinthus* became the dominant species and was combined with large colonies of *Porites* species and many other *Acropora* species at the edge of the reef back. The small reef patch indicated in figure 13 was separated from the main reef by 90 m of sand and coral rubble reaching a depth of 21 m. The eastern edge of the reef patch was mostly composed of *A. palifera* which was largely replaced by *A. humilis* toward the centre. Coral cover at the western edge reached 100 % and consisted mostly of ramose *Acropora* species. This zone ended abruptly at 10 m depth when the reef edge became nearly vertical. The ahermatypic coral, *Tubastrea micranthus* (Ehrenberg), was abundant to a depth of 27 m, at which point the reef face became submerged in sand and rubble.

The south side of dissected reef 1 sloped downward at a constant angle of about 45° to reach the sea floor at 32 m depth. The floor was mostly composed of *Acropora* and free-living fungiid corals, especially *Fungia fungites* (Linnaeus), *F. echinata* (Pallas) and *Halomitra* sp. The upper 10 m of the reef edge had a coral cover of up to 80%, with ramose *Acropora* species dominant.

The reef face of the north side was similar in appearance. A sandy floor was reached at 17 m. There were two elongate ridges of limestone elevated approximately 3 m between the small reef strip, shown in figure 13, and the main reef. The water depth near the reef strip reached 21 m.

The reef back of dissected reef 2 in the line of the transect is similar to that of reef 1. The reef face sloped steeply to 11 m where it became submerged in sand. Between 11 and 18 m depth, the floor consisted of very irregular patches of sand, rubble and eroded limestone. An almost horizontal sea floor of sand and rubble occurred at 18 m.

The reef back margin at the centre of the northern wing sloped steeply to a sand and rubble sea floor at 37 m. Coralline algae and *A. palifera* dominated the upper 11 m, at which depth the coral cover rapidly decreased to less than 1%. The northern side of the reef was essentially similar except that the reef face was vertical to 37 m. The southern side of the reef was very similar to the southern side of dissected reef 1 with sand and rubble being encountered at 18 m.

DISCUSSION

The differences between the ribbon reefs described in the first paper of this series and the reef systems described in this paper appear to have both a bathymetric and a hydrodynamic origin. The deltaic and dissected reefs are basically extensions of the shelf edge reef system occurring beyond the influence of the Queensland Trench. They do not form the western rim of the trench as do the ribbon reefs and hence their outer faces do not plunge to great depth. Perhaps as a result they do not have the deep openings and passes which separate most successive ribbon reefs and which permit the passage of the tidal currents to and from the continental shelf. They form an effective barrier 131 km long to tidal movement. However, tidal range, which is uniformly near minimal for the Great Barrier Reef (G.B.R.) throughout the whole distance of the ribbon reef system, increases rapidly towards the Torres Strait (Maxwell 1968); so does the volume of water involved, as the continental shelf widens over the area of the Torres Straight and the western Gulf of Papua. These factors combine to produce the very strong tidal currents characteristic of the whole Torres Strait region.

Clearly, the northern barrier reefs, which provide such a formidable obstacle to tidal movements, have their present morphologies largely determined by them. This applies especially to the deltaic reefs, where the barrier continues both to the north and south.

Reefs of the Pompey Complex in the Southern Region of the G.B.R. have been described by Maxwell (1970) as being 'deltaic', and certainly that term can be applied equally to both reef systems. Both are characterized by the presence of well defined channels containing very strong tidal currents. As Maxwell stated, the currents serve to scour the passages while the reef body serves to localize depositions which provide bathymetric elevations suitable for reef colonization.

In other respects, however, the Pompey Complex differs greatly from the northern reefs, primarily in being composed of a multitude of enclosed and semi-enclosed lagoons interspersed through a matrix of channels, reef zones of varying elevations and sand zones. Brief personal observation indicates that the major channels are not as steep-sided as those of the northern reefs; those investigated were asymmetrical, having western walls steeper than eastern ones, with currents running obliquely to them, over the reef flat. The single channel sounded had a relatively uniform depth of 98–109 m throughout its central portion (i.e. about three times the depth of the northern reefs), and a very wide outer region rising to a relatively uniform 32 m. Soundings perpendicular to, and parallel with, the outer reef face revealed a broad continental slope scoured to about twice normal depth seaward of the openings of major channels.

No other reef systems within the G.B.R. province appear to have much in common with those described in this paper. None have similar bathymetric or hydrodynamic situations, nor have other reefs had a comparable evolutionary history. It is these three all-important factors, in combination, which have given the northern barrier reefs their present distinctive morphologies. The third factor, the evolution of the reefs, is discussed in the third paper (Veron 1978, this volume).

Particular thanks are due to Mr L. D. Zell for assisting in all phases of this project. Diagrams were prepared by Miss B. Harker and Mr L. D. Zell and photography undertaken by Mr L. Brady. Field work was greatly assisted by several people, especially Mr J. Barnett, Professor M. Pichon and Mr R. A. Birtles.

This project was supported by the Australian Research Grants Committee and the Australian Institute of Marine Science.

REFERENCES (Veron)

Blackwood, F. P. 1844 *Naut. Mag.* **13**, 537–541.
Maxwell, W. G. H. 1968 *Atlas of the Great Barrier Reef.* (258 pages.) Amsterdam, London and New York: Elsevier.
Maxwell, W. G. H. 1970 *Deep Sea Res.* **17**, 1005–1018
Veron, J. E. N. 1978 *Phil. Trans. R. Soc. Lond.* B **284**, 123–127 (this volume).
Veron, J. E. N. & Hudson, R. C. L. 1978 *Phil. Trans. R. Soc. Lond.* B **284**, 3–21 (this volume).

Phil. Trans. R. Soc. Lond. B. **284**, 39–61 (1978)   [39]

Printed in Great Britain

Geomorphology of reef islands, northern Great Barrier Reef

By D. R. Stoddart,† R. F. McLean‡ and D. Hopley§

†*Department of Geography, University of Cambridge, Downing Place, Cambridge, U.K.*

‡*Department of Biogeography and Geomorphology, Research School of Pacific Studies, Australian National University, Canberra, Australia* 2600

§*Department of Geography, James Cook University of North Queensland, P.O. Box* 999, *Townsville, Queensland, Australia* 4810

During the 1973 Great Barrier Reef Expedition, 67 reef islands were mapped between latitudes 11° 30′ S and 17° S on the Great Barrier Reef. During the mapping, the major topographic, lithological, sedimentological and vegetational features of the islands were distinguished, and their elevations relative to a sea level datum established. The islands themselves were categorized in terms of topographic and vegetational complexity. Previous classifications by Steers, Spender, Fairbridge and others are reviewed in the light of these findings. Some of the islands had been previously mapped by Steers in 1928–29 or 1936; on others, changes could be identified from the evidence of shoreline advance or retreat and from vegetation patterns. The floristics and vegetation units of the islands are briefly described, on the basis of the field mapping and a large collection of flowering plants. Vegetation is influenced by stage in island development, latitudinal variation in rainfall, effects of ground-nesting seabirds, and probably also by disturbance by aboriginal man. Development of mangroves on reef flats is related to stage of reef flat and island development, and relation to tidal levels. This study of the geomorphology of the islands raises questions over the nature, origin and history of specific features (ramparts, beach ridges, boulder tracts, exposed limestones) which the Expedition attempted both to define and to answer.

Introduction

Islands, in spite of their limited size, provide a key to Holocene geomorphic history in the coral reef seas. The information they contain is a function of their topographic complexity, and no reef islands in the world approach those of the northern Great Barrier Reef in this respect. Hence much of the effort of the 1973 Expedition was devoted to investigating reef islands between 17° and 12° S. This paper is concerned with describing the main types of islands and their geomorphic features; later papers investigate these in more detail and present conclusions on their evolution.

The islands were mapped by Stoddart using compass-and-pacing methods; maps were constructed at scales of 1:1000 to 1:5000, with paces converted to metres. Low Isles was mapped with compass and metric tape. The accuracy of linear measures on most islands is thus constrained by the method used, but closure errors were generally small. Dimensions can be compared within and between reefs, since the error attached to the measuring method is likely to be uniform throughout. Most of the maps were constructed during day-time low spring tides. The maps have been supplemented by monochrome vertical aerial photographs made available by the Division of National Mapping, Canberra, and by vertical and oblique monochrome and colour aerial photographs subsequently taken by Hopley. Levelling traverses were carried out mainly by Stoddart and McLean during the first part of the Expedition, and by

Hopley and A. L. Bloom during the second. Heights are referred to the Queensland datum of mean lower low water springs; tide tables were used to predict tidal heights at the time of survey, and local water levels were then used to relate the profiles to datum. Uncertainties in such determinations are discussed by Scoffin & Stoddart (1978) and by McLean, Stoddart, Hopley & Polach (1978, part A of this Discussion). During this investigation, McLean was mainly concerned with island sediments, McLean and T. P. Scoffin with platform conglomerates, Hopley with beach-rock, Scoffin and P. G. Flood with reef-top sediments, and P. E. Gibbs with soft-bottom communities. B. G. Thom and others assisted with problems of interpretation during the early phase of the Expedition, as did J. E. N. Veron during phases I and III. Stoddart also collected about 5000 sheets of vascular plants, representing over 1100 numbers, from many of the islands mapped. Some of these additional studies are reported elsewhere in this volume, but is should be emphasized that the island studies were largely cooperative projects in which the above and other members of the Expedition all participated.

TABLE 1. ISLANDS EXAMINED IN THE NORTHERN PROVINCE OF THE GREAT BARRIER REEF IN 1973 AND ON PREVIOUS OCCASIONS

island	1770[1]	1819–21[2]	1843[3]	1848[4]	1910[5]	1929[6]	1936[7]	1973[8]
Arlington	—	—	—	—	—	—	×	×
Ashmore	—	—	—	—	—	—	—	×
Beesley	—	—	—	—	—	—	—	×
Bewick	—	—	—	×	—	—	×	×
Binstead	—	—	—	—	—	—	×	×
Bird	×	—	—	×	×	—	—	×
Burkitt	—	—	—	—	—	—	×	—
Chapman	—	×	—	×	—	—	×	×
Combe	—	—	—	—	—	×	—	×
Coquet	—	—	—	—	—	—	×	×
Eagle	×	—	—	×	—	—	—	×
Ellis	—	—	—	—	—	—	—	×
Fife	—	—	—	×	×	—	×	×
Green	×	—	—	—	—	×	—	×
Hampton	—	—	—	—	—	—	—	×
Hope, East	×	—	—	—	—	—	×	×
Hope, West	×	—	—	—	—	—	×	×
Houghton	—	—	—	—	—	×	×	×
Howick	—	—	—	—	—	—	×	×
Ingram–Beanley	—	—	—	—	—	—	×	×
Kay	—	—	—	—	—	—	—	×
King	—	—	—	—	—	—	×	—
Leggatt	—	—	—	—	—	—	—	×
Low	×	×	—	×	—	×	×	×
Low Wooded	—	—	—	—	—	—	×	×
Lowrie	—	—	—	—	—	—	×	×
Mackay	—	—	—	—	—	×	×	×
Magra	—	—	—	—	—	—	—	×
Michaelmas	—	—	—	—	—	×	×	×
Morris	—	—	—	×	×	—	—	×
Newton	—	—	—	—	—	—	×	×
Night	—	—	—	—	—	—	×	—
Nymph (= Enn)	—	—	—	—	—	—	×	×
Pelican	—	×	—	×	×	—	×	×
Pethebridge (= Kew), East	—	—	—	—	—	—	—	×
Pethebridge (= Kew), West	—	—	—	—	—	—	×	×
Pickard	—	—	—	—	—	—	—	×
Pickersgill, North	—	—	—	—	—	—	×	×

TABLE 1 (cont.)

island	1770[1]	1819–21[2]	1843[3]	1848[4]	1910[5]	1929[6]	1936[7]	1973[8]
Pickersgill, South	—	—	—	—	—	—	×	×
Piper (Farmer–Fisher)	—	—	—	×	×	—	—	×
Pipon	—	—	—	—	—	—	×	×
Raine	—	—	×	×	×	—	—	×
Sand	—	—	—	—	—	—	—	×
Saunders	—	—	—	—	—	—	—	×
Sherrard	—	—	—	—	—	—	×	×
Sinclair–Morris	—	—	—	—	—	—	—	×
Stainer	—	—	—	—	—	—	—	×
Stapleton	—	—	—	—	—	×	—	×
Sudbury	—	—	—	—	—	×	×	×
Three	×	—	—	×	—	×	×	×
Turtle I	—	—	—	—	—	×	×	×
Turtle II	—	—	—	—	—	—	×	×
Turtle III	—	—	—	—	—	—	×	×
Turtle IV	—	—	—	—	—	—	×	×
Turtle V	—	—	—	—	—	—	×	×
Turtle VI	—	—	—	—	—	—	×	×
Turtle Mid-reef	—	—	—	—	—	—	×	×
Two	—	—	—	×	—	—	×	×
Undine	—	—	—	—	—	—	×	×
Upolu	—	—	—	—	—	—	—	×
Waterwitch	—	—	—	—	—	—	—	×
Watson	—	—	—	—	—	—	—	×
Wilkie	—	—	—	—	—	—	×	—

References: (1) Cook, in Beaglehole (1955); (2) King (1827); (3) Jukes (1847); (4) MacGillivray (1852); (5) MacGillivray (1910); (6) Steers (1929); (7) Steers (1938); (8) this expedition.

ISLAND TYPES

The reef islands of the Great Barrier Reef were first extensively studied by Steers (1929, 1937, 1938), who distinguished sand cays, shingle cays, and what he termed 'low wooded islands' (1929, pp. 20–27), comprising an assemblage of windward shingle rampart, leeward sand cay, and intervening shallow reef flat with mangrove swamp, characteristically developed on small reefs of the inner shelf north of about 16° S. Two of these low wooded islands, Low Isles and Three Isles, were mapped in great detail by Spender (1930; T. A. Stephenson, Stephenson, Tandy & Spender 1931) during the 1928–29 Expedition, and Steers (1938) later mapped a further 16, proposing Bewick and Nymph (= Enn) Islands as type examples. Spender (1930, pp. 277, 285–286) preferred the term 'island-reef' to low wooded island, and Fairbridge & Teichert (1947) compromised with 'low wooded island-reef'. None of these terms is perfect, but Steers's 'low wooded island' is now well established in the literature and is used here. Umbgrove (1928) described rather similar islands from Djakarta Bay, Java, Steers (1940, pp. 32–35) comparable forms in the Pigeon and Salt Cays, Jamaica, and Stoddart (1965) analogous islands in the British Honduras barrier reef lagoon. None of these examples, however, exhibits the topographic complexity of the Queensland low wooded islands, and all of them are, moreover, in areas of low tidal range.

These low wooded islands and the other types recognized by Steers (1929) were incorporated in a general classification of reefs based on the nature and distribution of superficial sediments by Spender (1930). Fairbridge (1950, pp. 347–349) revised the classification to include unvegetated sand cays, vegetated sand cays, shingle cays, sand cays with separate shingle ramparts (= low wooded islands), and islands with a core of older reef material. While this scheme needs to be extended to serve as a general classification of reef islands of the world (Stoddart & Steers 1977), it serves as a useful framework for discussion of the islands of the northern Great Barrier Reef.

Table 1 lists all the islands surveyed during the 1973 Expedition, with notes on previous surveys. Detailed accounts of individual islands will be given in other publications.

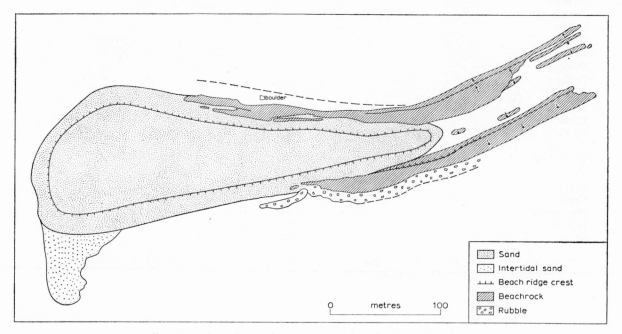

Sand
Intertidal sand
Beach ridge crest
Beachrock
Rubble

FIGURE 1. A sand cay with relict beachrock: Waterwitch, 1973.

UNVEGETATED SAND CAYS

These are usually located on the leeward sides of reef tops; their location and size are controlled by patterns of wave refraction. Three types may be distinguished:

(a) Small ephemeral cays. These are either intertidal sandbores less than 0.1 ha in area, or steeper, rather larger islands which are not overtopped by swash at all high tides. The former group includes the sand cays of several low wooded islands (Binstead, Chapman, Turtle II. Turtle IV, Watson), and the latter such islands as North and South Pickersgill, Pickard and Undine. None of these cays possesses beach-rock.

(b) Large, generally oval-shaped islands up to 300 m long and 100 m wide, with areas of 1.0–1.4 ha. These have steep marginal beaches and pronounced swash ridges; some have central depressions. Arlington, Mackay, Sudbury and Upolu fall in this class. Beach-rock was not found in the examples mapped.

(c) Other cays of variable dimensions (up to 400 m long and 120 m wide) which are surrounded by extensive relict beach-rock. The present form and size of the cay may differ from that outlined by the beach rock. Presumably in these cases an earlier phase of cay development has been terminated by storm activity, and a new cay has re-formed. Examples are Ashmore, Ellis and Waterwitch (figure 1).

Unvegetated cays are generally small (for 19 examples mapped, mean length is 140 m, mean width 40 m, and mean area 0.5 ha), and also variable in form and size over time. Five of those surveyed in 1973 were also mapped by Steers in 1936: table 3 gives comparable dimensions and shows substantial changes. In all cases the islands have decreased in size: Arlington in 1973 was 0.2 of its size in 1936, Sudbury 0.8, Undine 0.4, Mackay 0.4, North Pickersgill 0.3, and South Pickersgill 0.15.

TABLE 2. DIMENSIONS OF UNVEGETATED SAND CAYS

island	latitude S	longitude E	cay area m^2	maximum length m	maximum width m	intertidal sand and rubble area m^2	beach-rock area m^2	conglomerate platform area m^2
Arlington	16° 39½′	145° 59½′	4 600	120	50	—	—	—
Ashmore	11° 53′	143° 37′	12 400	325	60	—	1400	—
Minstead West Cay	13° 13′	149° 34′	920	60	22	—	—	—
Chapman West Cay 1	12° 53′	143° 36′	140	17	12	—	—	—
Chapman West Cay 2	12° 53′	143° 36′	350	32	19	—	—	—
Ellis	13° 22′	143° 41½′	4 000	175	30	5500	—	1960
Mackay	16° 03′	145° 39′	9 350	190	63	—	—	—
North Pickersgill	15° 51′	154° 33½′	3 800	120	40	310	—	—
Pickard	12° 14′	143° 09′	4 450	250	30	—	—	—
Sand	14° 31′	144° 51′	1 290	115	27	—	—	—
South Pickersgill	15° 51′	154° 33½′	490	53	12	340	—	—
Sudbury	16° 57′	146° 09′	13 950	205	105	480	—	—
Turtle II Cay	14° 44′	145° 12′	940	70	20	—	—	—
Turtle IV Cay	14° 43½′	145° 12′	3 280	110	40	—	1205	—
Turtle Midreef Islet	14° 43′	145° 11′	450	65	18	7200	—	—
Turtle Reef Cay	14° 43′	145° 10′	210	19	10	570	—	—
Undine	16° 07′	145° 38½′	4 610	220	26	—	—	—
Waterwitch	14° 12′	144° 53′	27 800	380	120	—	8600	—
Watson North Cay	14° 28′	144° 49′	2 570	120	35	—	85	—

TABLE 3. CHANGES IN UNVEGETATED CAYS 1936–1973

island	year	total area ha	area above high tide level ha	area above high tide as percentage of total area	vegetated area ha	vegetated area as percentage of total area	maximum length m	maximum width m
Arlington	1936	2.05	0.97	39	0.05	2.6	295	90
	1973	0.46	0.13	27	0	0	120	50
Sudbury	1936	1.72	—	—	0	0	230	105
	1973	1.40	0.73	52	0	0	205	105
Undine	1936†	1.12	0.13	11	0	0	275	50
	1973	0.46	0.03	7	0	0	220	26
Mackay	1936†	2.39	0.82	34	0.18	7.4	385	105
	1973	0.93	0.28	29	0	0	190	63
North Pickersgill	1936†	1.51	0.33	22	0	0	170	140
	1973	0.41	0.01	2	0	0	120	40
South Pickersgill	1936†	0.53	0.10	18	0	0	75	38
	1973	0.08	0.05	59	0	0	53	13

Source of data: calculated from maps given by Steers (1938) and from 1973 maps.

† Original measurements given in paces, converted at 1 pace = 1.2 m.

What governs the transition between cays of type (a) and larger islands of type (b) is not immediately apparent, nor what governs the transition of type (b) islands to vegetated sand cays. Some of the larger islands here termed unvegetated do in fact possess vascular plants, though only as scattered individuals which are clearly ephemeral. Thus in 1929 on Sudbury, Steers (1929, p. 257) noted 'seven small seedlings, one of *Ipomoea* (?) and six of *Sesuvium portu-lacastrum*'; in 1936 there were no plants at all (Steers 1938, pp. 67 and 68); and in 1973 there were three coconut seedlings and a small patch of *Sesuvium*. Similarly Mackay in 1929 'was well covered in its higher parts by grasses and creeping plants' (Steers 1929, p. 257); by 1936 (following a cyclone in 1934) the continuous vegetation had disappeared, being replaced by two or three clumps of grass, a single *Ipomoea* and a few other plants (Steers 1938, p. 70); and in 1973 there were four drift coconut seedlings, but no other plants present. In June 1936 Arlington had a vegetated area of 0.05 ha, of grasses and creepers, which was being eroded on all sides (Steers 1938, p. 68); in 1973 it had neither plants nor drift seeds. Spender (1930, p. 285) also noted that Pickersgill, which had previously been described as slightly vegetated, had no plants on it in 1929, and this was also the case in 1973.

TABLE 4. CHARACTERISTICS OF VEGETATED SAND CAYS

island	latitude S	longitude E	cay area m²	maximum length m	maximum width m	vegetated area m²	vegetated area as percentage total area	number of species of vascular plants	beach-rock area m²	intertidal sand and rubble area m²
Beesley	12° 11½'	143° 12'	6950	420	30	720	10	11†	7250	—
Combe	14° 24'	144° 54'	45700	545	155	27190	59	24	6840	—
Eagle	14° 42'	145° 23'	12530	430	150	8230	66	34	4630	—
East Hope	15° 44'	145° 28'	35530	270	240	21900	62	40	1445	—
Fife	13° 39'	143° 43'	71650	580	230	58130	81	31	6610	—
Green	16° 45½'	145° 58½'	139100	690	300	117420	84	114	8540	—
Kay	12° 14'	143° 16'	4300	185	43	285	7	3	2680	5430
Magra	11° 51½'	143° 17'	33470	450	130	20950	63	16†	3060	—
Michaelmas	16° 36½'	145° 59'	29030	385	70	7580	26	5	910	2412
Morris	13° 30'	143° 43'	65380	595	170	47350	72	29	980	—
Pelican	13° 55'	143° 50'	80530	430	250	57100	71	17	1840	—
Raine	11° 36'	144° 01'	273000	860	420	163300	60	12	6800	—
Saunders	11° 42'	143° 11'	97200	610	215	64115	66	30	470	—
Stainer	13° 57'	143° 50'	15336	235	115	5130	33	8	860	6150
Stapleton	14° 19'	144° 51'	46800	620	125	26820	57	12	5520	—
Turtle III	14° 44'	145° 11'	14610	200	150	6350‡	44	32	13230	9400
Upolu	16° 41'	145° 56'	13800	300	65	735	5	2	0	—

† Collection not complete.   ‡ Excludes 620 m² mangrove.

VEGETATED SAND CAYS

Vegetated sand cays (excluding the discrete sand cays of low wooded islands) mapped in 1973 are listed in table 4. The mean length of 17 such islands is 460 m and the mean width 170 m; the mean area (5.8 ha) is ten times that of unvegetated cays. As a class, however, these islands are highly diverse: Upolu with two species of vascular plants is scarcely distinguishable from Sudbury with none, whereas Green Island is ten times larger and has over 100 species of plants. However, all the vegetated cays mapped possess beach-rock, with the single exception of Upolu, an association noted also by Steers (1929, p. 20; 1937, p. 16).

Vegetated cays may be further distinguished by topography and vegetation type. One group comprises elongate, narrow islands, with steep beaches often surmounted by dunes which reach maximum altitudes of 7 m. These islands include Beesley, Combe, Eagle, Kay, Stapleton and Upolu, with areas ranging from 0.7 to 4.7 ha. Three of them have vegetated areas less than 10 % of the total, the others from 57 to 66 %. Numbers of plant species range from 2 to 11 in the first group and from 12 to 34 in the second. The vegetation consists of herbs, grasses and low scrub.

A second group comprises larger oval-shaped islands, averaging 530 m in length, 200 m in width, and 9.3 ha in area. Most of these islands are rather flat topped and featureless, though levelling within the woodland on Green Island showed terraces at 3.5–4.0 and at 4.3 m. The vegetation varies from the dense broadleaf woodland of Green Island, to scrub on Morris, dwarf scrub, herbs and grasses on Magra, Michaelmas, Raine, Saunders and Stainer (all of which have large seabird populations). The vegetated areas are more than 60 % of the total on all islands except Michaelmas and Stainer. Beach-rock is extensive at each island, including Michaelmas where it was not mapped by Steers; in all cases it is fairly closely associated with modern beaches and usually records retreats of 10–30 m. Raine Island has extensive phosphorites forming low cliffs round much of the island; Green and Bird Islands have superficial broken phosphorites under woodland in their centres.

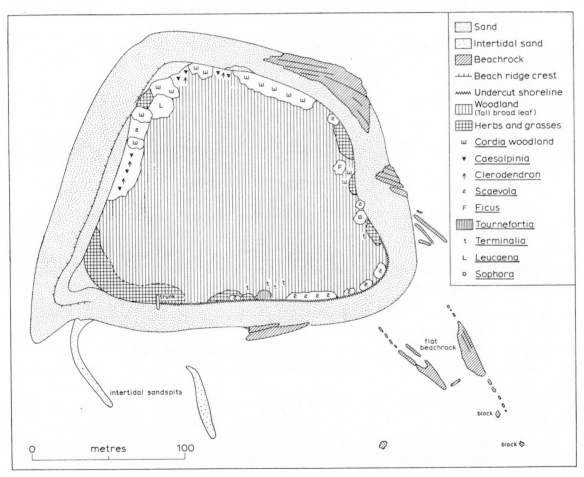

FIGURE 2. A vegetated sand cay: East Hope, 1973.

A third group consists of more equidimensional islands such as East Hope (figure 2), Fife and Pelican, 270–580 m long, 200–250 m wide, and 3.5–8.0 ha in area. The smallest of these, East Hope, is densely wooded, but the others are seabird islands with low scrub, herbs and grasses. In size, character of vegetation, and extent of beach-rock, these islands lie within the range of variation of the oval group. As at Green, East Hope has two distinct surface levels: a lower terrace at 3.4–4.4 m, and a higher at 4.9–5.5 m.

The orientation of the vegetated cays is more variable than Steers (1929, p. 19) supposed. Taking all islands together (unvegetated, vegetated, and cays of low wooded islands), the modal 30° orientation class is 080–110°, with 40 %. South of Cape Melville, orientation tends to be more nearly E–W, and north of that point NW–SE.

The vegetated cays of the northern Barrier Reef may be compared with the Bunker and Capricorn Groups in the south (Steers 1929, 1938; Flood 1977). The Bunker and Capricorn cays are substantially larger than those in the north. They range in length from 510 to 1880 m (mean 1020 m), in width from 130 to 960 m (mean 365 m), and in area from 5.3 to 116.7 ha (mean 28.0 ha). This mean area is more than three times greater than the mean area of vegetated cays in the northern province. Several of the southern islands are very densely wooded with *Pisonia* and *Pandanus* forest of very different aspect to that of the north, but in spite of their greater size their floristic diversity is no greater than on the northern islands.

TABLE 5. CHANGES IN VEGETATED SAND CAYS

island	year	total area ha	vegetated area ha	vegetated area as percentage of total area	beach-rock area ha	maximum length m	maximum width m
Combe	1929	4.93	3.92	79.5	0.41	530	155
	1973	4.57	2.72	59	0.68	545	155
Michaelmas	1936	3.13	1.46	47	n.r.	415	95
	1973	2.90	0.76	26	0.09	385	70
Stapleton	1929	3.96	1.69	43	0.40	810	96
	1973	4.67	2.68	57	0.55	620	125

Source of data: calculated from maps given by Steers (1929, 1938) and from 1973 maps.

Detailed comparisons of change over time can only be made for three islands (Combe, Michaelmas, Stapleton), by using Steers's maps of 1929 or 1936 (table 5). Combe shows a decrease in total area and in vegetated area, and a corresponding increase in area of exposed beach-rock; Michaelmas has also decreased in size; but Stapleton has increased and its vegetation cover expanded.

Low wooded islands

This group, comprising types III–V of Spender's classification of reefs (1930, pp. 285 and 286) and type 4 of Fairbridge (1950, pp. 347–349), is, through the complexity of forms represented, the most informative but in many respects ambiguous of Great Barrier Reef island types. As previous workers have recognized, it includes islands of differing characteristics, and it is perhaps unfortunate that the best known member of the class, Low Isles (which is also the southernmost), lacks many of the typical features common to the rest.

Low Isles and Three Isles are known in detail from the maps of Spender (1930), and Low Isles in particular from the ecological surveys of T. A. Stephenson *et al.* (1931), W. Stephenson, Endean & Bennett (1958), and the geomorphological observations of Moorhouse (1933, 1936) and Fairbridge & Teichert (1947, 1948). Both islands were re-surveyed in detail in 1973, and form the basis of a separate discussion (Stoddart, McLean, Scoffin & Gibbs 1978, this volume). In addition to Low Isles and Three Isles, Steers (1938) also mapped 16 more islands which could be included in the class of low wooded island. All of these except King, Burkitt, Wilkie and Night were remapped in 1973, when a further sixteen low wooded islands were also mapped for the first time. We therefore now have geomorphic information on 34 such islands, including comparative data over the last 37–45 years on 14 of them. Table 6 lists the main attributes of the low wooded islands mapped in 1973.

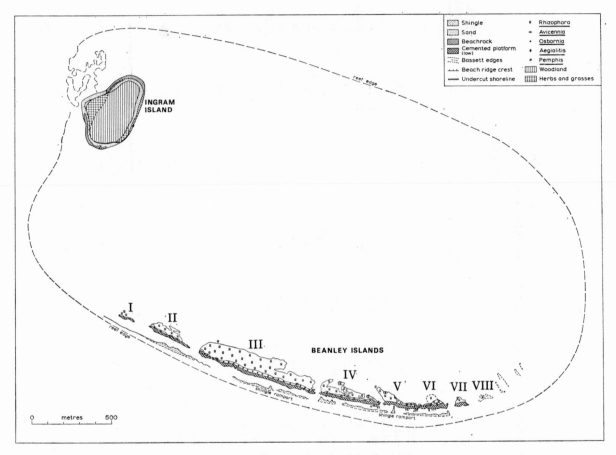

FIGURE 3. Ingram–Beanley Islands, 1973.

These extend over more than 4° of latitude, and are located on reef tops varying in extent from less than 10 to more than 500 ha. The total land area (including mangrove swamps) exposed at low water varies from 1.1 to nearly 150 ha. It is not surprising, therefore, that there is considerable variety in form, nor that many features are not matched by the idealized forms suggested from the Low Isles type example.

In this section we first consider some of the characteristic features of low wooded islands, and then categorize the islands themselves in terms of the occurrence of these features.

Characteristic features

These are discussed as they occur in transverse section from the seaward (windward) to leeward sides of the reef.

TABLE 6. CHARACTERISTICS OF LOW WOODED ISLANDS

island	lat. S	long. E	area of reef top ha	total land area ha	total land as percentage of reef top	cay area ha	promenade and old shingle ridge area ha	mangrove area ha	enclosed pool area ha	new rampart area ha	number of species of vascular plants
Bewick	14° 26'	144° 49'	185.2	147.3	79.5	10.46	11.42	125.44	0	—	30†
Binstead	13° 13'	149° 34'	29.6	0.9	2.9	0.09	0.87	0	0	0	20
Chapman	12° 53'	143° 36'	n.d.	2.2	n.d.	0.05	1.76	0.34	0	0	17
Coquet	14° 32½'	144° 59'	61.8	26.3	42.6	12.07	8.45	5.79	4.20	—	11†
Hampton	14° 34'	144° 53'	90.5	29.4	32.4	0.43	0	28.92	0	—	2†
Houghton	14° 31½'	144° 58'	135.8	50.0	36.8	12.46	11.54	26.00	5.25	—	13½
Ingram–Beanley	14° 25'	145° 53'	540.0	26.2	4.8	11.09	4.27	8.90	0	1.92	47
Leggatt	14° 33'	144° 40'	54.9	8.5	15.5	2.08	0.23	2.86	0	3.34	12†
Low	16° 23'	145° 34'	136.2	37.5	27.5	2.25	2.94	32.29	0	3.65	58
Low Wooded	15° 05'	145° 23'	87.9	51.8	58.9	4.16	12.01	31.96	0	0.75	62
Lowrie	13° 17'	143° 36'	n.d.	3.4	n.d.	0.24	—	2.37	0	—	13
Newton	14° 30'	144° 55'	72.3	31.3	43.3	2.58	3.05	25.70	0	4.00	33†
Nymph	14° 39½'	145° 15'	135.3	35.8	26.5	—	24.17	7.62	21.29	—	25†
Piper (Farmer–Fisher)	12° 14½'	143° 13½'	n.d.	8.8	n.d.	6.16	1.21	1.43	0	—	18†
Pipon	14° 07½'	144° 31'	334.0	72.0	21.6	0.74	4.86	62.12	—	4.29	37
Sherrard	12° 59'	143° 34'	n.d.	1.8	n.d.	1.39	0.36	0	0	—	17
Sinclair–Morris	14° 33'	144° 54'	31.7	8.1	25.6	2.23	0.95	1.77	0	3.18	30†
Three	15° 07'	145° 17'	133.5	47.1	35.2	15.81	11.50	8.18	0	11.56	59
Turtle I	14° 44'	145° 11'	42.3	18.2	43.1	—	9.32	5.36	0	3.55	53
Turtle II	14° 44'	145° 12'	61.0	23.0	37.8	—	8.82	12.00	3.57	2.21	15†
Turtle IV	14° 43½'	145° 12'	29.0	3.8	13.1	0.33	1.63	—	0	1.83	16†
Turtle V	14° 42'	145° 12'	14.8	4.7	31.8	—	4.68	0	0	0.02	14†
Turtle VI	14° 43'	145° 10½'	9.0	1.1	12.3	—	0.59	0.20	0	0.32	20†
Two	15° 01'	145° 27'	128.7	33.1	25.7	19.46	4.79	1.48	0	7.39	65
Watson	14° 28'	144° 49'	57.2	10.3	18.0	0.26	3.55	5.33	—	1.14	26†
West Hope	15° 45'	145° 27'	315.0	31.3	9.9	0	7.90	17.49	—	0.55	47

† Collection not complete.

(a) Reef flat

The reef flat forms a rocky platform, drying at low water springs, round the windward margins of reefs. It is laterally continuous, usually 60–100 m wide, and encloses the central reef top which on small reefs may resemble the reef flat or on large reefs may form an enclosed depression (called a pseudolagoon by Hedley & Taylor (1907) and Steers (1929), though this term was rejected by Spender (1930) and Steers (1937, p. 254)). The nature of these enclosed reef tops and their sediment covers is discussed by Flood & Scoffin (1978, part A of this Discussion). Generally the reef flats are covered with a felt of green and brown algae and corals are uncommon. At Three Isles in particular the surface of the flat consists of lineations of low relief a few metres apart and parallel with the reef edge (the 'honeycomb rock' of Stephenson *et al.* 1931), apparently the result of erosional planing of reef flat structures. The surface of the flat slopes gradually seawards (average slope 1 : 80 or 0° 40′), to levels at which corals can survive tidal exposure. The highest level of living reef flat corals appears to be 1.0 m (mean low water springs (m.l.w.s.) 0.5 m), but is frequently lower. Seaward of this point the slope increases rapidly and emersion becomes less frequent and coral cover greater. No algal ridges *sensu stricto* were found on low wooded island reefs, even in the most exposed situations, but shallow algal-rimmed pools and terraces do occur near the reef edge on some flats. Steers found them at Night, Houghton, Bewick, Pipon and Beanley (1937, pp. 122–124; 1938 pp. 85 and 86), and we also found them at West Hope and Leggatt.

(b) Rampart

Ramparts are asymmetric ridges of coral shingle with a steep inward face, locally reaching 80°, and a gentle seaward slope of less than 10°. Their outer margin is a feather edge of shingle on the reef flat and is often too indistinct to map; in plan it roughly parallels the edge of the reef. The inner edge is arcuate, with occasional shingle tongues which on the windward side are at right angles to the reef edge but elsewhere are at an angle to it. Though the term rampart is used by many American workers as synonymous with beach ridge, it is useful to retain it for these shingle accumulations of the reef flat which are wholly or almost wholly submerged at high water (Teichert 1947, pp. 154 and 155).

The mean maximum width of ramparts on the low wooded islands mapped is 45 m; the widest reach 65 m. The mean maximum elevation of the rampart crest is 1.8 m (mean high water springs (m.h.w.s.) 2.3 m), but the highest found, at Watson, reached 3.12 m. The mean maximum elevation of the rampart crest above the reef flat or moat floor immediately to landward is 0.8 m (maximum 2.9 m).

Shingle tongues are common round the entire inner perimeters of ramparts but are particularly well developed at the windward point, where, as Spender (1930, 206) showed, they suggest a basic homology of form between different reefs. Not all are recent. At Low Isles a shingle tongue extends far into the mangrove, and at Three Isles the main tongue is lithified and vegetated. Fresher tongues on other islands are often more than 100 m long, and at Houghton one reaches 340 m; these may have invaded pre-existing mangrove woodland during storm events.

In general, ramparts consist of clean white shingle, dominantly composed of *Acropora* sticks, and these are the components usually exposed in the landward face. On some islands with mangroves, close to the mainland, however, shingle on the backslope is embedded in a muddy matrix which inhibits mobility and reduces permeability; this is well seen at Leggatt and Sinclair–Morris.

Bassett edges form a particular and distinctive feature of eroding older ramparts. These are lithified foreset beds forming projecting steeply dipping ridges, straight or more often arcuate in plan, recording former locations of the inner edges of unconsolidated ramparts (Steers 1937, p. 26). Patches of bassett edges on otherwise bare reef flats indicate where ramparts formerly existed. Their elevations at Low Isles and West Hope range from 1.1 to 1.9 m (range at neaps 1.2–1.6 m).

Spender (1930) in his survey of Low Isles drew attention to the existence of inner and outer ramparts. The inner consisted of old, blackened, partially compacted shingle, the outer of fresh, white, loose shingle. Fairbridge & Teichert (1947, 1948) identified an older innermost rampart at Low Isles, and a more recent new rampart, formed since 1929. Most low wooded islands have multiple ramparts, though only the outer has the simple form described above; they are distinguished by colour, degree of weathering of the clasts, extent of colonization by vegetation, and relative location. In particular the older ridges coalesce and overlap to form shingle islands, margined at their outer edge by a ridge or breastwork of fresh white shingle (Moorhouse 1933, 1936). At Low Isles in 1973 new breastworks were found in places where the 1929 ramparts had disappeared.

Ramparts may be colonized by a low scrub of *Aegialitis annulata* and *Avicennia marina*. Older shingle sheets characteristically carry succulent mats of *Sesuvium portulacastrum*, *Salicornia quinqueflora*, *Arthrocnemum* spp., and *Suaeda australis*.

(c) Moat

Shingle ramparts may enclose shallow ponds on inner reef flats, which fail to drain at low water. These are ephemeral features subject to rapid changes in extent and water level with changes in the size and location of the enclosing ramparts. Corals grow in these moats at levels well above those of the outer reef flat, commonly reaching 0.9–1.0 m and expectionally 1.8 m (mean low water neaps (m.l.w.n.) 1.2 m). The corals include carpets of *Montipora* in shallow moats, and large microatolls and giant clams in deeper moats. Fairbridge & Teichert (1947, p. 4) noted that growth of corals in such ponded situations could lead to difficulties in the use of corals to reconstruct past sea levels.

Elsewhere, these shallow water bodies are colonized by mangroves, notably *Aegialitis annulata*, *Avicennia marina* and *Rhizophora stylosa*, and these in their turn serve as a baffle limiting the inward migration of shingle ramparts.

(d) Platforms or promenades

Conglomerate platforms often fringe the inner margins of moats and form the seaward edge of the larger land bodies of low wooded islands. They are usually confined to the seaward side of the mangrove–shingle islands, but are occasionally also found on the sand cays (e.g. Leggatt). Platforms are absent at Low Isles, but were mapped at Three Isles by Spender and interpreted as 'a cemented and conglomerated inner rampart now being eroded' (1930, pp. 207–210), an explanation followed by Steers (1930, p. 5). Spender observed that the level of the platform at Three Isles (2.1 m) was less than that of the highest ramparts at Low Isles (2.3 m) so that relative elevation of the platform was not necessarily implied; he also drew attention to the apparent absence of corals in growth position in the material of the platforms.

In his map of Three Isles, Spender (1930) clearly identified an upper and a lower platform, but he did not make this distinction in his discussion. Subsequently, Steers (1937, pp. 16–28) emphasized the existence of two discrete features, and equated them with benches on mainland and high-island coasts. The lower platform he stated was 'awash at high water springs' (m.h.w.s. 2.3 m), the upper (and less widespread) 'never exceeds' 3.3 m. Steers (1938, pp. 75 and 78) argued from their topography, structure, location and composition that both platforms were lithified shingle ramparts truncated and cliffed on their seaward sides by erosion. Fairbridge & Teichert, who worked only at Low Isles where these features do not exist, added nothing to Steers's account, and interpretation of the platforms plays little part in their model of low wooded island evolution. Steers himself (1931) favoured a eustatic explanation for the difference in elevation of the two platforms, and he also (1938) clearly felt that the lower as well as the higher had been slightly raised with respect to present sea level. At Turtle I and Nymph (= Enn) he estimated the level of the lower platform at 1.0–1.2 m above the reef flat, and of the high platform at 1.0 m above this (Steers 1938, pp. 77, 79 and 81).

Platforms were identified on most low wooded islands mapped in 1973. In some cases they could be clearly assigned to 'upper' or 'lower' categories, but often their relative position was unclear, and frequently the morphological distinction between platform and beach-rock was ambiguous. The use of terms such as 'upper' and 'lower', therefore, while carrying a precise connotation on individual reefs, does not necessarily imply accordance in height or similarity of origin and history between reefs.

Lower platforms on 14 reefs have a mean width of 30 m and a maximum width of 68 m (on Watson). Often, however, particularly beneath *Aegialitis* and *Avicennia* scrub, the outer edge of the lower platform is difficult to distinguish for irregular basset edges of eroding ramparts, with which they are indeed probably genetically linked. Mean maximum elevations of lower platforms on eight reefs (18 profiles) is 2.3 m, identical to the level of m.h.w.s. (as Steers observed); several examples reach between 2.6 and 3.0 m.

Upper platforms are less widespread. Mean maximum width on seven islands is 30 m, and the greatest width (on Coquet) is 40 m. Mean maximum elevation on nine islands (19 profiles) is 2.9 m, 0.6 m above m.h.w.s. and the mean maximum elevation of the lower platform. This level approximates to that of extreme h.w.s. However, several cases have maximum elevations of 3.2–3.5 m. The upper platform usually has an abrupt vertical or undercut seaward slope. Its surface is highly variable. In many cases the outermost few metres consists of jagged basset edges, similar to but higher than those of the reef flat. Elsewhere the surface may be dissected by large circular potholes 1 m or more deep; usually, however, where basset edges are absent the surface is horizontal in transverse profile, and in at least some cases this horizontality results from the presence of superficial flat-bedded shingle deposits on top of the main conglomerate, a feature first noted by Steers (1929, p. 255) at Houghton. Frequently, lower platform deposits lap up against and cover pinnacles of upper platform, and residuals of the latter can be found protruding through the former. Small pocket beaches may occur in gaps eroded in the face of the upper platform, and these generally have extensive beach-rock.

At several localities, platform conglomerate was found to overlie fossil microatolls in the position of growth (e.g. at Turtle I and Three Isles). These are interpreted as having grown in a former moat ponded by a shingle rampart. The rampart then advanced, overriding and killing the corals, was lithified, and subsequently eroded to form the present platform. The mechanism is described in more detail by Scoffin & Stoddart (1978, this volume). The genetic link between modern ramparts and the platforms is considered so close by Scoffin & McLean (1978, part A of this Discussion) that they term the platform conglomerate 'rampart rock'. Nevertheless, the platforms raise important questions, as Steers and Spender realized. Are there really two discrete levels? Are they accordant between reefs? If so, do they have time-significance? What external events (such as sea level change or storm activity) led to their formation?

(e) Shingle island

Platforms are usually surmounted by a series of old shingle ridges, now stabilized and vege-tated. In plan these resemble modern breastworks and probably had a similar origin. There are usually two or three such ridges, but in places there are much wider sequences of ridges of variable width and height. These are well seen on West Hope, Watson, Low Wooded Island, and several of the Turtle group. Their maximum elevation varies from 3.5 to 4.9 m. Some of these ridges are misleadingly called dunes in the older literature. Thus Green Ant Island at Low Isles is described as an 'accumulation of shingle, sand and pumice forming a dune-like bank about 50 yds [45.7 m] wide and probably as much as 20 ft [6.1 m] above datum at the summit' (Stephenson *et al.* 1931, p. 28; also Spender 1930, p. 207). The term dune here refers to morphology and not to composition or origin; true dunes are found on some sand cays, but they are absent from the windward sides of low wooded islands.

(f) Mangrove swamp

The mangrove swamp at Low Isles was studied in some detail in 1928–29, and has been re-examined by Macnae (1966). Mangroves on low wooded islands are clearly opportunistic in the sense that their distribution depends on the location and dynamics of shingle ramparts, platforms and ridges, and also on variations in reef-top topography. Steers (1937, p. 133), however, considered that the extent of mangrove could be used to place the low wooded islands in a sequence of development, from initial colonization to a stage where most of the reef top is covered by mangroves and the sand cay partly surrounded (as at Bewick and Nymph). This view was accepted by Fairbridge & Teichert (1948, p. 85) and Stephenson *et al.* (1958, p. 309), who found evidence of rapid mangrove colonization at Low Isles since 1929; Macnae (1966, p. 88), on the other hand, found no such evidence. The general characteristics of low wooded island mangroves will be considered elsewhere and the detailed record of change at Low Isles and Three Isles is discussed by Stoddart *et al.* (1978, this volume). Here we simply note the main attributes of the mangrove zone.

Mangrove areas vary from less than 1 ha to a maximum of 125 ha (on Bewick); the mean mangrove area on 22 low wooded islands is 19 ha. There is no relation between reef-top area and percentage covered by mangroves: some large reefs have very small areas of mangrove (Two, Three, Pipon), others, such as Bewick, a large proportion.

Aegialitis annulata and *Avicennia marina* are characteristic of shingle ramparts, often in very exposed situations, and moats. *Rhizophora stylosa* is the main colonizer of reef tops, with occa-sional tall trees of *Sonneratia alba*. At higher levels, *Rhizophora* is replaced by *Ceriops tagal*, several species of *Bruguiera* and *Xylocarpus*, and at the highest levels by *Osbornia octodonta* (especially on shingle substrates) and *Excoecaria agallocha*. The mean maximum elevation reached by man-groves, usually in the lee of windward shingle ridges, based on 40 profiles from 15 islands, is 2.0 m, with some examples reaching 2.4 m (m.h.w.s. 2.3 m).

A key to the history of low wooded island mangroves is given by the discovery, first at Houghton and later elsewhere, of fields of fossil microatolls in the position of growth within mangrove woodland. Such corals had been noted at Low Isles and Two Isles by Spender (1930, p. 207) but their significance had not been realized. On both Houghton and Leggatt these microatolls reach elevations of 1.35 m (m.l.w.n. 1.2 m); this may be compared with elevations of up to 2.0 m for fossil microatolls beneath conglomerate platforms, and of 0.9–1.0 m for living corals in rampart-ponded moats. On Houghton the microatolls form an extensive field, fortuitously revealed by hurricane damage, and similar microatolls have also been found within the swamps at Hampton and Bewick. Their interpretation in the context of sea level change and island history is discussed by Scoffin & Stoddart (1978, this volume) and McLean et al. (1978, part A of this Discussion).

(g) Sand cay

Leeward dry-land sediment accumulations are characteristic of most low wooded islands, but they differ in nature even more widely than ordinary sand cays (table 7). In a first group, the sand cay is discrete and separate, and may be compared with an ordinary sand cay. Some are unvegetated ephemeral islets of small size (Binstead, Chapman, Sand, Turtle II, Watson: mean area 0.1 ha). Others are larger vegetated cays (Bird, Farmer, Ingram, Low, Lowrie, Pipon, Sherrard, Sinclair, Three, Two: mean area 6.3 ha). Most of these are covered with woodland or scrub, and the number of species of vascular plants on each is generally higher than on the simple sand cays.

A second group consists of recognizably discrete cays partly or largely surrounded by mangrove, in some cases forming a single land unit with the windward shingle ridges and platforms. These include Bewick, Howick, Leggatt and Newton (mean area 4.9 ha). Of these, Howick and Newton are partly wooded, but the others are covered with grassland.

Taken together, the vegetated discrete cays of low wooded islands have a mean length of 420 m, a mean width of 185 m, and a mean area of 12.0 ha. They are, as a group, almost exactly twice the area of ordinary vegetated sand cays, but still less than half as large as the vegetated cays of the Bunker and Capricorn Groups.

In one important respect the vegetated sand cays of low wooded islands are more complex than most simple vegetated sand cays. This is in the presence of a low terrace round a central higher and more extensive core, a situation similar to that noted by Steers (1929, p. 347), following Stanley, on Middle Island. Sample terrace levels are given in table 8. Terraces at comparable levels, with similar vertical separation, have already been noted on Green Island. The two levels are distinguished not only by elevation but also by soil development and sedimentological characteristics (McLean & Stoddart 1978, part A of this Discussion) and vegetation: the higher terrace usually carries woodland or dense scrub, the lower a more open community of shrubs and herbs. Where the lower terrace is absent, cay beaches are steep and may have extensive arrays of well cemented beach-rock; where it is present the beaches are lower and beach-rock is patchily developed, often at lower intertidal levels, and less well cemented. There is no doubt that the lower terrace is an aggradation feature formed after the main formation of the cay; in two cases (Two Isles and Three Isles) it can be shown that part at least of its development has taken place over the last few decades (Stoddart et al. 1978, this volume).

The widespread existence of the two terrace levels on the cays has not previously been recognized, but it clearly needs to be discussed in the same context as the existence of upper and

TABLE 7. DISCRETE SAND CAYS OF LOW WOODED ISLANDS

island	lat. S	long. E	cay area m²	cay area as percentage of reef top	maximum length m	maximum width m	vegetated area m²	vegetated area as percentage of cay area	number of species of vascular plants	beach-rock area m²	mangrove area m²
Bewick	14° 26'	144° 49'	104 580	5.5	600	300	104 580	100	18	6176	0
Binstead	13° 13'	149° 34'	920	3.1	60	22	0	0	0	0	0
Bird	11° 46'	143° 05'	40 310	n.d.	380	150	29 250	73	8†	9030	0
Chapman 1 and 2	12° 53'	143° 36'	490	n.d.	17; 19	12; 19	0	0	0	0	0
Farmer (Piper)	12° 14½'	143° 13½'	61 590	n.d.	535	201	39 965	65	6†	9 565	0
Howick	14° 30'	144° 58'	45 715	n.d.	695	230	28 720	63	27	8280	0
Ingram	14° 25'	145° 53'	110 910	2.1	490	350	89 810	81	44	13 780	0
Leggatt	14° 33'	144° 40'	20 840	3.8	320	125	14 220	68	8†	2312	0
Low	16° 23'	145° 34'	22 500	1.7	240	130	14 060	63	34	6015	0
Lowrie	13° 17'	143° 36'	2380	n.d.	78	42	266	11	4	0	0
Newton	14° 30'	144° 55'	25 760	3.6	395	120	13 998	54	27	2898	0
Pipon	14° 07½'	144° 31'	7400	0.2	240	65	3850	52	32	5124	0
Sand	14° 31'	144° 51'	1290	n.d.	115	27	0	0	0	0	0
Sherrard	12° 59'	143° 34'	13 870	n.d.	220	110	7104	51	10	2468	0
Sinclair	14° 33'	144° 54'	19 260	6.1	300	112	11 570	60	24	0	3010
Three	15° 07'	145° 17'	158 130	11.8	715	285	131 400	83	46	16 210	0
Turtle II	14° 44'	145° 12'	940	0.2	70	20	0	0	0	0	0
Two	15° 01'	145° 27'	194 600	15.1	720	350	164 450	85	48	20 140	0
Watson	14° 28'	144° 29'	2570	0.4	120	35	0	0	0	85	0

† Collection not complete.

lower platforms and of sequences of shingle ridges. The difference in nature of beach-rock outcrops is also of interest. The mean maximum width of the older beach-rock is 2.4 m; in all cases it is associated with modern beaches, except at Sherrard, where relict beach-rock indicates a bodily translocation of the whole cay a distance of 150 m to the northwest. The average range of elevation on low wooded island cays is 1.2–2.4 m (m.l.w.n. 1.2 m; m.h.w.s. 2.3 m),

TABLE 8. ELEVATIONS OF TERRACES ON SAND CAYS OF LOW WOODED ISLANDS

island	height of lower terrace m	height of higher terrace m	height difference m	maximum elevation m
Leggatt	3.4–3.6	4.6–4.9	1.2–1.3	4.9
Two	3.4–3.7	6.0–6.3	2.6	6.6
Three	3.0–3.8	4.8–5.9	1.8–2.1	6.0
Howick	3.4–3.5	4.4–4.9	1.0–1.4	4.9
Ingram	2.7–4.0	5.0–6.2	2.2–3.3	6.9

TABLE 9. SUMMARY OF SIGNIFICANT ELEVATION DATA DERIVED FROM PROFILES

feature	mean elevation m	number of measurements	maximum elevation m	minimum elevation m
highest living corals	0.47	11	0.94	0.09
highest ramparts	1.77	14	3.12	0.80
living moat corals	1.04	24	1.78	0.42
highest lower platform	2.31	23	3.27	1.40
highest upper platform	2.87	19	3.53	2.08
highest mangrove	2.01	38	2.86	0.71
dead microatolls	1.47	18	2.30	0.98
height range of sand cay lower terrace	3.34–3.89	10	4.0–4.4	2.7–3.5
height range of sand cay higher terrace	5.01–5.63	10	6.0–7.2	4.3–4.9
highest land	5.74	17	8.99	4.41
top of beach-rock	2.29	21	3.30	1.25
bottom of beach-rock	1.09	21	2.20	0.40

Extreme h.w.s. 2.9; m.h.w.s. 2.3; m.h.w.n. 1.6; m.l.w.n. 1.2; m.l.w.s. 0.5; extreme l.w.s. 0.

but the highest examples reach 2.7–3.0 m (extreme h.w.s. 2.9 m). Erosion of the windward extremities of islands revealing complex arcuate bands of beach-rock, as at Three Isles and Newton, is quite common, even where the shore is now protected by mangroves; some of these beach-rock arrays are unusually high, and some are even vegetated. Steers (1938, pp. 78–86) also noted the existence of possibly raised beach-rock on low wooded islands, apparently associated with the lower platform at Bewick, Pipon and King, and with the higher platform at Nymph and Ingram. The higher more massive beach-rock consists either of inclined ledges with 10–15° seaward dip, often showing complex patterns of overlap, or horizontal platforms, often covered with *Sesuvium* and other succulent herbs. The less massive beach-rock on the shores of lower terraces stands at lower elevations; the height range is commonly 1.1–2.0 m.

(h) Boulder zone

This was defined on the northwest (leeward) side of Low Isles by Steers (1929) and Spender (1930, p. 201). It is a recurrent feature on low wooded islands, and occurs either close to the leeward reef edge (as at Low and Watson) or near the island shore (as at Bewick). The zone is up to 200 m long, with boulders reaching 3–4 m in greatest dimension. The constituents are mainly individual coral colonies, undoubtedly storm-deposited. Similar deposits have been found, also in leeward situations, cemented into platform rocks, forming a coarse boulder conglomerate, as at Howick.

Types of low wooded island

Taken together, the rather heterogeneous group of low wooded islands here described have a mean total land area at low tide of 28.5 ha (25 cases). Of this total, dry land (i.e. cay, shingle ridges and platforms) comprises a mean area of 9.8 ha, the mangrove zone 16.5 ha, and the fresh shingle ramparts 2.1 ha. However, these figures conceal wide variations, and it is helpful to distinguish four subgroups of low wooded islands, differentiated by location and nature of sedimentary deposits and the extent of mangroves on the reef top. Characteristics of individual islands have already been given in table 6.

(a) Mangroves of limited extent, sand cay separate

Low Isles forms the classic example of this type. The reef top is large; the cay and the mangrove–shingle island are well separated; and there are well defined fresh shingle ramparts on the windward reef. These features are repeated on Lowrie, Two Isles, Three Isles and Pipon; Sinclair–Morris is somewhat comparable but smaller. On some reefs the sand cay is embryonic (Chapman, Binstead, Watson, Turtle IV) or almost non-existent (West Hope). In others it is substantially larger than the mangrove–shingle island (Ingram–Beanley, Sherrard, Piper). Mangroves may be limited to a few seedlings on fresh shingle tongues (e.g. East Hope, which has therefore here been treated as a simple sand cay), or are restricted to species characteristic of higher levels (*Excoecaria, Osbornia, Xylocarpus*) on the shingle cay, with only a narrow fringe of *Rhizophora* (Chapman, Binstead, Sherrard). It is possible to speculate on the reasons for these differences, but here we simply draw attention to them.

(b) Mangroves extensive, joining the shingle and sand cays

Of all the islands, Bewick (figure 4) has the most continuous mangrove cover, though the outline of the sand cay is quite distinct. Both Newton and Nymph also have extensive reef-top cover, though with large enclosed lagoons. The sand cay at Newton is well defined, but the boundary between mangrove and dry-land vegetation at Nymph is much less distinct. Sinclair–Morris is an example of a smaller reef with cay, mangrove and rampart forming a continuous unit.

Three islands on more elongate reefs have rather different characteristics. Houghton and Coquet have large distinct sand cays and extensive mangroves, but the windward shingle area is linked to the cay by a continuous belt of conglomerate platform and shingle ridges; Low Wooded Island is similar, except that the sand cay has less well defined boundaries.

(c) Turtle-type islands

Steers (1929, p. 25) drew attention to the islands of the Turtle Group as representing an 'intermediate stage between the simple sand cay and the complex cay', lacking a central open flat, and with the mangrove–shingle cay 'closely wrapped round the sand cay'. Later (1937, pp. 128–130) he emphasized the lack of sharp boundaries of the 'cay-like area', the extensive high shingle ridges forming most of the dry-land area, and the 'old aspect' of the terrain by comparison with other islands. Vegetated shingle ridges, with pronounced surface topography, dominate the reef top; on Turtle VI (figure 5) there is a distinction, similar to that on some sand cays, between a higher central area and a lower surrounding terrace, with slabs of old beach-rock on the slope between the two. The sediments of these islands also differ from those of other low wooded islands (McLean & Stoddart 1978, part A of this Discussion).

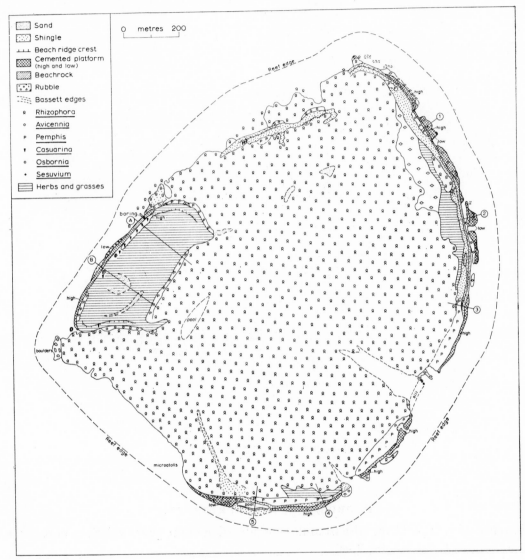

FIGURE 4. Bewick Island, 1973.

(d) Miscellaneous

Several islands of very variable characteristics were also mapped and cannot be assigned to the above categories. Hampton could be considered as wholly a mangrove island, but there are small low remnants of conglomerate platform near the windward reef edge and a dry cay area within the mangrove. Steers (1937, p. 128) similarly found a completely mangrove-encircled sand cay at Hannah. Sand Island has an embryonic sand cay, patches of conglomerate platform and much rubble, with mangrove scrub of *Avicennia* and *Aegialitis*; it was undoubtedly larger in the past. The two Pethebridge (= Kew) Islands consist of a dry shingle island, conglomerate platforms and rubble sheets to windward, with to leeward intertidal sand spits, ephemeral cays, and only embryonic mangroves.

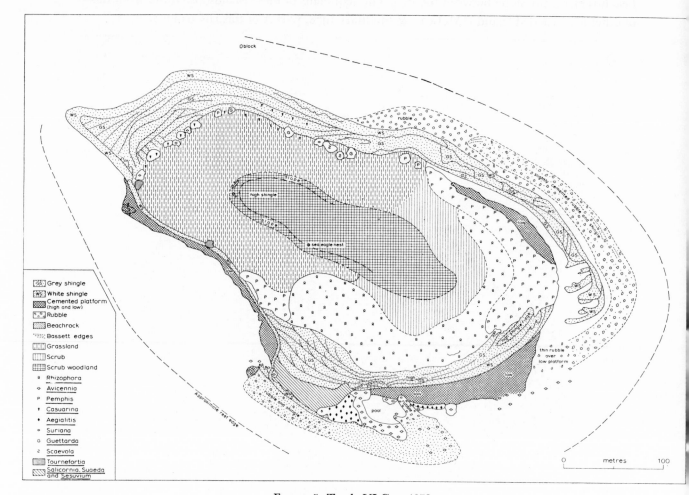

FIGURE 5. Turtle VI Cay, 1973.

PROBLEMS OF INTERPRETATION

Two main sets of problems arise from this consideration of the geomorphology of reef islands of the northern Great Barrier Reef.

(a) Spatial distribution

Both Steers (1929) and Spender (1930) emphasized differences in the distribution of islands on the reefs. The outer ribbon reefs have almost no islands at all. The Admiralty charts mark six between Cairns and Cape York, all sandbanks except for Waterwitch (shown by relict beach-rock to have formerly been larger) and Raine (which is on a detached reef, not a ribbon reef). The large platform reefs of the outer shelf, within the ribbon reefs, in general also lack islands, except for a few intertidal sand banks. Near the centre of the shelf, some large reefs carry simple sand cays (Eagle, Pickersgill, Sudbury, Combe, Stapleton, East Hope). Most of the islands, however, are concentrated on smaller reefs of the inner shelf. These are either simple sand cays formed by refraction round the reefs (Green, Michaelmas, Fife, Kay, Magra, Morris, Pelican, Saunders, Stainer), or, especially near the mainland coast, they are low wooded islands of varied size and form. Of the islands mapped in 1973, the mean distance of sand cays from the mainland was 24.5 km and from the shelf edge 36 km, and of low wooded islands 14 and 38 km respectively. Sand cays are found across the whole width of the shelf (minimum distance from mainland 4.5 km, maximum 104 km), whereas 94 % of low wooded islands are within 20 km and 26 % within 10 km of the mainland.

Spender (1930) suggested that this distribution resulted from a systematic difference in level of the reef tops, the outer reefs being too low for permanent sediment bodies to accumulate, and the inner reefs being progressively higher towards the mainland with sediment accumulations forming more readily on their tops. Steers (1930, 1931, 1937) maintained that exposure to wave action was a more important control of sediment accumulation. The outer reefs, exposed to ocean swell, were swept clear of debris. The outer platform reefs immediately in the lee of the ribbon reefs and intersected only by narrow channels are too protected for shingle ridges to form. The smaller, more widely separated inner shelf reefs, rising from shelf levels of 20–30 m, are exposed to the Southeast Trades blowing longitudinally along the shelf, affecting not only the growth direction of the reefs and the reef morphology but also patterns of wave refraction and storm-deposition of debris. The implications of these two general models for reef-island distribution in other parts of the world have been discussed by Stoddart (1965). Most workers have supported Steers's general view, but Spender's postulate of systematic variation in reef-top levels transverse to the mainland coast has received some support from theoretical considerations of hydroisostasy (Thom & Chappell 1978, part A of this Discussion).

(b) Temporal development

Crucial to the interpretation of the record of the reef-top features is the calibration of their response to relative sea level changes in the Holocene. Previous workers, lacking radiometric dating techniques, have relied on geological interpretations of topographic features, many of which were themselves ambiguous in origin. Steers (1929, 1937) used the presence of upper and lower platforms, the fairly consistent height difference between them, the possible existence of raised beach-rock on the cays, and the presence of high and low erosional benches (though at rather different levels) on high-island and mainland shores to suggest that sea level may have fallen eustatically and episodically during the period of reef-island formation. Spender (1930) argued from the accordance in the maximum elevations of platforms and modern ramparts that there was no necessary argument for such falls; he did not, however, discuss the problem presented by the presence of platforms at two distinct levels.

In addition to the platforms, the present investigation has introduced additional evidence: (a) the existence of upper and lower terraces on cays, which can be correlated between islands with fairly consistent vertical separation, and (b) the widespread identification of reef-top microatolls, either beneath platform deposits or within mangrove swamps, at intertidal levels which are unusually high by comparison with modern growing corals. Our investigation has therefore focused on relating these different features to each other (figure 6) and to an independent radiometric time scale.

The presence of the microatolls on many low wooded islands also has implications for a question raised by Steers (1937, pp. 135 and 136), denied by Spender (1930, p. 290), and discussed by Fairbridge & Teichert (1948): how far do the relative extents of shingle ridges. sand cays and especially mangrove swamps on low wooded islands record a temporal succession from a stage with no mangroves at all on the flat to one, like Bewick, where the cover is complete, and if so, what controls the rate and timing of stages in the developmental sequence (Fairbridge & Teichert 1948, p. 85)?

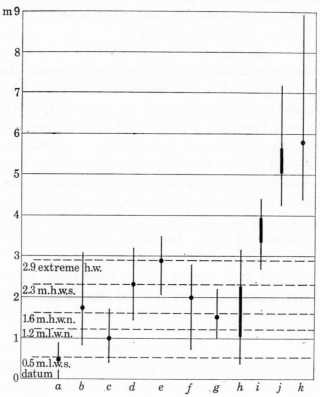

FIGURE 6. Elevations of significant topographic features of low wooded islands: (*a*) living corals; (*b*) ramparts; (*c*) moat corals; (*d*) lower platform; (*e*) upper platform; (*f*) highest platform; (*g*) dead microatolls; (*h*) beach-rock; (*i*) lower shelf; (*j*) upper shelf; (*k*) highest point.

Steers (1937, pp. 135 and 136) and Fairbridge & Teichert (1948, p. 85) thought that extent of mangrove on the reef top was the main indicator of stage or age of low wooded islands. Spender (1930, p. 290) considered that at any one time the relative extent of reef-top features was determined by reef level and environmental conditions, rather than stage of development. He even suggested that mangroves might be decreasing rather than increasing in area at Low and Three Isles (Spender 1937, p. 142), largely because of the removal by erosion of protecting ramparts and platforms or by the chemical erosion of reef-top substrates by the mangroves themselves. Steers (1937, p. 145) suggested that perhaps the number and size of the shingle ridges would be a better index of topographic development than the extent of mangroves. Detailed surveys over nearly 50 years have shown extensive spread of mangroves at Low Isles, and no spread at all at Three Isles (Stoddart *et al.* 1978, this volume); and in general over this period of time there appears to have been remarkably little geomorphic change. There is, however, considerable evidence of catastrophic damage to mangroves as a result of cyclones. This was noted by Steers (1938) at Houghton, Wilkie and Night Islands, and again at Houghton and Newton by us. These effects seem to be local rather than regional in their incidence. If there is a progressive extension of mangroves in the manner suggested by Steers and Fairbridge & Teichert, it clearly takes place at different rates on different reefs and with frequent random interruptions because of storms.

The morphological evidence alone, therefore, though suggestive, is ultimately ambiguous in terms of origin and genesis, and reliance on this evidence alone would place severe constraints on our reconstruction of Holocene reef history in the northern Great Barrier Reef area. As other papers in this Discussion will show, however, the purely morphological evidence can be supplemented by detailed sedimentological, lithological and ecological studies, and calibrated by reference to a radiometric time scale, so that the area of speculation in the choice of alternative models of the development of islands can be reduced.

REFERENCES (Stoddart *et al.*)

Beaglehole, J. C. (ed.) 1955 *The journals of Captain James Cook on his voyages of discovery*, vol. 1. Cambridge: Hakluyt Society.

Fairbridge, R. W. 1950 *J. Geol.* **58**, 330–401.

Fairbridge, R. W. & Teichert, C. 1947 *Rep. Gt Barrier Reef Comm.* **6**, 1–16.

Fairbridge, R. W. & Teichert, C. 1948 *Geogrl J.* **111**, 67–88.

Flood, P. G. 1977 *Atoll Res. Bull.* **195**, 1–7.

Flood, P. G. & Scoffin, T. P. 1978 *Phil. Trans. R. Soc. Lond.* A **291**, 55–71 (part A of this Discussion).

Hedley, C. & Taylor, T. G. 1908 *Rep. Aust. Ass. Adv. Sci.* **1907**, 394–413.

Jukes, J. B. 1847 *Narrative of the surveying voyage of H.M.S. Fly.* (Two volumes, 423 and 362 pages.) London: T. and W. Boone.

King, P. P. 1827 *Narrative of a survey of the intertropical and western coasts of Australia.* (Two volumes, 451 and 673 pages.) London: J. Murray.

MacGillivray, J. 1852 *Narrative of the voyage of H.M.S. Rattlesnake.* (Two volumes, 402 and 395 pages.) London: T. and W. Boone.

MacGillivray, W. 1910 *Emu* **10**, 216–233.

Macnae, W. 1966 *Aust. J. Bot.* **14**, 67–104.

McLean, R. F. & Stoddart, D. R. 1978 *Phil. Trans. R. Soc. Lond.* A **291**, 101–117 (part A of this Discussion).

McLean, R. F., Stoddart, D. R., Hopley, D. & Polach, H. 1978 *Phil. Trans. R. Soc. Lond.* A **291**, 167–186 (part A of this Discussion).

Moorhouse, F. W. 1933 *Rep. Gt Barrier Reef Comm.* **4**, 35–36.

Moorhouse, F. W. 1936 *Rep. Gt Barrier Reef Comm.* **4**, 37–44.

Scoffin, T. P. & McLean, R. F. 1978 *Phil. Trans. R. Soc. Lond.* A **291**, 119–138 (part A of this Discussion).

Scoffin, T. P. & Stoddart, D. R. 1978 *Phil. Trans. R. Soc. Lond.* B **284**, 99–122 (this volume).

Spender, M. A. 1930 *Geogrl J.* **76**, 194–214 and 273–297.

Spender, M. A. 1937 *Geogrl J.* **89**, 141–142.

Steers, J. A. 1929 *Geogrl J.* **74**, 232–257 and 341–367.

Steers, J. A. 1930 *Scient. Rep. Gt Barrier Reef Exped. 1928–29,* **3**, 1–15.

Steers, J. A. 1937 *Geogrl J.* **89**, 1–28 and 119–139.

Steers, J. A. 1938 *Rep. Gt Barrier Reef Comm.* **4**, 51–96.

Steers, J. A. 1940 *Geogrl J.* **95**, 30–42.

Stephenson, T. A., Stephenson, A., Tandy, G. & Spender, M. A. 1931 *Scient. Rep. Gt Barrier Reef Exped. 1928–29,* **3**, 17–112.

Stephenson, W., Endean, R. & Bennett, I. 1958 *Aust. J. mar. Freshwat. Res.* **9**, 261–318.

Stoddart, D. R. 1965 *Trans. Inst. Br. Geog.* **36**, 131–147.

Stoddart, D. R., McLean, R. F., Scoffin, T. P. & Gibbs, P. E. 1978 *Phil. Trans. R. Soc. Lond.* B **284**, 63–80 (this volume).

Stoddart, D. R. & Steers, J. A. 1977 In *Biology and geology of coral reefs* (eds O. A. Jones & R. Endean), vol. 4, pp. 59–105. New York: Academic Press.

Teichert, C 1947 *Proc. Linn. Soc. N.S.W.* **71**, 145–196.

Thom, B. G. & Chappell, J. 1978 *Phil. Trans. R. Soc. Lond.* A **291**, 187–194 (part A of this Discussion).

Umbgrove, J. H. F. 1928 *Wet. Meded. Dienst Mijnb. Ned.-Oost-Indië* **7**, 1–68.

Phil. Trans. R. Soc. Lond. B. **284**, 63–80 (1978)   [63]

Printed in Great Britain

Forty-five years of change on low wooded islands, Great Barrier Reef

By D. R. Stoddart,† R. F. McLean,‡ T. P. Scoffin§ and P. E. Gibbs‖

†*Department of Geography, Cambridge University, Downing Place, Cambridge, U.K.*

‡*Department of Biogeography and Geomorphology, Research School of Pacific Studies, Australian National University, Canberra, Australia 2600*

§*Grant Institute of Geology, University of Edinburgh, West Mains Road, Edinburgh EH9 3JW, U.K.*

‖*The Plymouth Laboratory, Marine Biological Association of the U.K., Plymouth PL1 2PB, U.K.*

During the 1928–29 Expedition, centred at Low Isles, Spender mapped the 'low wooded islands' or 'island-reefs' of Low Isles and Three Isles in detail, and additional information was published by Steers, T. A. Stephenson and others. From this work, two different models of the evolution of low wooded islands were proposed, Spender holding that the islands were in a state of equilibrium resulting from their location on the reef, Steers that they could be placed in an evolutionary sequence. Moorhouse described the results of cyclones at Low Isles in 1931 and 1934, and Fairbridge & Teichert reconsidered the general issues following aerial reconnaissance and a brief visit to Low Isles in 1945. Subsequently, aspects of change since 1928–29 have been studied at Low Isles by W. Stephenson, Endean & Bennett in 1954 and by W. Macnae in 1965. Maps produced since 1929, however, have all been based on Spender's surveys. In 1973, Low Isles and Three Isles were remapped in detail, and a direct comparison can now be made over an interval of 45 years. This shows changes in island topography, and substantial alteration in the size and location of shingle ramparts which has affected conditions for coral growth on reef flats. Mangroves have extended greatly at Low Isles, but not at all at Three Isles. The implications of these findings for the general models of Steers and Spender will be discussed and related to the Holocene history of the Great Barrier Reefs.

Introduction

No coral island in the world has been so intensively studied over so long a time as Low Isles, the southernmost low wooded island of the Great Barrier Reef (figure 1). Even before it became the headquarters for the Great Barrier Reef Expedition 1928–29 it had been noticed and described, first by Cook in 1770 (Beaglehole 1955, p. 343), then by King in 1819 (1827, vol. I, p. 207), and by the *Rattlesnake's* naturalist MacGillivray in 1848 (1852, vol. I, pp. 101–103). In 1928–29 it served as the base for Steers's extensive studies of reef islands, and it was mapped in great detail at a scale of 1:5000 by Spender as a basis for the surveys of the biologists (Steers 1929, pp. 251–253; Spender 1930; Stephenson, Stephenson, Tandy & Spender 1931). The first air photographs available were flown in September 1928 at a scale of 1:2400 and were used in these studies.

Subsequently, note was taken of minor geomorphic changes during storms in 1931 and 1934 (Moorhouse 1933, 1936), chiefly by annotating Spender's map. Steers briefly revisited the island in 1936. New air photographs (1:3000) were flown in January 1945, and these, together with a visit on 30 January–4 February 1945, led to a substantial rediscussion of many features by Fairbridge & Teichert (1947, 1948). This reinterpretation rested rather heavily on the air photographs, for their visit was during neap tides and coincided with bad weather during which rainfall totalled 280 mm. Again the changes were annotated on Spender's map.

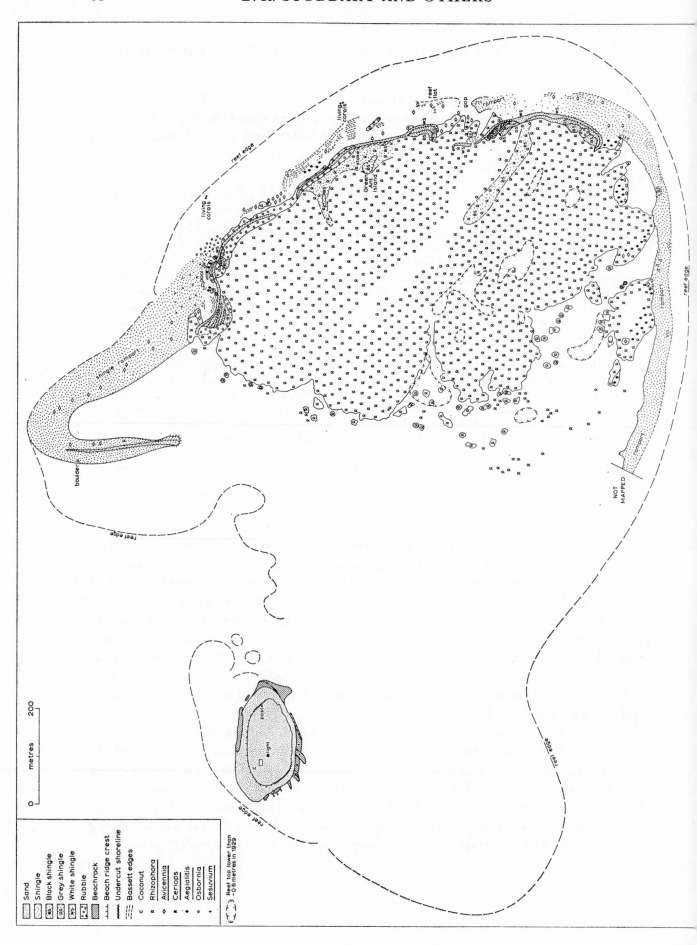

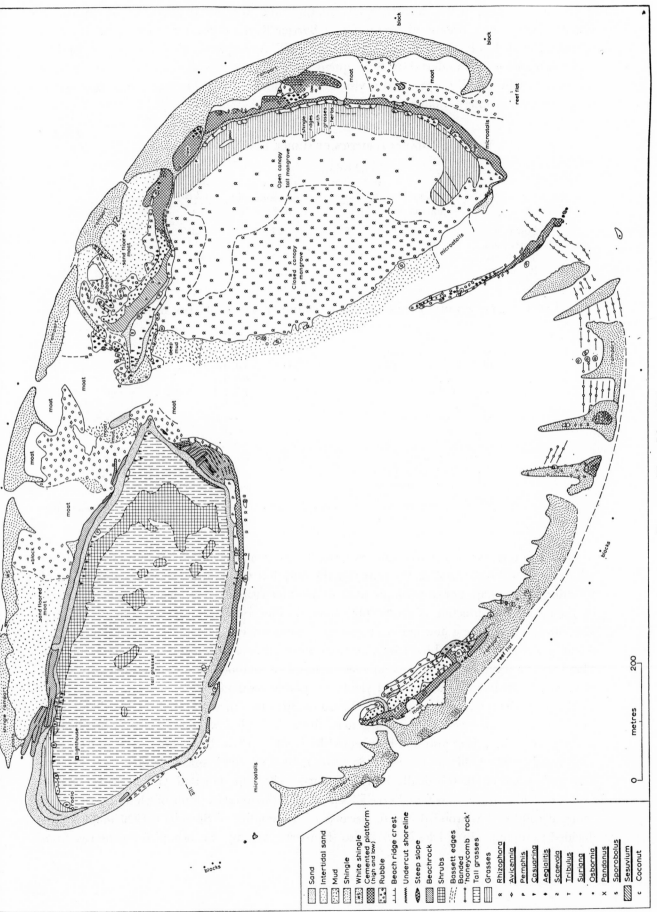

FIGURE 2. Three Isles, 1973.

The next major study was that of Stephenson, Endean & Bennett (1958), during an expedition (12–26 August 1954) organized by the Great Barrier Reef Committee following the cyclone of 1950. Detailed notes were made of changes since T. A. Stephenson's work in 1928–29, but no fresh mapping was undertaken. Subsequent studies included visits to collect marine algae in May 1963 by Cribb (1973), and a visit of a few hours by W. Macnae (1966, pp. 86–88) in March 1965 to observe mangroves.

TABLE 1. CHARACTERISTICS OF SIMPLE LOW WOODED ISLANDS

	area of reef top/ha	Total land area at low water springs ha	Land area as percentage of reef top	Area of sand cay ha	Mangrove area/ha
Low Isles	136	37.5	27.5	2.25	32.3
Three Isles	134	47.1	35.2	15.81	8.2
Two Isles	129	33.1	25.7	19.46	1.5
Pipon Isles	334	72.0	21.6	0.74	62.1

TABLE 2. ELEVATIONS OF SELECTED GEOMORPHIC FEATURES OF LOW WOODED ISLANDS

	Low	Three	Two
highest open-water corals	0.6	0.9	0.4
highest shingle rampart	—	1.1–1.6	1.7
highest living moat corals	—	0.9–1.1	—
lower platform	—	1.7–2.1	1.3–2.1
upper platform	—	2.7–3.1	—
highest old shingle ridge	3.8	—	—
highest mangrove	2.2	1.1	1.8
low terrace on sand cay	—	3.0–3.8	3.4–3.7
high surface on cay	4.2–5.5	4.8–5.9	6.0–6.3
maximum height of cay	5.5	6.0	6.6
height range of beach-rock	1.7–2.7	0.7–2.8	0.6–2.7

Extreme h.w.s. 3.9; m.h.w.s., 2.3; m.h.w.n., 1.6; m.l.w.n., 1.2; m.l.w.s., 0.5; extreme l.w.s., 0. All elevations in metres.

In spite of this level of activity, however, resulting in comprehensive inventories of the biota (Endean 1956; Stephenson & Wells 1956), the only full topographic survey remains that of Spender, and perhaps not surprisingly some of the later visits have yielded contradictory views on the nature and direction of geomorphic changes. The re-survey undertaken in 1973, therefore, aimed to produce a new map from which changes since 1929 could be directly measured. A camp was occupied on Low Isles Cay from 24 to 29 August, during spring tides, and the island was surveyed by tape-and-compass traverse. This method is less rigorous than Spender's original theodolite triangulation, but yields acceptable results for comparison.

Three Isles Reef was also mapped at 1:5000 by Spender (1930), though in less detail than Low Isles. His results were used in the general discussion of low wooded islands by Stephenson et al. (1931). Three Isles also had been noted by Cook in 1770 (Beaglehole 1955, p. 371) and was visited by MacGillivray in 1848 (1852, vol. i, 105 and 106). Agassiz landed there in 1898 and made some characteristically hasty inferences (1898 pp. 113 and 114, pl. 7–11). Steers (1929, p. 256) took the view that Three Isles is perhaps the finest example of what he termed the 'normal case' of low wooded island. Because of the quality of Spender's 1929 map, it was decided to re-map Three Isles in 1973 also (figure 2). A camp was occupied on the cay from

7 to 27 September, though during this time periods were also spent working on Two Isles and Low Wooded Island. The topography was mapped by pace-and-compass traverse.

This paper is confined to an analysis of the geomorphic and ecological changes which have taken place between the detailed surveys at Low Isles and Three Isles in 1928–29 and 1973, together with some references to the very similar features at Two Isles and Pipon Isles, mapped by Steers in 1936 (1938, pp. 75 and 76) and also in 1973. Comparative data are also available on many other low wooded islands, mapped by Steers in 1928–29 and 1936 and again by us in 1973, but these surveys are less detailed and they will be considered elsewhere. Table 1 gives dimensions of the four low wooded islands considered here, and table 2 further geomorphic data; elevations are referred to the Queensland datum (mean lower low water springs).

Problems of interpretation

The results of the re-survey bear directly on the problems of interpretation of the development and history of low wooded islands first identified by Steers (1929, 1930, 1937) and Spender (1930, 1937). Low wooded islands comprise three main elements on the tops of small patch reefs of the Great Barrier Reef Inner Shelf. These are (1) an array of broadly arcuate shingle ramparts round the windward reef rim; (2) a mangrove swamp immediately in the lee of the ramparts; and (3) a leeward sand cay. Low wooded islands differ considerably in the relative extent of these components, the proportion of the reef top occupied by them, and the complexity of the rampart systems. Most low wooded islands have high conglomerate platforms associated with the ramparts; these are present at Three Isles, where they were mapped by Spender, but not at Low Isles, which is to that extent unrepresentative of the class. Steers (1930, 1931, 1937) believed that the conglomerate platforms were derived from lithified shingle ramparts, but owed their present elevation to a slight eustatic fall in sea level. Spender drew no such conclusion; he argued that ramparts advanced as waves of shingle from the reef edge, each reaching a limiting distance before a new one began to form, and that older ramparts then became consolidated. Thus for Steers, though not for Spender, the sequence of ramparts and platforms had historical significance. Steers also believed that the relative extent of mangroves on the reef top (as well as the complexity of the ramparts) indicated the stage of development of the low wooded island complex; Spender thought this view oversimplified, and argued that in at least some cases the postulated sequence might well have been reversed.

These conflicting views were generalized into widely differing interpretations of reef island development and history in the Great Barrier Reef Province. If they can be tested by the detailed remapping of Low Isles, Three Isles, Two Isles and Pipon Islands, then the conclusions drawn at these two locations will have wider significance. We discuss the two reefs together, under the general heads of ramparts, platforms, mangrove swamp, sand cay, and faunistic changes. Figure 1 shows the 1973 survey of Low Isles and figure 2 that of Three Isles. Low has a reef-top area of 136 ha and maximum dimensions of 1.8 and 1.5 km; Three an area of 134 ha and dimensions of 1.7 and 1.2 km; Two an area of 129 ha and dimensions of 1.7 and 0.95 km. At low water, 27.5 % of the reef top is occupied by land (including mangrove) at Low, 35 % at Three, and 26 % at Two.

RAMPARTS

Spender (1930) mapped an almost continuous zone of shingle ramparts round the windward side of the Low Isles Reef. He distinguished an outer rampart, forming 'a low wall, rising at its inner edge perpendicularly from the flat . . . about 2 to 3 feet [0.6–1 m] high' with 'an even, gradual slope' on its seaward side (Spender 1930, p. 197; Stephenson et al. 1931, p. 24). The crest of this outer rampart reached 1.5–2.1 m above datum at Low Isles, compared with less than 1.5 m at Three Isles. In both cases the outer rampart consisted of fresh white shingle. The inner rampart had the same form as the outer, but consisted of blackened shingle with interstices filled with sand and mud; it was colonized by Avicennia, Aegialitis and Sesuvium. Spender interpreted the ramparts as successive waves of shingle driven across the flat by wave action; the rampart moved until it had reached a limiting distance defined by wave strength, and then a new rampart formed to seaward and began migrating inwards. Spender appeared to ascribe rampart formation to ordinary wave conditions rather than to storm events. The average width of the outer rampart at Low Isles in 1929 was about 60 m and the maximum 80 m. At Three Isles the corresponding features was rather less continuous than at Low Isles, especially in the southeast; it also had an average width of 60 m and a maximum width of 70 m. The inner rampart was not readily distinguishable as a distinct entity at Three Isles; at Low Isles it was also about 60 m wide.

In February 1931 heavy southeast winds led to the erosion of the ramparts on the south and southwest sides of Low Isles, with deposition in the southeast. New banks of coral shingle 22 and 30 m long were formed, with steep landward and gentle seaward slopes (Moorhouse 1933). On 12 March 1934 a cyclone pushed the ramparts inwards 23–27.5 m, especially in the north-east. The new banks built in 1931 were flattened, but were rebuilt within two months. These changes had marked effects on coral growth in moats ponded on the landward sides of the ramparts (Moorhouse 1936).

Fairbridge & Teichert (1947, 1948) made a full re-analysis of the Low Isles ramparts in 1945. They distinguished four ramparts. The first (innermost and oldest) was represented by shingle tongues mapped within the mangroves by Spender, but not recognized as a separate rampart by him. The second was equivalent to Spender's inner rampart, and the third to his outer; both in 1945 consisted of blackened and corroded shingle (Fairbridge & Teichert 1948, p. 79). The fourth resembled the 1929 outer rampart in that it consisted of 'fresh, unblackened . . . very light coloured' shingle. It had formed since 1929, in part from the new ridges described in 1931 by Moorhouse. Spender's inner rampart 'had not moved at all' since 1929. The old outer rampart retained much of its original form, but had moved inwards by up to 40 m; on the east side the new fourth rampart was superimposed upon it in two banks each about 250 m long. Taken as a whole the ramparts in 1945 retained their original lateral continuity.

By 1954, however, Stephenson et al. (1958, p. 289) drew attention to the absence of shingle on much of the eastern and southeastern sides of the reef, from the level of high water neaps to below low water springs: the former presence of ramparts was indicated only by consolidated bassett edges. Mapping in 1973 confirmed this change, which probably resulted from erosion during the 1950 cyclone. The reef-flat ramparts have disappeared from most of the eastern side of the mangrove island; extensive bassett edges indicate their former locations. The edge of the mangrove itself is lined by a largely symmetrical breastwork of white shingle, overlapping the landward edges of Spender's old inner rampart. The total length of this breastwork is about 1100 m, divided into three main sections. Ordinary ramparts similar to those described by Spender still exist in the south and extreme north, though the northern spit has changed its shape and location; these ramparts are 25–40 m wide. Seaward of the ramparts is a gently sloping reef flat, covered with an algal turf, averaging 40–50 m in width, and with patches of algally bonded shingle. The surviving old dark shingle ridges inside the breastwork are generally

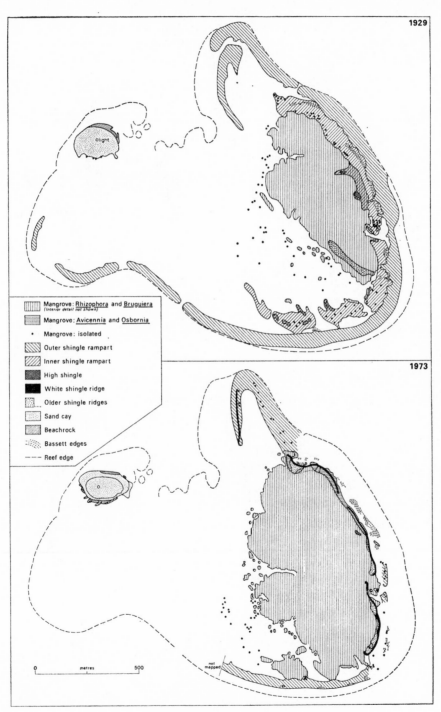

FIGURE 3. Topographic change at Low Isles, 1929–1973.

20–35 m wide. This major change in rampart form and location presumably results from storm activity, and indicates that rampart evolution is a less regular process than Spender thought. It also shows that superimposition of ridges occurs where their forward movement is checked by the presence of dense mangrove woodland.

Re-mapping at Three Isles (figure 4) shows a very similar situation. A large area of outer rampart has disappeared at the southeast point, leaving bare reef flat. The remainder of the

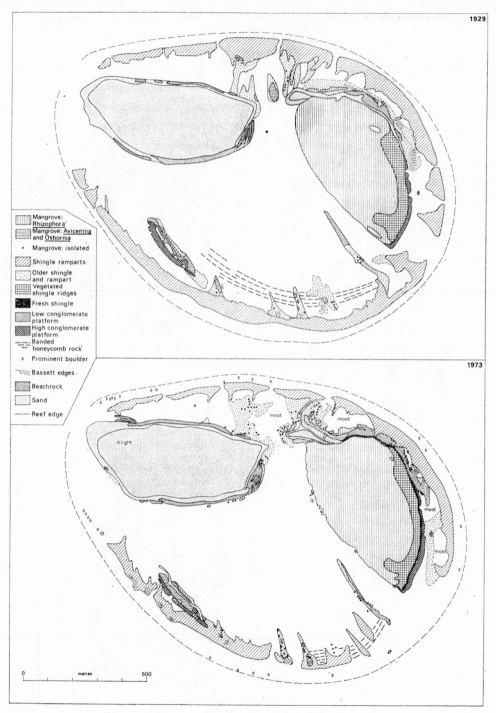

FIGURE 4. Topographic change at Three Isles, 1929–1973.

rampart retains its general form, though with many more gaps (resulting in draining of moats), but it has also moved considerably forward. This is particularly noticeable at Third Island, where in 1929 the rampart was 60–90 m seaward of the shore and now impinges directly on it. These changes in the inner edge of the rampart are detailed in figure 5. There is no indication at Three Isles of the kind of cyclic development suggested by Spender, but rather a continuing process of redistribution of shingle, interrupted by major storm events which locally strip the reef flats.

Re-mapping at Two Isles also shows major changes in the location of the inner edge of the rampart between 1936 and 1973, especially in the north.

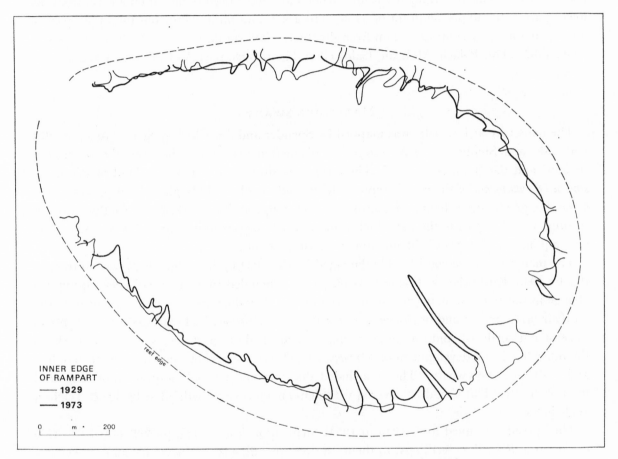

INNER EDGE
OF RAMPART
—— 1929
—— 1973

0   m ·   200

FIGURE 5. Changes in the Three Isles ramparts, 1929–1973.

PLATFORM

Spender mapped two conglomerate platforms at Three Isles, a high and a low, with a height difference between them of 1.7 m; the high platform ('promenade') was nearly 400 m long (Stephenson *et al.* 1931, p. 30). It was surmounted by 'an accumulation of shingle, sand and pumice forming a dune-like bank about 50 yds wide and probably as much as 20 ft [6 m] above datum', which had no real counterpart at Low Isles. Some dissection of this upper platform had occurred by 1973, especially on Third Island, but otherwise the features remained as mapped by Spender. The height range of the lower platform was determined as 1.7–2.1 m, and of the upper 2.7–3.1 m. Where levelled, the shingle bank reached a maximum elevation of 1.6 m above the upper platform, i.e. to 4.7 m. A less well marked platform at Two Isles ranged between 1.3 and 2.1 m; the rock outcrop is 13–25 m wide, and the band of shingle ridges above it is up to 100 m wide. In the case of Three Isles, both Agassiz (1898, pp. 113 and 114) and Spender (1930, pp. 207–210) thought the platform had formerly been more extensive round the windward side of the reefs.

Spender argued from the concordant heights of the Three Isles upper platform and the crest of the Low Isles shingle ramparts that the former was a lithified rampart eroded and cliffed on its seaward side. The lower platform, which frequently breaks down into bassett edges, is more clearly a cemented rampart. Radiocarbon dates have been obtained on a coral from the lower part of the upper platform of 3750 ± 110 a B.P. and on a clam from the upper part of 3050 ± 70 a B.P.; by contrast a clam from the lower platform dates at 1460 ± 70 a B.P. (ANU-1380, 1382, 1475: Polach, McLean, Caldwell & Thom 1978, part A of this Discussion).

Mangrove swamp

The mangrove at Low Isles was mapped by Spender and described by Steers (1929, pp. 254 and 255) and Stephenson et al. (1931). A distinction was drawn between the 'mangrove swamp' and the 'mangrove park'. The former consisted of a dense woodland of *Rhizophora stylosa* (*R. mucronata* of the earlier reports) with sandy pools, muddy glades and passages, and some shingle tongues extending inwards from the ramparts. The park, to the southwest of the swamp, was 'that part of the flat which is more or less successfully colonised by outlying trees or samplings of *Rhizophora*' (Stephenson et al. 1931, p. 50).

The mangroves were considerably damaged by the 1934 cyclone, but there was no trace of this by 1945. Fairbridge & Teichert (1948, p. 76) concluded that there had been no marked change in the extent of mangrove in the swamp, but they documented the spread of *Bruguiera* on their first rampart and of *Avicennia* on their second. *Avicennia* had extended considerably in the east and northeast, and was also growing on their third rampart. In the mangrove park, on the other hand, the development of *Rhizophora* was 'very noticeable', patches of it 'expanding and coalescing everywhere'. They concluded that it would be 'only a question of time until most of the Reef Flat would be covered with dense mangrove growth' (Fairbridge & Teichert 1948, p. 85).

This spread continued from 1945 to 1954: Stephenson et al. (1958, p. 309) noted the wide extension of *Rhizophora* seedlings: to them, 'a slow spreading seems inevitable and . . . eventually the entire flat may be converted into a mangrove swamp'. Nevertheless, in spite of these remarks, Macnae (1966, p. 88) took the view that no significant changes had taken place between 1929 and his visit in 1964; however, he carried out no surveys and made no measurements.

The 1973 map confirms the trends identified in 1945 and 1954 and disproves Macnae's contention. Planimetric measurement (figure 6) gives a total mangrove area of 21.9 ha in 1929 and 36.5 ha in 1973, an increase of 67 %. The 1929 swamp area remains as it was, with the same identifiable pools and channels. Much of the mangrove park, however, has become continuous woodland, though lower and less dense than that of the 1929 swamp. Expansion has been particularly marked along the old ramparts in the south. The present mangrove margin closely follows the shoal areas contoured by Spender in 1929. Spender had suggested that the deeper reef-top pools were erosional and resulted from chemical rotting during the former presence of mangroves (Spender 1930, pp. 8 and 290; also Fairbridge & Teichert 1948, p. 82); he also suggested (1937) that mangroves were retreating from a former wider extent on the southern ramparts. Such retreat on the ramparts could perhaps have resulted from storm activity; but the evidence of change since 1929 clearly shows expansion, not contraction, of the mangrove cover, and also indicates that the pattern of expansion is a function of pre-existing reef-top topography.

The position at Three Isles is quite different: within the limits of the mapping method there has been no measurable change in the extent of the mangrove swamp since 1929; it has remained constant at 8.2 ha. No reason can be given for this difference, though it might be noted that in

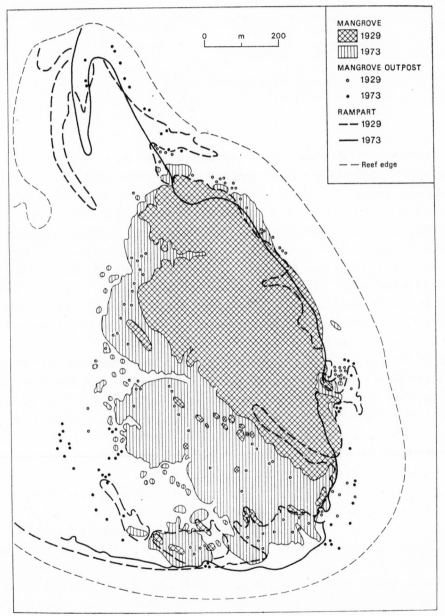

FIGURE 6. Change in the mangrove swamp and ramparts, Low Isles, 1929–1973.

1929 the edge of the swamp was a wall of tall mangrove, as in 1973, with no spread of seedlings or 'mangrove park' on the flat. Spender (1930, p. 207) himself remarked on this contrast between Low Isles and Three Isles, and suggested that mangrove growth might be limited by factors intrinsic to each particular low wooded island.

Studies in 1973 gave the most complete collection to date of Low and Three Isles mangroves, with information on their distribution. Several species now common have not previously attracted attention. Table 3 lists the species collected at these islands.

TABLE 3. MANGROVES RECORDED ON SELECTED LOW WOODED ISLANDS IN 1973

	Low	Three	Two	Pipon
Aegialitis annulata R.Br.	×	×	×	×
Aegiceras corniculatum (L.) Blanco	×	—	—	—
Avicennia marina (Forst.) Vierh.	×	×	×	×
Bruguiera exaristata Ding Hou	—	×	×	—
Bruguiera gymnorrhiza Lam.	×	—	×	—
Ceriops tagal (Perr.) C.B.Rob.	×	×	×	×
Excoecaria agallocha L.	×	—	—	—
Lumnitzera racemosa Willd.	—	—	—	×
Osbornia octodonta F.v.M.	×	×	×	×
Rhizophora stylosa Griff.	×	×	×	×
Sonneratia alba J.E.Sm.	—	×	×	×

Rhizophora stylosa is the dominant species, forming a closed forest up to 20 m tall. It is also the pioneer species on the sheltered reef top, and large numbers of seedlings grow in shallow moats enclosed by ramparts; because of the large tidal range these are often more than 1 m tall. This is the species referred to as *R. mucronata* by Spender (1930), Stephenson *et al.* (1931) and Stephenson *et al.* (1958). At higher levels immediately in the lee of the ramparts, especially near Green Ant Island at Low Isles, *Rhizophora* gives way to tall woodland of *Bruguiera gymnor-rhiza* (the *B. rheedi* of earlier reports) and *Ceriops tagal*, as described by Fairbridge & Teichert (1948) and Macnae (1966). Macnae also records *Excoecaria agallocha* for this community at Low Isles but this was not collected in 1973. Neither *Sonneratia* nor *Xylocarpus* have been found at Low Isles, though *Aegiceras corniculatum* is present on Green Ant Island, the only location of this species on a reef island of the northern Great Barrier Reef.

The mangrove vegetation of the ramparts is quite different. Most species form dwarf trees and shrubs, except for a tall and extremely dense woodland of *Osbornia octodonta* on the north-eastern flank of the forest. *Osbornia* was not mentioned at Low Isles in 1928–29, and it may possibly have colonized the reef since then; it was recorded at that time at Three Isles. Macnae referred to it in 1964 as a colonizer, with *Pemphis* and *Thespesia*, on rampart crests. In 1973 the rampart had either been removed by erosion or had been pushed inland through the *Osbornia* fringe, so that the trees were growing directly on the reef flat.

The lower ramparts have a characteristic mangrove vegetation dominated by *Aegialitis annulata* and *Avicennia marina* (the *A. officinalis* and *A. eucalyptifolia* of Spender and T. A. Stephenson). Macnae (1966, p. 88) stated that *Aegialitis* had colonized Low Isles since 1929, but this is in error; it was mentioned there by Steers (1929, p. 254), Spender (1930, p. 199) and Stephenson *et al.* (1931, p. 60). Though it can form tall trees, it usually occurs at discrete clumps of shrubs 1–1½ m tall on older ramparts and bassett edges. *Avicennia marina* is found in similar situations, and rarely exceeds 2 m in height on outer ramparts. Both species can be exposed to severe wave action during high tides and rough weather.

The mangroves of Three Isles are similar, except that *Osbornia* is more extensive and there are large rampart areas of *Aegialitis* and *Avicennia*. Both *Sonneratia alba* and *Bruguiera exaristata* were found at Three but not at Low Isles.

The mangrove associates *Pemphis acidula*, common on conglomerate platforms of low wooded islands, is inconspicuous at Low but extensive on the eastern side of Three Isles. Similarly, presumably because of the absence of suitable substrate, succulent halophytes such as *Suaeda*, *Arthrocnemum* and *Salicornia* are much less extensive at Low Isles than at Three Isles and other low wooded islands.

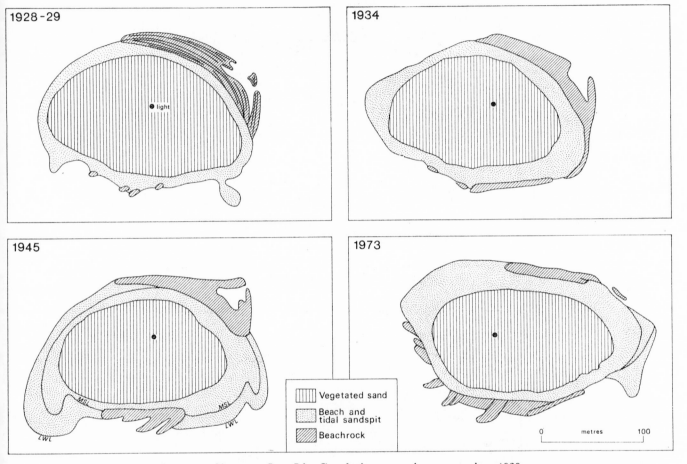

FIGURE 7. Change at Low Isles Cay during successive surveys since 1929.

SAND CAY

The sand cays of the two reefs differ markedly in size: that on Low covers 2.25 ha compared with 15.8 ha on Three. Spender's map of 1929 may be compared with Moorhouse's annotations in 1934, Fairbridge & Teichert's air photograph interpretation in 1945, and the 1973 survey (figure 7). Relative location is established with reference to the lighthouse and the distinctive beach-rock outcrops. In 1928–29 the main beach-rock outcropped in the north, in 1973 in the south. If these maps are superimposed (figure 8) a picture of the range in variation of cay form and the total extent of the beach-rock is obtained (for comments on changes in beach-rock outcrops at other times, see Steers (1937, p. 134), Fairbridge & Teichert (1947, 1948), and Stephenson *et al.* (1958, p. 273)). The changes appear to be oscillatory. In part they are doubtless seasonal, in part responses to random storm events. Part of the response may result from changes in the vegetation cover: MacGillivray (1852, vol. I, p. 101) found the cay 'well wooded' in 1848, but it has long been settled and the vegetation much disturbed.

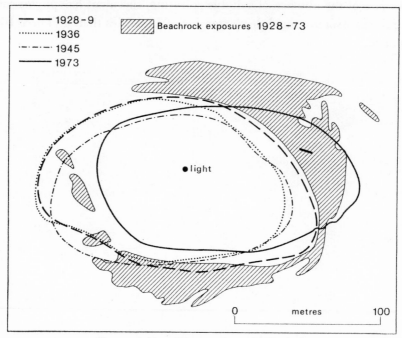

FIGURE 8. Low Isles sand cay, 1929–1973.

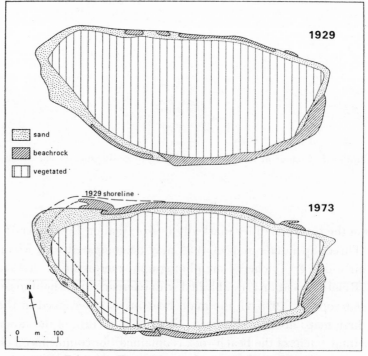

FIGURE 9. Three Isles sand cay, 1929 and 1973.

Changes at Three Isles are less obvious but more interesting. By 1929 there had been considerable erosion at the east end of the cay, exposing a complex array of beach-rock in front of a tall receding cliff. Since then, more beach-rock has become exposed along the north shore. However, there has also been considerable aggradation on the western end. The 1929 shore can be traced as the backslope of an aggradation terrace covered with pioneer vegetation. Where this backslope reaches the south shore, massive old beach-rock is exposed; it is probably continuous beneath the terrace to the north shore (figure 9). In contrast the beach-rock on the terrace shores is weakly developed and at lower elevations in the intertidal. It is not clear what has triggered this phase of aggradation, nor to what extent it may in part pre-date the 1929 survey.

Similarly, at Two Isles there has been considerable aggradation in the form of low terraces at the east and west end of the cay since 1936, and beach-rock patterns have changed correspondingly (Steers 1938, pp. 75–76): the steep slopes separating the low terrace from the higher densely wooded part of the cay in 1973 coincide almost exactly with the shoreline as mapped by Steers in 1936.

Faunistic changes on the sand flat

Two major faunistic surveys have been carried out at Low Isles: by Stephenson *et al.* (1931) and by Stephenson *et al.* (1958). In view of the geomorphic changes since 1929, we were therefore concerned to identify changes in the sand flat fauna during this period. The sand flat is an extensive area of muddy sand mixed with coral shingle (Flood & Scoffin 1978, part A of this Discussion), extending to a depth of 1.5 m (Marshall & Orr, 1931) and colonized by various marine angiosperms, notably *Thalassia hemprichii* (Ehren.) Aschers., together with *Halodule uninervis* (Forssk.) Aschers., *Halophila ovalis* (R. Br.) Hook., and *Enhalus acoroides* (L.f.) Royle. An account of the fauna collected in 1973 is given elsewhere (Gibbs 1978, this volume). A comparison of the relative abundance and species composition found during this survey with those described in the earlier surveys is difficult for a number of reasons (cf. Stephenson *et al.* 1958), but primarily because the fauna is very diverse in species, many of which occur only as scattered individuals, relatively few being abundant. Also, the earlier accounts are rather brief, the sand flat fauna being investigated as part of broad surveys of the whole reef flat. However, confining comments to the more conspicuous species of the macrofauna, it may be noted that the most common of the surface-burrowing gastropods are still *Cerithium vertagus* L. and *Nassarius coronatus* (Bruguière) together with *Rhinoclavis asper* (L.), *Strombus gibberulus* L. and *Polinices pyriformis* (Recluz). Other nassariids, strombids and naticids are present in fewer numbers along with *Terebra* spp. and *Oliva* spp. Ubiquitous echinoderms such as *Archaster typicus* Müller and Troschel and *Holothuria atra* Jaeger remain common.

The burrowing species composing the infauna are more varied. The presence of certain species can be seen from characteristic disturbances of the sand surface; these species include the enteropneusts *Balanoglossus carnosus* (Willey) and *B. australiensis* Hill (= *Ptychodera flava* Eschscholtz of earlier accounts?) which seem to be as locally abundant as formerly reported, as are the large stomatopod *Lysiosquilla maculata* (Fabricius) and the burrowing holothurian *Holothuria arenicola* Semper. Curiously, another burrowing crustacean, *Axius* (*Neaxius*) sp. cf. *plectrorhynchus* Strahl, which is quite common on the flat, does not appear in the earlier collections (see McNeill 1968). Numerically, the infauna is still dominated by the actiniarian *Edwardsia* (chiefly *E. gilbertensis* Carlgren) and chaetopterid polychaetes. Of the latter, only *Mesochaetopterus sagittarius* Claparède (= *M. minutus* Potts) was found in the earlier surveys, but in 1973 two other genera, *Phyllochaetopterus* and *Spiochaetopterus*, were also very common, as was the terebellid *Pista typha* Grube. On the other hand, sipunculans and *Phyllodoce malmgreni* Gravier were reported as being very plentiful in sand in 1928, but few of the former and none of the latter could be found in 1973.

Although these differences in the relative abundance of certain of the more conspicuous species become apparent when comparing the results of the several surveys, they may reflect short-term fluctuations in populations rather than major changes, and in general the fauna of the Low Isles 'sand flat' appears to have altered little since 1928. However, it should be pointed out that a common sublittoral spatangoid, *Maretia planulata* (Lam.), may have colonized the intertidal deposits adjacent to the Anchorage on Low Isles between 1928–29 and 1954 (Endean 1956), and was still common in 1973 in this same area, along with the clypeastroid *Laganum depressum* Lesson. Spatangoids are noted for their particular substrate preferences (Nichols 1959), and the colonization by *M. planulata* of the tidal flat adjacent to the Anchorage between 1928 and 1954 possibly reflects a subtle change in the character of the sediments in this area during this period.

IMPLICATIONS

These new data indicate that geomorphic processes at Low Isles are less regular and predictable than either Steers or Spender proposed. So far as unconsolidated features such as the inner and outer ramparts first described at Low Isles are concerned, it is clear (*a*) that they do not have historical significance in the sense of indicating specific sea level events, and (*b*) that their formation and destruction takes place irregularly but in geological terms rapidly, in response to random storm events. Mangrove growth on the reef top may indicate a process of extension over time, but the time scale which calibrates it varies widely from reef to reef: in this sense the mangroves are opportunistic colonizers of a substrate of varying suitability. Hence while we agree with Fairbridge & Teichert's conclusion that 'these various reef types present stages in an evolutionary process which led inevitably from the submerged reef stage, through the formation of a sand-cay and the establishment of a mangrove forest on the rampart-protected Reef Flat, to the end stage at which the entire reef platform is covered with vegetation' (1948, p. 85), this concept has little explanatory value in comparing low wooded islands because of the variations in physical conditions between them and the differences in rate parameters that these imply.

Certain basic questions raised by Steers and Spender at Low Isles deserve comment. Which forms first, the sand cay or the ramparts? Is the development of the swamp contingent on the existence of the ramparts? It is likely that the formation of a *Rhizophora* swamp close to the windward edge of a reef could not take place without the protection of the ramparts, as Spender (1930, p. 201) concluded: on islands such as Murdoch and Hampton, where there is *Rhizophora* without ramparts, the swamp is on the lee end of the reef. The only mangroves normally found without the protection of ramparts are *Aegialitis* and *Avicennia*, and these are usually rooted in old rampart conglomerate and rubble. However, the changes at Low Isles show that the existence of the swamp, especially at the high levels to which it can build, checks the inward movement of shingle and concentrates it in a zone of banded breastworks. These usually support a dry-land scrub of *Thespesia*, *Pandanus* and other trees and shrubs. Without the physical barrier of the swamp it is likely that the shingle would form more diffuse and ephemeral constructional features.

It is difficult to find compelling evidence for Spender's suggestion that reef-top mangroves might be decreasing rather than increasing in area, and that 'this might have happened at Low Isles and has happened at Three Isles' (1937, p. 142). At Low Isles the suggestion comes from King's statement (1827, vol. I, p. 207) that in 1819 there were 'distinctly three' islands, confirmed by MacGillivray (1852, vol I, p. 101), two of them being 'merely groves of mangrove on the reef'. Spender suggested that one of these groves, now disappeared, formerly existed on the inner rampart at its southernmost location; this is in fact one of the areas extensively colonized since 1929. The deeper reef-top holes, rather than being eroded by organic acids associated with former mangroves, more probably result from inequalities of reef formation which themselves control the location of mangrove growth. Spender's suggestion at Three Isles comes from his inferred fragmentation of previously more continuous ramparts, giving greater exposure on the flat and presumed regression of the mangroves; there seems to be no other direct evidence for former mangrove extension.

Whether the sand cay forms before or after the ramparts is probably not a question to which there is a general answer. On any reef the formation of ramparts and other features must affect patterns of wave refraction and water movement over the reef top, and the cay presumably adjusts to such changes. More important, as a greater proportion of the reef top is occupied by ramparts and mangroves, production of calcium carbonate sand must decline (McLean & Stoddart 1978, part A of this Discussion). Radiometric dates on many islands indicate that most cay sands are about 3000 years old; one from the Three Isles cay gives an age of 3220 ± 80 a B.P. (ANU-1414). It seems that cays formed rapidly in leeward situations once the level of the reef top and of sea level coincided, but that sediment supply subsequently diminished. The existence of low terraces round the Three Isles cay could possibly partly result from this decrease in rate of sand supply as well as from changes in wave pattern and possible slight negative shifts in sea level.

The differing degree of development of conglomerate platforms on different reefs requires comment. There is a well developed 'upper platform' reaching 3.1 m at Three Isles, and an extensive 'lower platform' reaching 2.1 m. There is a platform at Two Isles within the height range of the Three Isles lower platform. There are no platforms at all on Low Isles. If the platforms indicate successively lower stands of sea level, as Steers (1931) suggested, it is difficult to see why there should be these differences on such otherwise similar reef tops.

CONCLUSION

The sequence of changes outlined here illustrates the complexity of the recent history of low wooded islands and the difficulties of proposing simple models of their evolution. The Queensland examples differ considerably from those of Djakarta Bay (Umbgrove 1929) where the sand cay occupies a much greater proportion of the flat and where the ramparts are much more mobile features (Vertappen 1954). In this Indonesian case Umbgrove suggested that the size of the cay might provide an index of the stage of topographic development of the complex. We have already discussed reasons why stage of mangrove development does not provide a satisfactory index of development. Steers (1937, p. 145) suggested that the width and extent of ramparts might prove a more useful indicator, but our evidence suggests that in this stage of our knowledge such indexes would be simplistic. Continued monitoring at Low Isles and Three Isles will hopefully continue to throw light on present processes and rates of change, which which ultimately could be used to calibrate the recent geomorphic record.

REFERENCES (Stoddart *et al.*)

Agassiz, A. 1898 *Bull. Mus. comp. Zool. Harvard Coll.* **28** (4), 95–148.

Beaglehole, J. C., ed. 1955 *The journals of Captain James Cook on his voyages of discovery*, vol. 1. Cambridge: Hakluyt Society.

Cribb, A. B. 1973 In *Biology and geology of coral reefs* (eds O. A. Jones & R. Endean) vol. 2 (Biology 1), pp. 47–75. New York: Academic Press.

Endean, R. 1956 *Pap. Dep. Zool. Univ. Qd* **1**, 121–140.

Fairbridge, R. W. & Teichert, C. 1947 *Rep. Gt Barrier Reef Comm.* **6**, 1–16.

Fairbridge, R. W. & Teichert, C. 1948 *Geogrl. J.* **111**, 67–88.

Flood, P. G. & Scoffin, T. P. 1978 *Phil. Trans. R. Soc. Lond.* A **291**, 55–71 (part A of this Discussion).

Gibbs, P. E. 1978 *Phil. Trans. R. Soc. Lond.* B **284**, 81–97 (this volume).

King, P. P. 1827 *Narrative of a survey of the intertropical and western coasts of Australia. Performed between the years 1818 and 1822.* (2 volumes, 451 and 637 pages.) London: J. Murray.

MacGillivray, J. 1852 *Narrative of the voyage of H.M.S.* Rattlesnake ..., *during the years 1846–1850.* (2 volumes, 402 and 395 pages.) London: T. and W. Boone.

Macnae, W. 1966 *Aust. J. Bot.* **14**, 67–104.

Marshall, S. M. & Orr, A. P. 1931 *Scient. Rep. Gt Barrier Reef Exped. 1928–29* **1**, 93–133.

McLean, R. F. & Stoddart, D. R. 1978 *Phil. Trans. R. Soc. Lond.* A **291**, 101–117 (part A of this Discussion).

McNeill, F. A. 1968 *Scient. Rep. Gt Barrier Reef Exped. 1928–29*, 1–98.

Moorhouse, F. W. 1933 *Rep. Gt Barrier Reef Comm.* **4**, 35–36.

Moorhouse, F. W. 1936 *Rep. Gt Barrier Reef Comm.* **4**, 37–44.

Nichols, D. 1959 *Phil. Trans. R. Soc. Lond.* B **242**, 347–437.

Polach, H. A., McLean, R. F., Caldwell, J. R. & Thom, B. G. 1978 *Phil. Trans. R. Soc. Lond.* A **291**, 139–158 (part A of this Discussion).

Spender, M. A. 1930 *Geogrl J.* **76**, 194–214 and 273–297.

Spender, M. A. 1937 *Geogrl J.* **89**, 141–142.

Steers, J. A. 1929 *Geogrl J.* **74**, 232–257 and 341–367; discussion, 367–370.

Steers, J. A. 1930 *Scient Rep. Gt Barrier Reef Exped. 1928–29* **3**, 1–15.

Steers, J. A. 1931 *C. r. Congr. géog. internat.* **2**, 164–173.

Steers, J. A. 1937 *Geogrl J.* **89**, 1–28 and 119–139; discussion, 140–146.

Steers, J. A. 1938 *Rep. Gt Barrier Reef Comm.* **4**, 51–96.

Stephenson, T. A., Stephenson, A., Tandy, G. & Spender, M. A. 1931 *Scient. Rep. Gt Barrier Reef Exped. 1928–29* **3**, 17–112.

Stephenson, W., Endean, R. & Bennett, I. 1958 *Aust. J. mar. Freshw. Res.* **9**, 261–318.

Stephenson, W. & Wells, J. W. 1956 *Pap. Dept. Zool. Univ. Qd.* **1** (4), 1–59.

Umbgrove, J. H. F. 1928 *Wet. Meded. Dienst Mijnb. Ned.-Oost-Indië* **7**, 1–68.

Verstappen, H. T. 1954 *Am. J. Sci.* **252**, 428–435.

Phil. Trans. R. Soc. Lond. B. **284**, 81–97 (1978)   [81]
Printed in Great Britain

Macrofauna of the intertidal sand flats on low wooded islands, northern Great Barrier Reef

By P. E. Gibbs

The Plymouth Laboratory, Marine Biological Association of the U.K., Plymouth PL1 2PB, U.K.

[Plates 1 and 2]

An account of the macrofauna inhabiting the mobile substrata of the intertidal zone on low wooded islands in the Northern Region of the Great Barrier Reef Province is given. Excluding mangrove habitats, three main sediment–fauna types can be recognized: (i) well sorted sands forming the sloping beaches of sand cays, colonized by *Ocypode*, mesodesmatids and hippids; (ii) muddy sand flats supporting a very diverse fauna dominated in most areas by *Edwardsia* and chaetopterids; and (iii) areas of fine deposit, characterized by *Uca*, *Gafrarium* and *Marphysa*.

Some associations of animals, chiefly involving commensal polychaetes and bivalves, are described.

Introduction

A considerable amount of taxonomic and faunistic work has been carried out on the marine fauna of the Great Barrier Reef Province (early studies are reviewed by Dall & Stephenson 1953). Zoogeographically, the province, extending from about 9° 30' S to 25 °S, is important due to its proximity to the Indo-Malayan area which is regarded as the faunistic centre of the Indo–West Pacific region (Ekman 1953). While our knowledge of the coral fauna and associated reef assemblages of such groups as the molluscs, echinoderms and fishes of the province is fairly extensive, other groups, such as the polychaetes, are poorly known. There is also a major gap in information concerning the detailed composition of the communities inhabiting the mobile substrata of both the littoral and sublittoral zones. The 1928–29 and 1954 Expeditions to Low Isles in the Northern Region (see Yonge 1930; Stephenson, Stephenson, Tandy & Spender 1931; Stephenson, Endean & Bennett 1958) carried out wide investigations of the reef's habitats and the resulting observations on the fauna of the 'sand flat' provide a useful basis for further ecological and faunistic studies.

During the second half of 1973, the Royal Society – Universities of Queensland Great Barrier Reef Expedition carried out geomorphological surveys of the low wooded islands north of 17° S, the latitude of Cairns (see Stoddart 1978). As a complementary study the author surveyed the communities inhabiting the intertidal sedimentary deposits on these islands. Although other islands and reefs were visited, this account is based chiefly on the observations made on low wooded islands between latitudes 14° 30' S and 17° 00' S, namely Low Isles, 16° 23' S, 145° 34' E (period of visit 23–30 August); East Hope Island, 15° 44' S, 145° 27' E (30 August–6 September); Three Isles, 15° 07' S, 145° 25' E (6–28 September); Two Isles, 15° 01' S, 145° 27' E (20–22 September); Nymph Island, 14° 39' S, 145° 15' E (9–10 October) and Ingram Island, 14° 30' S, 144° 53' E (22–23 October). In this area the tidal range is about 3.0 m at springs and 1.0 m at neaps (values for Cairns, Tide Tables, Department of Harbours and Marine, Brisbane).

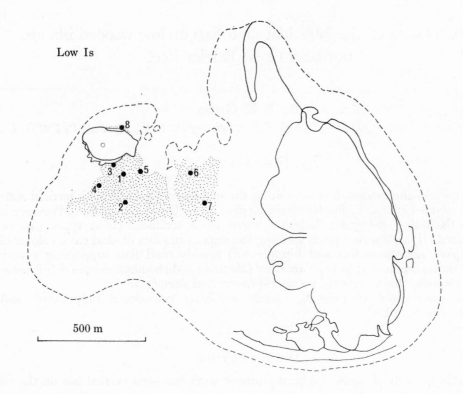

Low Is

500 m

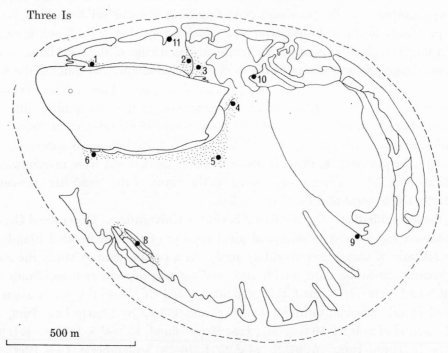

Three Is

500 m

FIGURE 1. Maps showing locations of stations on Low Is (L1–L8) and Three Is (3.1–3.11). Outline features are based on the surveys of D. R. Stoddart.

The geomorphological features of low wooded islands have been described by Steers (1929, 1930), Spender (1930) and Stephenson et al. (1931) (see also papers in this volume and part A of this Discussion). Typically, the central part of the reef complex, usually termed the reef flat (or 'pseudolagoon'), is covered by sediments composed chiefly of calcareous skeletal detritus derived from various reef organisms, notably corals, forams and algae (see Maxwell 1973; Flood & Scoffin 1978, part A of this Discussion), often with an admixture of fine material of indeterminate origin but including plant remains. The grading of the surface deposits is very variable, but generally black muds occur around the mangroves, poorly sorted muddy sands, often with a high proportion of coarse coral fragments, cover the reef flat, and well sorted, medium to coarse sands form the beaches of the sand cay. The depth of sediment on the reef flat varies from a few centimetres in hollows between areas of exposed rock, to 5 m or more (see Marshall & Orr 1931). On most low wooded islands, sea grasses colonize the sand flats, chiefly *Thalassia hemprichii* (Ehrenb.) Aschers. with *Enhalus acoroides* (L.f.) Royle, *Halophila* spp. and *Halodule* spp. (see Den Hartog 1970).

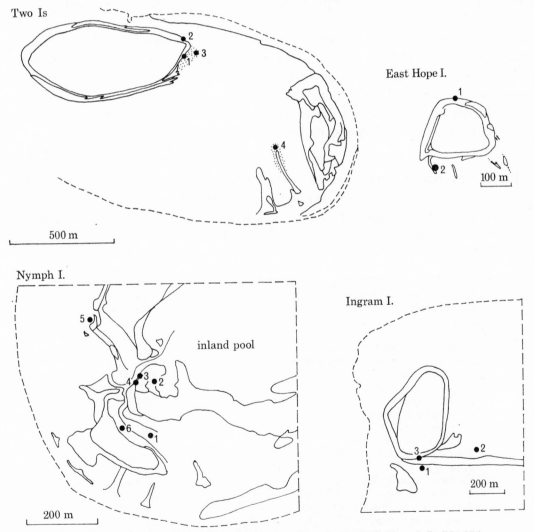

FIGURE 2. Maps showing locations of stations on Two Is (2.1–2.4),Nymph I. (N1–N6), East Hope I. (EH1–EH2) and Ingram I. (IN1–IN3).

Qualitative samples of the macrofauna were collected in selected areas on the sand flats by digging, specimens being picked out by hand. Although extraction by sieving with 1.0 or 2.0 mm diameter mesh was possible in some areas, generally the presence of coarse coral debris in the deposits precluded this method. Foraminifera, which often compose a significant proportion of the deposit, were not collected since the reef forms are generally well known (see Collins 1958; Maxwell 1968). Drainage of the central reef flat area is often limited by the underlying rock formation and surface pools remain throughout the period of exposure. The resulting high water table makes excavation of the sediment to any appreciable depth impossible, and thus it is probable that some of the deeper burrowing species are not represented in the collections. However, the presence of some of these species can be recognized by their characteristic burrows, for example *Lysiosquilla maculata*. Where the sediment cover is thin, sampling is further complicated by the burrowing of many species into fissures in the underlying rock.

The positions of the infauna sampling stations are shown in figures 1 and 2. The surface burrowing forms, chiefly gastropods, were collected over wider areas, especially where their presence was indicated by surface tracks. Examination of material along the strand-lines on sand cays provided evidence of the presence of some rarer or inaccessible species, particularly deep-burrowing bivalves.

A systematic list of the species collected is given in table 1. Taxonomic work on the collections is still in progress and thus, at this stage, some names are provisional. Accounts of little-known species and groups are in the course of preparation. The major parts of the collections have been, or will be, deposited as follows: British Museum (Natural History) – Polychaeta, Crustacea, Echinodermata, Pisces; Australian Museum – Gastropoda, Galeommatidae; Western Australian Museum – Lamellibranchia; South Australian Museum – Sipuncula.

TABLE 1. SYSTEMATIC LIST OF SPECIES COLLECTED IN THE SURVEY AREA FROM INTERTIDAL SEDIMENTARY DEPOSITS

See figure 1 for positions of stations. Generally distributed species are designated gen; species newly recorded for the province are marked †; mollusc species marked * or stations [ ] indicate records based on shells only.

Coelenterata	
Anthozoa	
Actinodendron plumosum Haddon	gen
Edwardsia gilbertensis Carlgren	gen
Edwardsia stephensoni Carlgren	3.3 IN1
Annelida	
Polychaeta	
Gastrolepidia clavigera Schmarda	gen
† *Lepidasthenia* sp. cf. *terrareginae* Monro	L2
† *Sthenelais zeylanica* Willey	L1 L6 L7
† *Bhawania pottsiana* Horst	L6 IN1
Eurythoe complanata (Pallas)	3.2 3.3
† *Pseudeurythoe paucibranchiata* Fauvel	L6 IN1
† *Phyllodoce* sp. cf. *fristedti* Bergström	3.1
Leocrates chinensis Kinberg	L1 EH2 3.3
† *Ophiodromus* sp. cf. *berrisfordi* Day	gen
† Syllid spp.	gen
Ceratonereis mirabilis Kinberg	3.3 3.11
† *Nereis* (*Neanthes*) *caudata* (delle Chiaje)	3.5
Perinereis cultrifera helleri Grube	3.3 3.11
Perinereis nuntia brevicirris (Grube)	IN3
Glycera gigantea Quatrefages	L1 L5 3.1 3.5 IN3
† *Glycera lancadivae* Schmarda	L3 3.1 3.4

TABLE 1 (*cont.*)

?† *Diopatra* sp.	N1
† *Onuphis* (*Nothria*) *holobranchiata* (Marenzeller)	L4 N1
Marphysa mossambica (Peters)	3.8 3.10 N1 N2 N6
† *Nematonereis unicornis* (Grube)	L1 3.5 N1 IN1
† *Lumbrineris tetraura* (Schmarda)	gen
† *Lumbrineris* sp.	L4 IN1
† *Arabella iricolor* (Montagu)	EH1 EH2 IN1 IN2
† *Arabella mutans* (Chamberlin)	gen
† *Drilonereis major* Crossland	3.6 3.11
Malacoceros (*Malacoceros*) *indicus* (Fauvel)	gen
† *Malacoceros* (*Rhynchospio*) *glutaeus* (Ehlers)	3.5
† *Scolelepis aitutakii* Gibbs	L5 IN1
† *Scolelepis squamata mendanai* Gibbs	L3
† *Magelona* sp.	L1
† *Cirratulus* sp.	gen
† *Cirriformia punctata* (Grube)	3.3 2.4
† *Cirriformia tentaculata* (Montagu)	3.1 2.4
† *Poecilochaetus tropicus* Okuda	L3
† *Chaetopterus variopedatus* (Renier)	L6
† *Mesochaetopterus sagittarius sagittarius* (Claparède)	gen
Mesochaetopterus sagittarius subsp.	gen
† *Phyllochaetopterus elioti* Crossland	gen
† *Phyllochaetopterus* sp. cf. *brevitentaculata* Hartmann-Schröder	EH2
† *Spiochaetopterus* spp.	gen
† *Nainereis laevigata* (Grube)	3.3
† *Scoloplos* (*Leodamas*) *chevalieri* (Fauvel)	N6 IN1
Dasybranchus caducus (Grube)	gen
† *Notomastus* sp.	2.4
† Maldanid spp.	gen
† *Owenia fusiformis* delle Chiaje	3.5 2.4 N5 IN1
† *Myriochele* sp.	L1
† *Euthelepus kinsemboensis* Augener	2.4
Loimia medusa (Savigny)	gen
Pista typha Grube	gen
† *Sabella melanostigma* Schmarda	L2 L4 3.2
Sipuncula	
† *Aspidosiphon gracilis* Baird	L6
Aspidosiphon sp. cf. *tortus* (Selenka & de Man)	3.3
Paraspidosiphon klunzingeri (Selenka & de Man)	L6 3.4 3.10 2.2 IN1
† *Siphonosoma* (*Siphonosoma*) *rotumanum* (Shipley)	L8
Siphonosoma (*Damosiphon*) *cumanense* Keferstein	gen
Sipunculus robustus Keferstein	L2 3.6
Echiura	
Anelassorhynchus sp. cf. *porcellus* Fisher	IN1
Mollusca	
Gastropoda	
Chrysostoma paradoxum (Born)	L4
Cerithium aluco (L.)	L3
Cerithium vertagus L.	gen
Ischnocerithium echinatum (Lamarck)	Three
Rhinoclavis asper (L.)	gen
Rhinoclavis sinensis (Gmelin)	Low
Lambis truncata sebae (Kiener)	L4
Strombus campbelli Griffiths & Pidgeon	L2
Strombus gibberulus L.	gen
Strombus luhuanus L.	L4 2.4
Strombus variabilis Swainson	Low
Natica gualteriana Recluz	Low 3.2
Polinices aurantium (Roeding)	Low
Polinices pyriformis (Recluz)	Low Three
Sinum sp. cf. *planulatum* Recluz	2.2

TABLE 1 (cont.)

Cypraea annulus L.	Low
Cypraea errones L.	Three
Cypraea moneta L.	Low
Nassarius albescens (Dunker)	Three
Nassarius coronatus (Bruguière)	gen
Nassarius luridus (Gould)	Low Three
Oliva annulata Gmelin	2.4
Oliva miniacea Roeding	3.2
Mitra mitra (L.)	L5 3.3
Vexillum vulpeculum (L.)	3.2
Melo amphorus (Solander)	L4 3.5
Conus arenatus Hwass	gen
Conus coronatus Gmelin	Three 2.4
Conus ebraeus L.	2.4
Conus eburneus Hwass	L4 3.2
Conus flavidus Lamarck	2.4
Conus pulicarius Hwass	3.5
Conus suturatus Reeve	Low EH2 3.5
Terebra affinis Gray	3.2
Terebra subulata (L.)	L5 2.4
* *Atys cylindricus* (Helbling)	[L1] [3.5]
* *Pyramidella sulcata* (A. Adams)	L5
Haminoea sp.	L4
?Aglajid sp.	IN2
Lamellibranchia	
Pinna muricata L.	L7
Anodontia (*Anodontia*) *edentula* (L.)	L4 IN1 [3.8]
* *Codakia* (*Codakia*) *punctata* (L.)	[Three-clay]
Codakia (*Codakia*) *simplex* (Reeve)	L2 [2.4]
Codakia (*Codakia*) *tigerina* (L.)	IN1 [3.8] [Three-cay]
Epicodakia bella (Conrad)	IN1
'Devonia' sp.	L5
'Erycina' sp. 1	IN1
'Erycina' sp. 2	L2 3.6
'Erycina' sp. 3	L5 L6 L7 2.4 IN2
Fronsella sp.	L5 IN2
'Galeomma' sp.	L3 L5
'Leptonid' sp.	L3
Phlyctaenachlamys lysiosquillina Popham	L2 L3 L5
Fimbria fimbriata (L.)	gen
* *Acrosterigma elongatum* (Bruguière)	[Three-cay]
* *Acrosterigma rugosum* (Lamarck)	[3.8] [Three-cay]
* *Acrosterigma subrugosum* Sowerby	[Three-cay]
* *Fragum* (*Fragum*) *fragum* (L.)	[Three-cay]
* *Fragum* (*Fragum*) *unedo* (L.)	[Three-cay]
* *Fragum* (*Fragum*) sp. cf. *dioneoum* Brod. & Sow.	[EH1]
Fragum (*Fragum*) sp. cf. *whitleyi* Iredale	L2 IN1
* *Mactra* (*Mactra*) *maculata* Gmelin	[Three-cay]
Atactodea glabrata (Gmelin)	gen (cays)
Davila crassula Deshayes	gen (cays)
Tellina (*Tellinella*) *virgata* L.	L4 2.2 3.4
* *Tellina* (*Cyclotellina*) *remies* L.	[Three-cay]
Tellina (*Quidnipagus*) *palatum* Iredale	[3.8]
* *Asaphis dichotoma* (Anton)	[3.8] [IN1] [Three-cay]
Gafrarium tumidum Roeding	3.8 N6
* *Gafrarium* sp.	[Three-cay]
* *Lioconcha* (*Lioconcha*) *castrensis* (L.)	[Three-cay]
Periglypta puerpera (L.)	3.5 [Three-cay]
Crustacea	
Lysiosquilla maculata (Fabricius)	gen
Alpheidae spp.	3.6 3.11

TABLE 1 (*cont.*)

?† *Axius* (*Neaxius*) sp. nr. *plectorhynchus* Strahl	gen
Callianidea sp.	L2 2.4
Hippa celaeno (de Man)	gen (cays)
Hippa pacifica (Dana)	EH1 3.6 2.1
Paguridae spp.	gen
Porcellana sp. nr. *ornata* Stimpson	L3
Albunea symnista L.	N1
Calappa hepatica (L.)	3.1 3.3 3.5 IN1
Thalamita admete (Herbst)	3.6
Thalamita crenata Latreille	3.3
Portunus sp.	3.5
Scylla serrata (Forskal)	N6
Leptodius exaratus (H. M. Edwards)	3.6
Pilumnus vespertilio (Fabricius)	3.6
?† *Tetrias* sp. affin. *fischeri* (A. M. Edwards)	L6
Pinnotheres sp.	L7
Ocypode ceratophthalma (Pallas)	gen (cays)
Ocypode cordimana Desmarest	gen (cays)
Macrophthalmus telescopicus (Owen)	L4 L5 L7
Uca tetragonon Herbst	gen (mud)
Uca sp.	N2 N3 N4
Mictyris sp. affin. *livingstonei* McNeill	N3

Echinodermata
Asteroidea

Archaster typicus Müller & Troschel	gen

Ophiuroidea

† *Amphioplus* (*Lymanella*) *bocki* Koehler	L5 L6 L7 3.2
Amphiura diacritica H. L. Clark	L6
Macrophiothrix koehleri A. M. Clark	3.11

Echinoidea

Laganum depressum Lesson	L5 L6
Maretia planulata (Lamarck)	L5
Schizaster lacunosus (L.)	L2
Metalia spatagus (L.)	L6
Metalia sternalis (Lamarck)	L6 L7

Holothurioidea

† *Bohadschia bivittata* (Mitzukuri)	2.4
Bohadschia marmorata Jaeger	L3 2.4
Labidodemas semperianum Selenka	L3
Holothuria (*Halodeima*) *atra* Jaeger	gen
Holothuria (*Mertensiothuria*) *leucospilota* Brandt	gen
Holothuria (*Thymiosycia*) *arenicola* Semper	gen
Stichopus chloronotus Brandt	gen
Leptosynapta latipatina H. L. Clark	L1
Chiridota rigida Semper	L3 L7

Chordata
Enteropneusta

Balanoglossus australiensis Hill	gen
Balanoglossus carnosus (Willey)	gen
Balanoglossus studiosorum Horst	L7 N5

Pisces

Leiuranus semicinctus (Lay & Bennett)	2.3
Callogobius sp. cf. *slateri* Steindacher	3.3
Cryptocentrus sp. cf. *leptocephalus* Bleeker	L1
Istiblennius meleagris Cuvier & Valenciennes	3.6
Carapus homei Richardson	L4

Observations

The intertidal region of low wooded islands such as Low, Two and Three may be broadly divided into three main sedimentary areas; these are (1) the steeply sloping beaches of the sand cay composed generally of well sorted medium to coarse sands with coarser debris (see McLean & Stoddart 1978), (2) the flats of poorly sorted muddy sand and (3) areas of fine deposit accumulations (muds and silty fine sands) often colonized by mangroves, mainly *Rhizophora* and *Avicennia*. While the communities of the sand cay beaches and fine deposit areas consist of relatively few, often abundant, species, the sand flats support a very diverse fauna, but many of the species occur only as isolated individuals or appear to be very localized in their distribution. Although small areas are dominated by certain animals, such as *Mesochaetopterus* or *Balanoglossus*, in general there is no faunistic heterogeneity over large areas of the flats.

Sand cay beaches

The steeply sloping beaches of all sand cays appear to be colonized by the ghost crab *Ocypode*, its burrows occupying a zone from about the level of high water neaps to above high water springs. Random samples of *Ocypode*, mostly collected at night by torchlight, show that most populations are composed of the two widespread Indo–West Pacific species *Ocypode ceratophthalma* and *O. cordimana*, the former species being dominant.

The middle region of the slope is generally lacking in macrofauna species. Lower down, frequently in a narrow zone above the base of the slope at its junction with the flat, dense populations of suspension-feeding mesodesmatid bivalves are a common feature. Two species are represented, *Davila crassula* and *Atactodea glabrata*. Although the two species are often found separately, mixed populations occur in coarse sand and coral debris, and apparently their habitat preferences overlap. In the same situations, mole-crabs belonging to the anomuran genus *Hippa* are likewise abundant. These scavengers are principally *H. celaeno* (= *adactyla* Fabricius of earlier accounts; see Haig (1974)) but a few specimens of the larger *H. pacifica* are also to be encountered. This mesodesmatid–hippid association is particularly common on the northern shore of East Hope Island (Station EH1). Other macrofaunal species in this zone are few and include the eunicid *Arabella iricolor*. Occasionally, beneath beach-rock boulders embedded in the sand, *Perinereis* spp. can be found.

Sand flats

As stated above, the sand flat environment of the islands supports a very diverse assemblage of species, most of which are rare or only occasional. To illustrate this point, it might be mentioned that of the total of about 150 species recorded from the sand flats on the six islands of the survey (see table 3), about 70 are represented in the collections by one or two specimens only.

The surface burrowers are chiefly gastropods. The widely distributed forms *Cerithium vertagus* and *Nassarius coronatus* are the commonest everywhere, together with *Rhinoclavis asper*, *Strombus gibberulus* and *Polinices pyriformis*. Other cerithiids such as *Ischnocerithium echinatum* and *Rhinoclavis sinensis*, nassariids (*N. luridus* and *N. albescens*), naticids, such as *N. gualteriana*, and several *Terebra* spp. are less common and apparently more localized. A variety of both cypraeids and conids occur (see table 1); during daylight exposure periods, these are usually aggregated under coral boulders lying on the flats, as are various crustaceans such as portunids (*Thalamita* spp., *Portunus*), xanthids (*Leptodius*, *Pilumnus*), several pagurid spp. and a few alpheids. The ubiquitous polychaete *Eurythoe complanata* is also most frequently encountered under boulders but there is a notable absence of the large terebellid *Reteterebella queenslandia* Hartman, a conspicuous species in this habitat on southern reef flats such as Heron Island (Bennett, in Hartman 1963).

The deposit-feeding holothurians are the most conspicuous group of the surface fauna and *Holothuria atra* is ubiquitous on all islands, often mixed with fewer *Bohadschia marmorata*, *Stichopus chloronotus* and *Holothuria leucospilota*, the latter being much more abundant on the rocky areas of the reef flat. Another echinoderm, *Archaster typicus*, is also a common but less conspicuous inhabitant of the flats, its sandy coloration often partially concealing its presence as it lies half-buried in the sand during low tide.

Although specimens may be difficult to secure, the relative abundance of some species is clearly discernible from their characteristic disturbances of the sand surface or the mode of construction of burrows. The faecal casts of deposit-feeding enteropneusts are distinctive, recalling those of *Arenicola* on temperate shores (figure 3, plate 1). One of the largest species, *Balanoglossus carnosus*, is abundant on most sand flats, but is rarely taken intact, due to the soft, fragile nature of its trunk region. The volume and frequency of the casts of this species indicate considerable feeding activity during the ebb tide involving a significant turnover of the surface deposits. A smaller species, *Balanoglossus australiensis* (probably the *Ptychodera flava* Eschscholtz of earlier reports including Trewavas (1931)) is also common, especially in muddier patches, but the closely related *B. studiosorum* appears to be rare. A certain degree of particle selection in feeding has been demonstrated for some enteropneusts (cf. *Balanoglossus gigas* Fr. Müller; see Burdon-Jones 1962) and possibly the presence of these aggregations of *Balanoglossus* spp. causes local modifications of the sediment characteristics.

Where the depth of sediment is sufficient, the entrances to the extensive burrows constructed by the large stomatopod *Lysiosquilla maculata* can be found. As described by Yonge (in Popham 1939), the burrow entrance is partially sealed with a plug of mucus-bound sediment (figure 4, plate 1) which, when removed, reveals a circular, mud-lined hole, 6–8 cm in diameter extending vertically to a considerable depth. No specimens were captured nor even observed in the burrows at low tide (cf. McNeill 1968). Similar but smaller mud-lined burrows, 3 cm in diameter, often in depressions (figure 5, plate 1), are more frequent. These are excavated by the thalassinid *Axius* (*Neaxius*) sp. cf. *plectrorhynchus*. Single individuals can often be observed at the entrance to their burrow during low tide, and a total of eight specimens were caught by blocking their retreat with rapid thrusts of a fork. All of these animals (four males, four females) were taken from separate burrows and thus it is not known whether a pair of these shrimps occupies each burrow as in the case of *Neaxius* sp. on Aldabra (cf. Farrow 1971).

A further common and distinctive burrow, not identified, is illustrated in figure 6, plate 1; in this, the entrance, which is about 2 cm in diameter, is consolidated by coarse debris, mostly coral twigs. Often, gobies can be seen resting at the entrance at low tide, but these quickly withdraw on the slightest disturbance. Despite many attempts, only one goby was captured (*Cryptocentrus* sp. cf. *leptocephalus*). Possibly gobies construct these burrows or modify existing ones, as is done by *Cryptocentrus cristatus* (Macleay) (Ogilby, quoted in Marshall 1964), although other burrow-constructing fish, such as opisthognathids (see for example, Ogilby 1920; Eibl-Eibesfeldt & Klausewitz 1961) may be responsible. More probably these burrows may be constructed by alpheid shrimps since the accumulation of debris at the burrow entrance appears to be a typical habit of certain alpheids, and associations between these species and gobies have been described from widely scattered localities (Macnae & Kalk 1962; Farrow

1971; Bayer & Harry-Rofen 1957). However, such associations do not appear to have been recorded from the Great Barrier Reef region.

As found by the earlier surveys of Low Isles, the infauna of these sand flats is dominated almost everywhere by the actiniarian *Edwardsia* (chiefly *E. gilbertensis*) and chaetopterid polychaetes. Previous accounts record only *Mesochaetopterus sagittarius* (= *M. minutus* Potts), but in fact two other genera, comprising at least three species, are also abundant; these are the robust *Phyllochaetopterus elioti* and the more delicate *Spiochaetopterus* species. Each of the three genera is readily recognizable by the appearance of the chitinous tubes which, when massed, effectively bind the sediment, frequently forming pronounced hummocks. This is particularly evident where dense aggregations of small *Mesochaetopterus sagittarius* are present (figure 7, plate 2), its intertwined sand-covered tubes maintaining a stable habitat for many smaller organisms such as the polychaetes *Myriochele* and syllid species.

A high proportion of the remaining infaunal species are also polychaetes (see table 1). Other tubicolous forms which are frequently encountered as scattered individuals are *Loimia medusa*, *Owenia fusiformis*, various, as yet unidentified, maldanids and *Pista typha*, the latter being commonest on Low Isles. A further species, *Sabella melanostigma*, occurs as isolated colonies, each consisting of three to five individuals. The commoner burrowing species include sedentariate forms such as *Malacoceros indicus*, *Cirratulus* sp. and *Dasybranchus caducus*, in addition to the errantiates *Glycera gigantea*, *G. lancadivae*, *Arabella iricolor*, *A. mutans* and *Lumbrineris tetraura*. Sipunculans were never taken in any numbers; the commonest is *Siphonosoma cumanense*, but *Sipunculus robustus* (= *S. angasi* Baird in Edmonds 1955), *Paraspidosiphon klunzingeri* (*sensu* Edmonds, 1956) and several *Aspidosiphon* spp. are also to be found. Echiura are poorly represented, only one specimen of *Anelassorhynchus* sp. cf. *porcellus* being taken.

In contrast to the mesodesmatids of the cay beaches, bivalves are surprisingly scarce on the sand flats. Excluding commensal species (see below), bivalves found living total only 21 specimens, comprising nine species, all of which appear to occur only as very dispersed individuals. These include *Fimbria fimbriata*, *Tellina virgata* and small specimens of the lucinids *Codakia* spp. and *Epicodakia bella*. Other species not found living but recorded through complete shells or single valves are more numerous, principally cardiids (see table 1).

To a large extent, burrowing echinoderms as a group have been overlooked in the region, with the notable exception of the widespread species *Holothuria arenicola*, which is commonly found on most flats. The sand flat on Low Isles supports an interesting selection of species, most of which appear to occur elsewhere only rarely in the intertidal zone. As found by the earlier surveys, around the Anchorage on Low Isles (Station L5), a good low spring tide uncovers fairly numerous populations of the clypeastroid *Laganum depressum* (including *L. dyscritum* Clark, 1932), and the spatangoid *Maretia planulata* (see Endean 1956). The latter species, which is a common sublittoral form, was observed coming to the surface of the sand in late afternoon before immersion by the incoming tide (figure 8, plate 2), presumably a response to low oxygen conditions. Such behaviour has also been described for this species at Lindeman Island by Ward (1965, as *M. ovata*). Three other spatangoids can be dug, namely *Schizaster lacunosus*, *Metalia spatagus* and *M. sternalis*, but all are seemingly rare since in each case only one or two specimens were uncovered. Two burrowing ophiuroids collected on Low Isles are of particular interest, namely *Amphiura diacritica* and *Amphioplus* (*Lymanella*) *bocki* (see below). Both of these amphiurids burrow to a depth of 8–10 cm, with the body disk lying in a cavity and arms extended towards the surface in a manner similar to that described for other members of the family

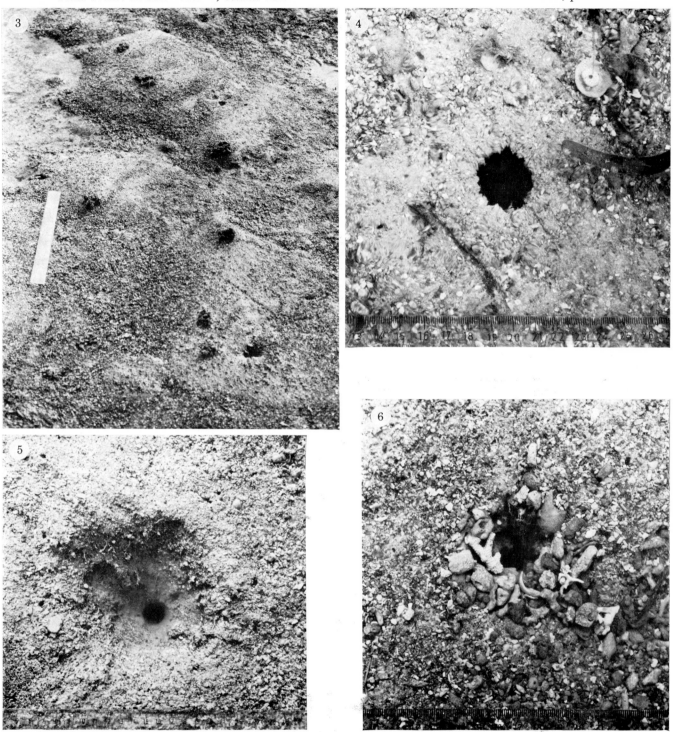

FIGURE 3. The faecal casts of the enteropneust *Balanoglossus carnosus* on the sand surface at low tide. Scale = 30 cm. Low Isles, Station L1, 27 August.

FIGURE 4. The entrance to the burrow of the stomatopod *Lysiosquilla maculata*. Low Isles, Station L1, 27 August.

FIGURE 5. The entrance to the burrow of *Axius* (*Neaxius*) sp. Low Isles, Station L1, 27 August.

FIGURE 6. The entrance to the burrow of ? alpheid sp. Low Isles, Station L2, 27 August.

FIGURES 7–10. For description see opposite.

(see Des Arts 1911; Mortensen 1927). The smaller burrowing holothurians, such as *Leptosynapta latipatina*, *Labidodemas semperianum* and *Chiridota rigida* all appear to be uncommon in the survey area although the latter two species are both widespread in Queensland waters (see Endean 1957).

Fine deposit areas

Areas of fine sediment (muddy and silty fine sands) were not surveyed extensively and no attempt was made to study the fauna of mangrove muds, although a comparison of the latter with that found on the mainland (see Macnae 1967) would prove instructive. However, where investigated (i.e. Stations 3.8, 3.9, 3.10, N6), fine deposits contain relatively few species characterized by the fiddler crab *Uca tetragonon*, the venerid *Gafrarium tumidum* and the eunicid polychaete *Marphysa mossambica*. Several other more widely distributed species also extend to these areas, namely *Dasybranchus caducus*, *Malacoceros indicus* and the occasional sipunculan (*S. cumanense* and *P. klunzingeri*). Complete shells of deep burrowing bivalves such as *Asaphis dichotoma*, *Tellina palatum* and *Anodontia edentula* were commonly uncovered, particularly at Station 3.8, but living specimens could not be found.

The shingle ramparts, a feature peculiar to these islands, provide a further habitat for infaunal species in places where the interstices have become infilled and consolidated with finer sediment. On Three Isles (Station 3.11) excavation at the base of the rampart yielded a variety of polychaetes, chiefly *Perinereis cultrifera helleri*, syllid sp. and *D. caducus*, with *S. cumanense*, several alpheid spp. and the large ophiuroid *Macrophiothrix koehleri*.

The inland tidal pools on some low wooded islands, for example Nymph, Turtle II and West Pethebridge, present an interesting environment for further study. Here it is worthwhile recording that on Nymph I. (Station N2) a very localized population of *Mictyris* sp. affin. *livingstonei* (see McNeill 1926) exists along the banks of the outflow channel, together with *Uca* sp.

Associations

Although our knowledge of the relations between partners is generally rather scant, associations between animals are common on mud and sand flats in many areas of the world, and the wide variety of associations to be found in localities within the Indo–West Pacific region is well exemplified by those described from Inhaca Island, Mozambique (see Macnae & Kalk 1962). Similar, but less well documented, examples occur in Queensland waters, and during the course of the present investigations of the sand flat fauna a number of associations were discovered, mostly instances of commensalism, that is, of a type where the partnership appears to be advantageous to one member without the other being inconvenienced or harmed (Dales 1957). These associations and their occurrences are summarized in table 2.

DESCRIPTION OF PLATE 2

FIGURE 7. A dense aggregation of the chaetopterid polychaete *Mesochaetopterus sagittarius sagittarius*. Scale is 5.5 cm in diameter. Ingram Island, Station IN1, 22 October.

FIGURE 8. A specimen of the spatangoid *Maretia planulata* emerging at the sand surface at low water mark in the late afternoon just before immersion by the incoming tide. Low Isles, Station L5, 26 August.

FIGURE 9. Three specimens (5.1–7.8 mm in length) of the commensal bivalve '*Erycina*' sp. 2 attached to the posterior end of the trunk of *Sipunculus robustus*. Three Isles, Station 3.6, 13 September.

FIGURE 10. Three specimens (1.3–5.4 mm in length) of the commensal bivalve '*Erycina*' sp. 3 attached to the dorsal side of *Fimbria fimbriata* (78.0 mm in length). Two Isles, Station 2.4, 22 September.

One of the most common and widespread of associations in the western Pacific is that of the polynoid *Gastrolepidia clavigera* living on holothurians (see Gibbs 1969, 1971). Although largely unrecorded in the Great Barrier Reef region (but see Whitley & Boardman 1929), this polychaete is associated with at least six holothurian species (Gibbs, Clark & Clark 1976) and on the sand flats individuals of *Stichopus chloronotus* and *Holothuria atra* often harbour this commensal. On Two Is and Three Is between 2 and 6 % of *H. atra* and between 6 and 40 % of *S. chloronotus* were found to have one or two *Gastrolepidia* clinging to their undersurfaces.

TABLE 2. LIST OF COMMENSAL SPECIES AND THEIR HOSTS FOUND ON SAND FLATS
WITHIN THE SURVEY AREA

Abbreviations: As, Asteroidea; De, Decapoda; En, Enteropneusta; Ho, Holothurioidea; La, Lamellibranchia; Pi, Pisces; Po, Polychaeta; Si, Sipuncula; St, Stomatopoda.

host		commensal		station
Chaetopterus variopedatus	(Po)	*Tetrias* sp. aff. *fischeri*	(De)	L6
Loimia medusa	(Po)	*Ophiodromus* sp. cf. *berrisfordi*	(Po)	L6
Paraspidosiphon klunzingeri	(Si)	'Erycina' sp. 1	(La)	IN1
Siphonosoma cumanense	(Si)	*Fronsella* sp.	(La)	L5 IN2
Sipunculus robustus	(Si)	'Erycina' sp. 2	(La)	L2 3.6
		'Devonia' sp.	(La)	L5
Lysiosquilla maculata	(St)	'Galeomma' sp	(La)	L3 L5
		Phlyctaenachlamys lysiosquillina	(La)	L2 L3 L5
?*Pinna muricata*	(La)	*Pinnotheres* sp.	(De)	L7
Fimbria fimbriata	(La)	'Erycina' sp. 3	(La)	L5 L6 L7 2.4 IN2
Archaster typicus	(As)	*Ophiodromus* sp.	(Po)	Hampton
Chiridota rigida	(Ho)	'Leptonid' sp.	(La)	L3
Holothuria arenicola	(Ho)	*Carapus homei*	(Pi)	L4
Holothuria atra	(Ho)	*Gastrolepidia clavigera*	(Po)	gen
Labidodemas semperianum	(Ho)	*Porcellana* sp.	(De)	L3
Stichopus chloronotus	(Ho)	*Gastrolepidia clavigera*	(Po)	gen
Balanoglossus australiensis	(En)	*Ophiodromus* sp. cf. *berrisfordi*	(Po)	gen
Balanoglossus carnosus	(En)	*Lepidasthenia* sp. cf. *terrareginae*	(Po)	L2
Balanoglossus studiosorum	(En)	*Ophiodromus* sp. cf. *berrisfordi*	(Po)	Turtle I

The abundance of enteropneusts on the sand flats has been referred to above, and it is not surprising that several commensals are to be found living in their burrows. The commonest of these is the hesionid polychaete *Ophiodromus*, which typically lives singly with *Balanoglossus australiensis*. This *Ophiodromus* sp. is deep brown in colour and thus is rather conspicuous on its dull yellow host. It is an active species and readily leaves the enteropneust when disturbed, a fact which may account for this polychaete being overlooked in the past. In some areas, for example at Station L3, 40–50 % of the *B. australiensis* population shelter this commensal. *Ophiodromus* was found on one occasion with *B. studiosorum* (on Turtle I), but is not confined to enteropneusts since it was also found cohabiting a tube of the terebellid *Loimia medusa*. The species is close to *O. berrisfordi* Day, but is probably new. A second *Ophiodromus* sp. inhabits the ambulacral groove of *Archaster typicus* (one example from Hampton I).

Another polychaete commensal of enteropneusts is the polynoid *Lepidasthenia*. In this case, only a single specimen was discovered living in the burrow of a *B. carnosus*. Although close to *L. terrareginae* (also described from Low Isles by Monro (1931)), the specimen appears to belong to an undescribed species.

Commensal crabs include the pinnotherid *Tetrias* sp. (affin. *fischeri*) and *Porcellana* sp. (nr *ornata*). A pair of the former (male and female) were cohabiting a tube of the polychaete

Chaetopterus variopedatus and one individual of the latter was found in the burrow of the holothurian *Labidodemas semperianum* clinging to the body of its host. A single specimen of a *Pinnotheres* sp. (female) was also retrieved from a collection of molluscs from the Low Isles sand flat which included a specimen of *Pinna muricata*, its probable host.

As a group, commensal bivalves belonging to the superfamily Galeommatacea are well represented in the present collections: a total of 30 specimens comprising eight species were taken, several of which will probably require description as new species. The taxonomy of this group requires revision and the names used here, kindly provided by Dr W. F. Ponder, are provisional. Of particular interest are those forms which utilize the burrows of the large stomatopod *Lysiosquilla maculata*, attaching themselves to the mud lining of the burrow, usually within a distance of about 15 cm of the entrance. Although many *Lysiosquilla* burrows were excavated throughout the survey area, curiously, commensal bivalves were detected in only three, all at stations on the Low Isles sand flat, in the following numbers:

Species	Station		
	L2	L3	L5
Phlyctaenachlamys lysiosquillina	1	3	1
'Galeomma' sp.	—	2	2
'Devonia' sp.	—	—	1

As shown, at least three species are present and all may be living in the same burrow. The best known of these species is *P. lysiosquillina*, described by Popham (1939) from material collected by the 1928–29 Expedition on Low Isles. However, it does not appear to have been subsequently found elsewhere, although the host is widely distributed in the Indo–Pacific region. 'Galeomma' sp. is of similar overall size to *P. lysiosquillina* with a shell length of 4–8 mm; in the field, its all-white mantle readily distinguishes it from *P. lysiosquillina*, which has orange coloured mantle papillae and tentacles. The third species, '*Devonia*', is represented by a single specimen 4.6 mm long.

Such commensal bivalves do not appear to frequent the 'vestibules' of *Axius* burrows which, although smaller, appear to present a similar niche as offered by the *Lysiosquilla* burrows. However, a more effective method of investigation other than excavation by digging might reveal bivalves inhabiting the deeper parts of the burrow system (cf. Farrow 1971). Elsewhere in Australian waters, *Axius* burrows harbour a number of bivalve species (see Cotton 1938).

Relatively few associations between bivalves and sipunculans have been recorded (see Stephen & Edmonds 1972), and thus it is perhaps surprising that associations involving three bivalves and three sipunculan species should be represented. Three individuals of the large sipunculan *Sipunculus robustus* were discovered with specimens of '*Erycina*' sp. 2, 5–10 mm in length, attached by byssus threads to the posterior end of the trunk. Two of these sipunculans had single commensals but the third had three (figure 9, plate 2). The latter was taken from a shallow burrow, about 1.0 cm below the sand surface beneath a beach-rock boulder (Station 3.6). A similar but unnamed bivalve living commensally on *S. robustus* is known from the Palau Islands (Satô 1935) and Banda Neira (Wesenberg-Lund 1954). Another species of '*Erycina*' (sp. 1) and *Fronsella* sp. occurred singly, attached to their respective hosts *P. klunzingeri* (one example) and *S. cumanense* (two examples); both of these bivalves are small, 2.5–3.2 mm in length. A further bivalve is of similar habit, namely 'Leptonid' sp., and this lives commensally on the holothurian *Chiridota rigida*, resembling *Devonia* spp. on synaptids (Anthony 1916; Ohshima 1930, 1931).

The association between the two bivalves '*Erycina*' sp. 3 and *Fimbria fimbriata* is of interest because of its frequency of occurrence in the survey area. Only eight specimens of the conspicuous white, surface-burrowing host were found, but five of these each had two or (in one case) three individuals of the '*Erycina*' sp. attached by byssus threads on the dorsal side to the posterior of the hinge line (figure 10, plate 2). The smallest of these '*Erycina*' is only 1.3 mm, the largest 5.4 mm in length.

The list of associations found on the sand flats is completed in mentioning *Carapus homei*, a well known associate of echinoderms (Mukerji 1932; Jangoux 1974) a specimen of which revealed itself in escaping from the cloaca of a *Holothuria arenicola* dug up on Low Isles.

DISCUSSION

The zonation of the fauna on the sloping beaches of the sand cays of low wooded islands is similar to that found in other Indo–West Pacific regions. As typical for most tropical sandy shores, *Ocypode* is dominant at high water level (see, for example, Dahl 1953; Hartnoll 1975) with the two widespread species *O. ceratophthalma* and *O. cordimana* generally occurring in mixed populations. The record of *O. kuhli* de Haan from Low Isles is considered doubtful and requiring verification (McNeill 1968). Between mid-tide level and low water neaps, mesodesmatid beds are a common entity and similar aggregations of *Atactodea* and *Davila* are widespread in the lagoon environments of Indo-Pacific atolls (Banner 1952; Cloud 1952; Beu 1972). In northern Queensland the same zone is often occupied by several hippid species, mainly *H. celaeno*; although feeding was not observed, these mole-crabs presumably perform a scavenging rôle in the same manner as described for related forms (see Mortensen 1922; Bonnet 1946; Matthews 1955).

Over the sand flat, bivalves are scarce and the surface-burrowing forms are mainly gastropods, many of which are widely distributed species found throughout the Indo–Pacific (cf. Morrison 1954; Taylor 1971; Gibbs 1975). The infauna, while consisting of numerous species, tends to be dominated numerically by relatively few forms with *Edwardsia*, chaetopterids, other tubicolous polychaetes and *Balanoglossus* spp. forming the bulk of the macrofauna. As a group, the polychaetes have been largely neglected in past surveys in the area; for example, less than 10 species were recorded for the Low Isles sand flat by the 1928–29 and 1954 Expeditions, whereas over 30 species were taken in 1973. In all, over 50 species occur on the sand flats, a high proportion of which, although Indo-Pacific in distribution, are here recorded for the first time for the Province as a whole. The burrowing echinoderms are equally interesting, particularly on Low Isles where four spatangoids, one clypeastroid, two ophiuroids and four or five holothurians exist. The specimens of *Amphioplus bocki* are the first recorded from Australian waters while *Amphiura diacritica* and *Leptosynapta latipatina* were previously known only from the holotypes collected respectively in the Whitsunday Group (Clark 1938) and at Friday Island (Clark 1921, 1946). *Maretia planulata* seems to have been absent from the Low Isles sand flat in 1928–29 but was common there in 1954 (Endean 1956). Since the species is frequently plentiful in sandy deposits in shallow depths, it is possible that the upper limit of the Anchorage population may fluctuate according to conditions at the level of low water.

An assessment of the changes which have occurred in the species composition and relative abundance of the Low Isles sand flat fauna between 1928–29, 1954 and 1973 proves difficult for several reasons (see Stoddart, McLean, Scoffin & Gibbs 1978, this volume) but chiefly because the earlier studies of this habitat were brief and the ecological data are rather fragmentary. However, in terms of the dominant species, *Edwardsia*, *Mesochaetopterus* and *Balanoglossus* remain abundant while such species as *Aspidosiphon cumingi* and *Phyllodoce malmgreni*, reported as plentiful in 1928, could not be found in 1973. Conversely, quite a number of the more conspicuous tubicolous polychaetes taken in 1973, including *Phyllochaetopterus*, *Spiochaetopterus*, *Loimia* and *Sabella*, do not appear in the 1928–29 reports (Monro 1931). In view of the present-day abundance of chaetopterids, the statement in Stephenson *et al.* (1931, p. 73) that 'the polychaetes as a whole do not appear to be among the dominant groups, except in rocky crevices . . .' is difficult to interpret.

TABLE 3. SPECIES DIVERSITY OF THE SAND FLAT FAUNA ON LOW ISLES AND OTHER ISLANDS IN THE SURVEY AREA

| group | number of species | | |
	Low Isles	other islands	total
Anthozoa	2	3	3
Polychaeta	30+	44+	50+
Sipuncula	5	4	6
Echiura	—	1	1
Crustacea	*ca.* 10	*ca.* 15	*ca.* 20
Gastropoda	24	26	36
Lamellibranchia	13	11	17
Echinodermata	16	8	17
Enteropneusta	3	3	3
total	**103**	**115**	**153**

Attention can also be focused on the anomuran *Axius* (*Neaxius*) sp., a fairly common and conspicuous animal in the survey area but one which is not mentioned in earlier accounts (see McNeill 1968). Possibly its burrows were misidentified as being constructed by small *Lysiosquilla*. The species may be very localized in its distribution since the genus *Axius* appears to have been recorded in the province only once (Port Molle), the single specimen being uncertainly referred to *Axius* (*Neaxius*) *plectrorhynchus* Strahl by Miers (1884) (see also Fulton & Grant 1902).

The species diversity of the fauna of the sand flats in the survey area is summarized in table 3. Based on the present collections, the macrofauna can be estimated at 150–160 species although undoubtedly this total could be substantially increased with further study of past collections. However, the figures illustrate the importance of Low Isles as a locality for further investigation, in that about two-thirds of the species in the total sand flat fauna of the area are recorded from there and about one-fifth of all species were taken only on Low Isles, notably some little-known echinoderms. Further, it can be noted that of the 19 commensal associations discovered during the survey, no less than 16 were found on Low Isles.

I am grateful to the Royal Society for the opportunity to participate in the Royal Society – Universities of Queensland Expedition. I thank Dr D. R. Stoddart, Expedition Leader, and other Expedition members, particularly Dr R. F. McLean, Dr T. P. Scoffin and Mr A. Smith for their assistance in the field. For much help and valuable advice with the identification of the collections I have to thank Professor C. E. Cutress, University of Puerto Rico (*Edwardsia*); Dr J. D. George, British Museum (Natural History) (Cirratulidae); Dr S. J. Edmonds, South Australian Museum (Sipuncula); Dr P. Colman, Dr W. F. Ponder, Australian Museum

Gastropoda and Galeommatidae); Mrs S. Slack-Smith, Western Australian Museum Lamellibranchia, excluding Galeommatidae); Dr R. W. Ingle, British Museum (Natural History) (Crustacea); Miss A. M. Clark, Mrs C. M. Clark, British Museum (Natural History) (Echinodermata); Professor C. Burdon-Jones, James Cook University (Enteropneusta); Dr A. Wheeler, British Museum (Natural History) (Pisces). I am most grateful also to the Council of the Marine Biological Association of the U.K. and the Natural Environment Research Council for leave of absence during the period of the Expedition.

References (Gibbs)

Anthony, R. 1916 *Archs Zool. exp. gén.* **55**, 375–391.

Banner, A. H. 1952 *Atoll Res. Bull.* **13**, 1–42.

Bayer, F. M. & Harry-Rofen, R. R. 1957 *Rep. Smithsonian Instn* **1956**, 481–508.

Beu, A. G. 1972 *J. malac. Soc. Aust.* **2**, 113–131.

Bonnet, D. D. 1946 *Science, N.Y.* **103**, 148–149.

Burdon-Jones, C. 1962 *Bolm Fac. Filos. Cienc. Univ. S. Paulo* (zool.) **24**, 255–280.

Clark, H. L. 1921 *Pap. Dep. mar. Biol. Carnegie Instn Wash.* **10**, 1–223.

Clark, H. L. 1932 *Scient. Rep. Gt Barrier Reef Exped. 1928–29* **4**, 197–239.

Clark, H. L. 1938 *Mem. Mus. comp. Zool. Harv.* **55**, 1–596.

Clark, H. L. 1946 *Publs Carnegie Instn*, No. 566, 1–567.

Cloud, P. E. 1952 *Atoll Res. Bull.* **12**, 1–73.

Collins, A. C. 1958 *Scient. Rep. Gt Barrier Reef Exped. 1928–29* **6**, 335–437.

Cotton, B. C. 1938 *Victorian Nat.* **55**, 58–61.

Dahl, E. 1954 *Oikos* **4**, 1–27.

Dales, R. P. 1957 *Mem. geol. Soc. Am.* **67**, 391–412.

Dall, W. & Stephenson, W. 1953 *Pap. Dep. Zool. Univ. Qd* **1**, 21–49.

Den Hartog, C. 1970 *Verh. K. ned. Akad. Wet.* **59**, 1–275.

Des Arts, L. 1911 *Bergens Mus. Årb.* **12**, 1–10.

Edmonds, S. J. 1955 *Aust. J. mar. Freshwat. Res.* **6**, 82–97.

Edmonds, S. J. 1956 *Aust. J. mar. Freshwat. Res.* **7**, 281–315.

Eibl-Eibesfeldt, I. & Klausewitz, W. 1961 *Senckenberg. biol.* **42**, 421–426.

Ekman, S. 1953 *Zoogeography of the sea.* (417 pages.) London: Sidgwick & Jackson.

Endean, R. 1956 *Pap. Dep. Zool. Univ. Qd* **1**, 123–140.

Endean, R. 1957 *Aust. J. mar. Freshwat. Res.* **8**, 233–273.

Farrow, G. E. 1971 In *Regional variation in Indian Ocean coral reefs* (eds D. R. Stoddart & C. M. Yonge) (Symp. zool. Soc. Lond. vol. 28), pp. 455–500.

Flood, P. G. & Scoffin, T. P. 1978 *Phil. Trans. R. Soc. Lond.* A **291**, 55–71 (part A of this Discussion).

Fulton, S. W. & Grant, F. E. 1902 *Proc. R. Soc. Vict.* N.S. **14**, 55–64.

Gibbs, P. E. 1969 *Phil. Trans. R. Soc. Lond.* B **255**, 443–458.

Gibbs, P. E. 1971 *Bull. Br. Mus. nat. Hist.* (*Zool.*) **21**, 99–211.

Gibbs, P. E. 1975 *Atoll Res. Bull.* **190**, 123–131.

Gibbs, P. E., Clark, A. M. & Clark, C. M. 1976 *Bull. Br. Mus. nat. Hist.* (*Zool.*) **30**, 103–144.

Haig, J. 1974 *Mem. Qd Mus.* **17**, 175–189.

Hartman, O. 1963 *Rec. Aust. Mus.* **25**, 355–358.

Hartnoll, R. G. 1975 *J. Zool. Lond.* **177**, 305–328.

Iredale, T. 1939 *Scient. Rep. Gt Barrier Reef Exped. 1928–29* **5**, 209–425.

Jangoux, M. 1974 *Revue Zool. afr.* **88**, 789–796.

McLean, R. F. & Stoddart, D. R. 1978 *Phil. Trans. R. Soc. Lond.* A **291**, 101–117 (part A of this Discussion).

Macnae, W. 1967 In *Estuaries* (ed. G. H. Lauff), pp. 432–441. Washington: American Association for the Advancement of Science.

Macnae, W. & Kalk, M. 1962 *J. Anim. Ecol.* **31**, 93–128.

McNeill, F. A. 1926 *Rec. Aust. Mus.* **15**, 100–128.

McNeill, F. A. 1968 *Scient. Rep. Gt Barrier Reef Exped. 1928–29* **7**, 1–98.

Marshall, S. M. & Orr, A. P. 1931 *Scient. Rep. Gt Barrier Reef Exped. 1928–29* **1**, 93–133.

Marshall, T. C. 1964 *Fishes of the Great Barrier Reef and coastal waters of Queensland.* Sydney: Angus & Robertson.

Matthews, D. C. 1955 *Pacif. Sci.* **9**, 382–386.

Maxwell, W. G. H. 1968 *Atlas of the Great Barrier Reef.* (258 pages.) Amsterdam: Elsevier.

Maxwell, W. G. H. 1973 In *Biology and geology of coral reefs* (eds O. A. Jones & R. Endean), vol. 1, pp. 299–345. New York: Academic Press.

Miers, E. J. 1884 In *Report on the zoological collections made in the Indo-Pacific Ocean during the voyage of H.M.S.* Alert 1881–2, pp. 178–322. London: British Museum.

Monro, C. C. A. 1931 *Scient. Rep. Gt Barrier Reef Exped.* **4**, 1–37.

Morrison, J. P. E. 1954 *Atoll Res. Bull.* **34**, 1–18.

Mortensen, T. 1922 *Vidensk. Meddr dansk naturh. Foren.* **74**, 23–56.

Mortensen, T. 1927 *Handbook of the echinoderms of the British Isles.* Oxford: University Press.

Mukerji, D. D. 1932 *Rec. Indian Mus.* **34**, 567–569.

Ogilby, J. D. 1920 *Mem. Qd Mus.* **7**, 1–30.

Ohshima, H. 1930 *Annotnes zool. jap.* **13**, 25–28.

Ohshima, H. 1931 *Venus* **2**, 161–177.

Popham, M. L. 1939 *Scient. Rep. Gt Barrier Reef Exped. 1928–29* **6**, 61–84.

Satô, H. 1935 *Sci. Rep. Tôhoku imp. Univ.* **10**, 299–329.

Spender, M. 1930 *Geogrl J.* **76**, 193–214.

Steers, J. A. 1929 *Geogrl J.* **74**, 232–257.

Steers, J. A. 1930 *Scient. Rep. Gt Barrier Reef Exped. 1928–29* **3**, 1–15.

Stephen, A. C. & Edmonds, S. J. 1972 *The Phyla Sipuncula and Echiura.* London: British Museum (Natural History).

Stephenson, T. A., Stephenson, A., Tandy, G. & Spender, M. 1931 *Scient. Rep. Gt Barrier Reef Exped. 1928–29* **3**, 17–112.

Stephenson, W., Endean, R. & Bennett, I. 1958 *Aust. J. mar. Freshwat. Res.* **9**, 260–318.

Stoddart, D. R. 1978 *Phil. Trans. R. Soc. Lond.* A **291**, 5–22 (part A of this Dissussion).

Stoddart, D. R., McLean, R. F., Scoffin, T. P. & Gibbs, P. E. 1978 *Phil. Trans. R. Soc. Lond.* B **284**, 63–80 (this volume).

Taylor, J. D. 1971 In *Regional variations in Indian Ocean coral reefs* (eds. D. R. Stoddart & C. M. Yonge) (Symp. zool. Soc. Lond., vol. **28**), pp. 501–534.

Trewavas, E. 1931 *Scient Rep. Gt Barrier Reef Exped.* **4**, 39–67.

Ward, M. 1965 *Aust. Zool.* **13**, 127–134.

Wesenberg-Lund, E. 1954 *Bull. Inst. r. Sci. natn Belge* **30** (16), 1–18.

Whitley, G. P. & Boardman, W. 1929 *Aust. Mus. Mag.* **3**, 330–336.

Yonge, C. M. 1930 *Scient. Rep. Gt Barrier Reef Exped. 1928–29* **1**, 1–11.

Phil. Trans. R. Soc. Lond. B. **284**, 99–122 (1978)   [99]

Printed in Great Britain

The nature and significance of microatolls

By T. P. Scoffin† and D. R. Stoddart‡

†*Grant Institute of Geology, University of Edinburgh, West Mains Road,*
Edinburgh EH9 3JW, U.K.

‡*Department of Geography, University of Cambridge, Downing Place,*
Cambridge, U.K.

With an appendix by B. R. Rosen

[Plates 1–4]

Microatolls, those coral colonies with dead, flat tops and living perimeters, result from a restriction of upward growth by the air/water interface. The principal growth direction is horizontal and is recorded in the internal structure, though fluctuations in water depth can influence the surface morphology producing a terraced effect. The morphology of the basal surface of the colony is controlled by the sand/water interface such that the thickness of the coral records the depth of water in which it lived.

In open water at the margin of reefs in the Northern Province of the Great Barrier Reef, tall-sided uneven-topped microatolls live, whereas, on the reef flats in rampart-bounded moats and ponds, thin flat-topped and terraced microatolls are abundant. Because water in moats can be ponded to levels as high as high water neaps (1.6 m above datum at Cairns) and still have daily water replenishment, microatolls on reef flats can grow to levels 1.1 m higher than open-water microatolls (which grow up to a maximum elevation of low water springs, i.e. 0.5 m above datum). This imposes a major constraint on the use of microatolls in establishing sea level history. The two factors controlling pond height during one sea stand (relative to the reef) are tidal range (which governs the height of high water neaps) and wave energy (which governs the height of ramparts which enclose moats).

Dating and levelling fossil microatolls exposed on the reefs show that 4000 years (a) B.P., high water neaps was at least 0.7 m higher than it is at present.

1. Introduction

(a) Definition

Early descriptions of microatolls were given by Darwin (1842), Dana (1872), Semper (1880, 1899) and Guppy (1886), using general names such as coral head or coral block. Guppy (1886) spoke of 'miniature atolls', Agassiz (1895) of 'diminutive atolls' and Krempf (1927) of 'dwarf atolls' ('*atolls nains*'). The term *micro-atoll* was first used by Krempf (1927), but without concise definition. It was widely adopted and variously defined. Kuenen (1933) used it for 'a colony of corals' with 'a raised rim, more or less completely surrounding a lower, dead surface'. MacNeil (1954) used it for 'massive colonial corals growing peripherally in shallow areas and whose dead upper surface (sometimes made concave by solution) is exposed at low tide'. Several authors, for example Newell & Rigby (1957), Hoskin (1963), Kornicker & Boyd (1962), Garrett, Smith, Wilson & Patriquin (1971) have adopted the term, inconsistent with early usage, to refer to patch reefs consisting of many corals which develop a structure having a raised growing rim and a low, commonly dead or sand-filled, centre. Scheer (1972) suggests that the term *mini-atoll* is more appropriate for such patch reefs. The term *faro* is in common use for large ring-shaped patch reefs at atoll margins. Recently Stoddart & Scoffin (1978) have reviewed the literature on microatolls.

A typical microatoll is a single coral colony, commonly massive and circular in plan, with a dead, predominantly flat, upper surface and a living lateral margin. A fossil microatoll has the same morphology and though no living polyps remain, both the internal structure and commonly the preservation of the peripheral corallites indicate that the polyps at the lateral margins lived on after those at the centre died. For microatolls to indicate the former close proximity of sea surface to their upper surfaces, their internal structure should reveal evidence of predominantly horizontal growth.

(b) Corals having microatoll form

Those colonial corals (including Scleractinia, Hydrozoa and Alcyonaria) that would, if unimpeded, grow with a dome-shaped upper surface, develop a microatoll form when upward growth is restricted. Massive corals were the most common microatolls on the Great Barrier Reefs studied, especially *Porites lutea*, but some branching corals also take on this same configuration when they reach water level. Certain corals (notably *Pavona*) encrust the lateral margins of shallow water corals and thus enhance the microatoll nature of the foundation coral.

A list of all the corals found with microatoll form on the Great Barrier Reef patch reefs and barrier reefs examined between Cairns and Cape Melville is given in the appendix. These identifications were kindly made by B. R. Rosen of the British Museum.

Although an exhaustive collection was not made it is of interest to note that in this region of the Great Barrier Reef as many as 43 species of corals (representing 23 different genera) were found growing with microatoll form. That this very limited environment should support such large numbers of coral genera all with the same general morphology suggests that growth form might carry more weight than species composition in certain environmental reconstructions.

(c) Origin

Living microatolls are all found with their tops at the lowest level of sea water for the area in which they grow; they are not found growing below lowest water level of spring tides in the open sea. This observation coupled with the nature of microatoll growth forms indicates that the dead, flat upper surface of microatolls relates to the close proximity of the water/air interface. If corals that have developed a dead surface in relation to a certain level of water break (owing to instability) such that part of the living margin is now lower in the water, this lower part will grow up to the water level and develop a new microatoll surface. This indicates that the critical effect in producing the dead surface is not a factor unrelated to water level, such as the inability of the flat parts of corals to clear raining sediment from their broad surfaces as has been proposed by some authors (Wood-Jones 1912; Krempf 1927). Also, the excessive sediment theory would not account for the existence of open-branching microatolls. Exactly what finally kills the polyps at the coral top is not known; it may be the desiccation on exposure, excessive ultraviolet radiation, or in some cases even lack of nutrients at the coral centre.

On death of part of the coral its skeleton is soon attacked by boring, grazing and encrusting organisms. The activities of this epifauna and epiflora tend to retard the recuperation of the coral growth at this site of attack. Boring organisms (filamentous algae, sponges, worms, sipunculids, bivalves) and, to an extent, grazing organisms that feed on the epiflora, all lower the dead coral skeleton and may be partly responsible for the concave nature of the tops of many microatolls. However, it is important to note that the bioerosive activities of these animals and plants are themselves not the cause of the microatoll form as defined here. This is evident from examination of the internal structure of the Great Barrier Reef microatolls. Where the dead top results principally from the bioerosion of a formerly dome-shaped coral truncation of growth lines is apparent (figure 1 a), whereas those skeletons showing horizontal growth lines paralleling the dead surface result from purely lateral growth during most of the coral development (figure 1 b). Also, the corals whose upper surfaces have been hollowed out by bioerosion generally have an irregular top in comparison with the essentially planar surface of the typical microatoll. An example of a coral with a bioeroded upper surface is shown in figure 9† where a massive colony of *Montastrea annularis* has had its upper surface extensively grazed by *Diadema antillarum* sea urchins well below low water level on a fringing reef in Barbados.

2. GROWTH FORMS AND FACTORS INFLUENCING MORPHOLOGY

(a) Upper surface

(i) Steady state of lowest water level

The most common form of microatoll on the patch reefs of the northern Great Barrier Reef is shown in figure 1 c. Initial growth was as a hemisphere at times t_1 and t_2 but by the time t_3 water level was reached and lateral growth commenced and continued through to the present, t_9. Upward growth is restricted by the water/air interface and downward growth by the sand/water interface. The tallness of the structure coated by living polyps and therefore the thickness of the coral skeleton are indicative of the depth of water in which the coral grew. Most reef flat microatolls are between 5 and 20 cm thick. The top dead surface is called the microatoll plane. Any bioerosion of the microatoll plane truncates the growth lines.

(ii) Deepening of water

Commonly the polyps at the margins grow slightly above (up to 2 cm) the microatoll plane and the lowest water level in the area in which these microatolls occur is just at the level of the uppermost living polyps (figure 1 d). The flexuring of the growth lines at the inner margin of the living annulus (figure 1 d) suggests that in these examples the corals have recently managed some upward growth indicating a recent slight increase in water depth. Of course the shape of that part of the skeleton that is covered by living polyps must result from growth. This shape cannot be accounted for by the persistent outward growth of a raised lip with concomitant erosion of the inner portion of the living annulus. Progressive development of this outer-growth/inner-erosion style would result in the structure shown in figure 1 e, and not the pattern observed of flexuring of growth lines inwards at the inner margin of the living annulus.

The width of the living annulus at water level was rarely more than a few centimetres before it too developed a dead centre (the secondary microatoll plane) shown in figure 1f. It can be seen that at this stage the polyps at the inner part of the annulus grow over, and in the opposite direction to, earlier skeletal growth that produced the primary microatoll plane (figures 1f, 2 and 10). Also, these polyps grow into an ever decreasing annulus and thus compete for space. As the upward growth of these two annuli also becomes restricted by water level they too

† Figures 2–7 appear on plate 1, figures 9–14 on plate 2, figures 15–20 on plate 3 and figures 22–27 on plate 4.

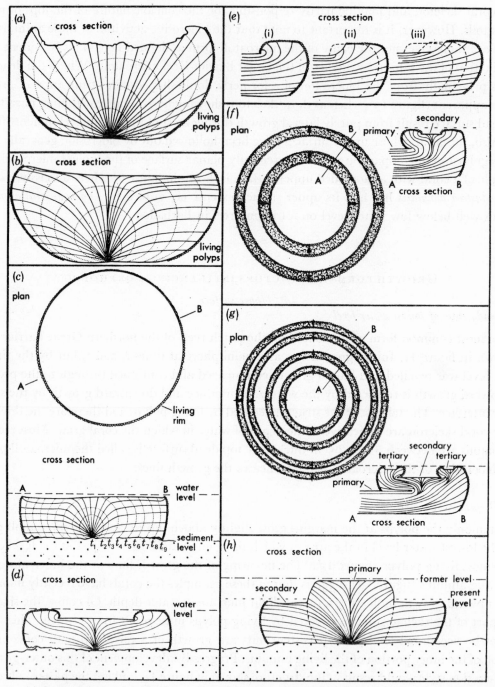

FIGURE 1. (a) Cross section of a coral colony with a living lateral margin and dead upper surface. The truncation
of the growth lines at the top are the result of surface bioerosion. (b) Cross section of a coral colony, a microatoll, with a living lateral margin and a dead upper surface. The growth lines parallel the top surface indicating horizontal growth consequent on an upward restriction. (c) Microatoll in plan and cross section showing
the general form and the stages of growth. Initial growth was as a hemisphere at times t_1 and t_2 but by time
t_3 water level was reached and lateral growth began and continued through to the present, t_9. (d) Cross
section of a microatoll whose living rim has grown a few centimetres higher than the dead central portion.
Note the flexuring of the growth lines at the inner margin of the living annulus. (e) Predicted structure of the
stages of development of a coral colony margin where bioerosion of the surface has progressed along with
outward growth of the rim, causing the gradual removal of the flexured growth lines. (f) Microatoll in plan
and cross section showing the development of the secondary microatoll plane. (g) Microatoll in plan and
cross section showing the development of the tertiary microatoll plane. The arrows on the plan view illustrate
the present direction of growth of the living polyps. (h) Cross section of a microatoll whose secondary microatoll plane is below the primary, indicating a shallowing of water.

bifurcate, producing a tertiary microatoll plane (figures 1g and 10). The maximum difference in heights recorded between a primary microatoll plane and a tertiary microatoll plane on one colony was 20 cm.

Three microatoll planes on one coral colony was the maximum number found, though theoretically, if the water is periodically deepened, there is no reason why more microatoll surfaces should not develop on the one coral colony. The surface structure becomes more convoluted with continued development of these surfaces because eventually the polyps of different annuli growing in opposing directions start to interfere with one another (figure 3).

(iii) Shallowing of water

The examples of microatolls with evidence of growth above the primary microatoll plane all grew in progressively deepening water. If the water shallows a new lower microatoll plane develops around the periphery of the coral (figure 1h). Exposure of the central part of the skeleton normally results in rapid bioerosion obscuring the earlier shape of the coral.

(b) Lower surface

Just as fluctuations in the level of the water/air interface are reflected in the morphology of the upper part of the colony and the adjacent internal structure, changes in the sand/water interface level affect the outline of the lower part of the coral and the adjacent internal structure, for the polyps tend to grow down close to the substrate level. The undersurfaces of microatolls commonly have a saw-tooth profile (figures 8a and 11). This shape indicates abrupt killing of lower polyps, followed, after a while, by further downward and outward growth starting from a level above the earlier lowest level. This growth pattern most probably results from the periodic piling around the microatoll base of sediment which has later been removed. Some shapes indicate that the sediment level was raised permanently (figures 8b and 11). The large microatoll shown in figure 11 indicates a marked raising of sand level at about the time the top reached water level. This raised sand level was maintained throughout the remaining growth with only minor fluctuations to produce the toothed base.

On reef flats normally only the small microatolls that have recently reached water surface are still connected to a solid substrate, the broad microatolls lie loosely on sand. The crevices under the lower surface provide a habitat for a range of cryptic animals, both sessile (the most common are sponges, bivalves, worms, foraminiferans, bryozoans) and motile (gastropods, fish, crustaceans, ophiuroids). These creatures help to disperse sand beneath the corals and may even bring about a lowering in the sediment under the solid skeleton. Microatoll shapes indicative of deepening water do not necessarily indicate a raising of water level, they may result from a lowering of the coral in a soft substrate.

(c) Regrowth after movement or breakage

Broad, thin microatolls are fairly easily lifted by waves; they can be moved, tilted, overturned or broken. When a microatoll is tilted, growth occurs along a new plane as shown in figure 8c.

Growth can continue after complete overturning (figure 8d) though some polyps may show a 180° change in growth direction.

If the two fragments separate when a microatoll breaks then growth becomes recurved at the edge of the break (figure 4). Lateral growth into deeper water can bring about instability of a microatoll. If the sandy substrate is removed from the periphery of a large microatoll, skeletal growth alone can be sufficient to bring about a disequilibrium causing breakage of the skeleton into segments. The margins tilt down and the polyps grow back up to the water level (figure 5).

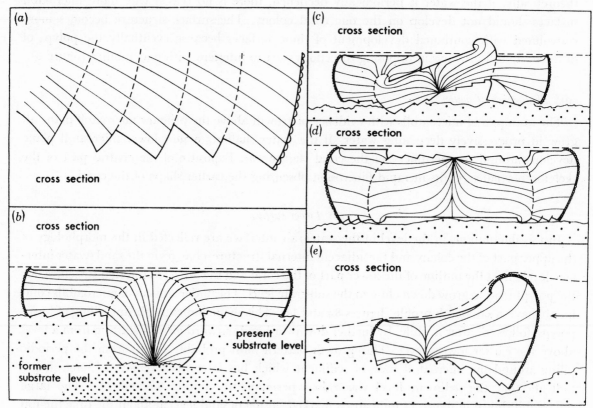

FIGURE 8. (a) Saw-tooth profile of the undersurface of a microatoll. The growth lines indicate the periodic death of the low polyps. (b) Cross section of a microatoll whose undersurface reveals a sharp, permanent raising of the sediment level after about a third of its development. (c) Cross section of a microatoll that has undergone tilting and regrowth. (d) Cross section of a microatoll that has been overturned and continued growth. (e) Cross section of an asymmetric microatoll. Dominant direction of flow of water is from right to left; this current scours the sediment away from the up-current side, piles it on the lee and at the same time, banks water high on the up-current side. Coral growth develops accordingly.

DESCRIPTION OF PLATE 1

FIGURE 2. Cross section of the margin of a *Favites* microatoll revealing the internal growth pattern in the development of primary (a) and secondary (b) microatoll planes. Width of photograph 35 cm. Reef flat, Two Isles, Great Barrier Reef.

FIGURE 3. Convoluted growth pattern on the surface of a large *Porites* colony. Reef flat, Turtle IV, Great Barrier Reef. Hammer 30 cm long.

FIGURE 4. Plan view of a *Porites* microatoll that has broken along a diameter and growth at the lateral margin has continued around the broken surface. Reef flat, Two Isles, Great Barrier Reef. Hammer is 30 cm long.

FIGURE 5. Large colonies of *Porites* that have split into segments on subsidence of the margins. Reef flat, Nymph Isle, Great Barrier Reef. Width of foreground in photograph 3 m.

FIGURE 6. Coalescing colonies of *Porites*. Note that the line of contact is at the same level as the surrounding microatolls planes suggesting a bevelling down of the former rims. Reef flat, Turtle IV, Great Barrier Reef. Width of foreground 1 m.

FIGURE 7. Development of a pavement of coalesced microatolls (predominantly *Porites*). Reef flat, Three Isles, Great Barrier Reef. Hammer is 30 cm long.

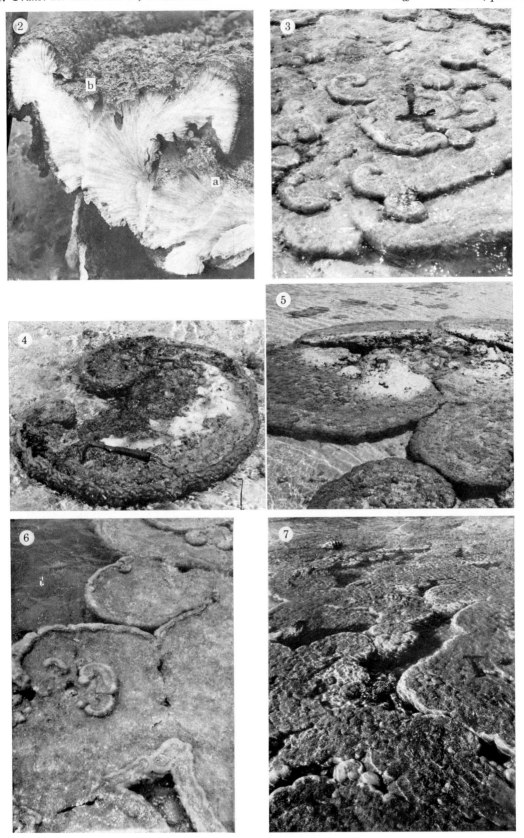

FIGURES 2–7. For description see opposite.

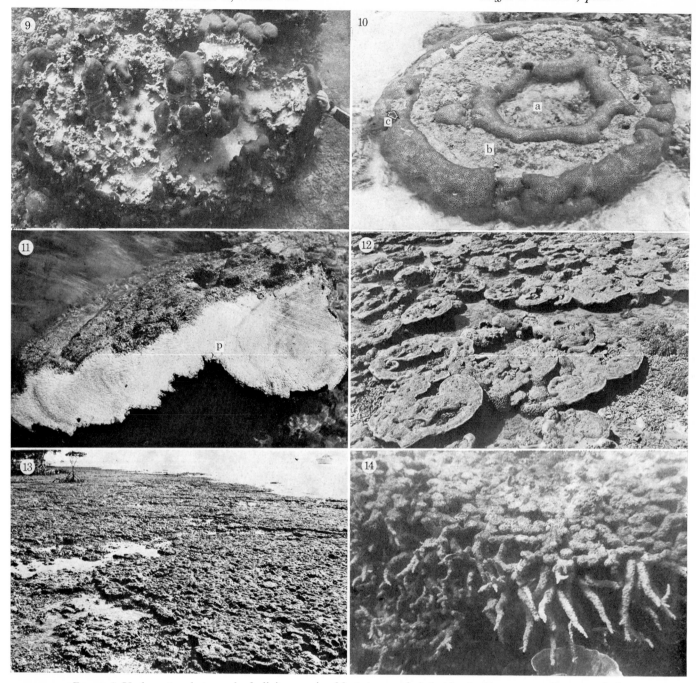

FIGURE 9. Underwater photograph of a living massive *Montastrea annularis* that has had its upper surface extensively grazed by *Diadema antillarum* sea urchins. The upper surface of this coral is at a depth of 2 m below low water level. Bellairs reef, Barbados, W.I.

FIGURE 10. *Platygyra* microatoll 1.5 m in diameter showing development of primary (a), secondary (b) and incipient tertiary (c) microatoll planes. Reef flat, Two Isles, Great Barrier Reef.

FIGURE 11. Sawed vertical section of half of a microatoll. The right side of the photograph shows the region of early development of the colony, the extreme left of the coral is the living margin. Arcuate growth bands can just be detected. The coral started growth in water about 20 cm deep but after a time (at position p) the sand level was raised to about 10 cm below the water surface. Subsequently, sand level fluctuated slightly to produce the saw-tooth profile at the base of the coral. The water level was roughly constant throughout. Reef flat, Nymph Isle, Great Barrier Reef.

FIGURE 12. Microatolls (dominantly *Goniastrea*) with inclined upper surfaces resulting from prevailing flow of water from right to left. Photograph taken at extreme low tide leeward margin of reef flat. Low Isles, Great Barrier Reef. Foreground 3 m wide.

FIGURE 13. Northeast margin of reef flat at Hampton Isle, Howick Group, Great Barrier Reef, showing the broad flat pavement consisting chiefly of the surfaces of dead and living microatolls. Foreground of photograph is 5 m wide.

FIGURE 14. Underwater photograph of colony of branching *Acropora* coral that grows in open water up to the level of low water springs. At the maximum elevation for growth, branches become flattened, stubby and dense. Leeward side of East Hope Island, Great Barrier Reef.

FIGURES 15–20. For description see page 105.

FIGURE 22. Windward margin of Watson Reef, Great Barrier Reef. The open water, at the right of the photograph, is at a lower level than the water in the moat retained by the rampart of loose coral shingle. Width of moat is about 6 m.

FIGURE 23. Two levels of ponded water. The higher pond, in the background, is retained by a wall of cemented shingle (the jagged projections represent eroded lithified foresets), the lower pond, in the foreground is retained by a rampart of loose coral shingle. Microatolls currently live up to the water surface in both these ponds. Windward flanks Watson Reef, Great Barrier Reef.

FIGURE 24. In-place dead microatolls in a drained moat. The rampart (at top of photograph) that formerly ponded water here has recently been breached. Windward margin of reef flat, Nymph, Great Barrier Reef. Foreground is 4 m wide.

FIGURE 25. In-place fossil microatolls exposed on the reef flat on Nymph Isle, Great Barrier Reef.

FIGURE 26. In-place fossil microatoll (left of the hammer) exposed in section in a cliff of cemented shingle. Upper platform, Nymph Isle, Great Barrier Reef. Hammer is 30 cm long.

FIGURE 27. Large dead microatoll at the leeward flanks of Low Isles. The open sea to the left was at the level of low water springs at the time of photography.

(d) Asymmetric growth

Most microatolls on reef flats show an obvious radial symmetry, but in some cases growth is uneven. The lip of the microatoll can be higher on one side than the other (figures 8e and 12). Microatolls of this shape grow in areas where, for the major part of the tidal cycle, water is flowing rapidly in one direction. The leeward sides of reef flats are lower than the windward. Consequently, on the flanks of the leeward sides of many reef flats, water that is draining from the windward flows steadily throughout the whole ebbing of the tide. This flow brings about scouring of sand from under the windward side of the microatoll (and some piling up of sand on the leeward) and also a banking up of water on the windward side allowing this side to grow both slightly higher and lower than it can on the leeward.

The living lip of microatolls is commonly wider on one side than the other. Though this uneven growth was commonly related to flowing water (wider side up-current, that is on the windward) this was not found to be invariably the case. In some instances it perhaps relates to slight tilting such that one side reached water level earlier than the other and therefore spread laterally earlier producing the asymmetric form.

(e) Size

Microatolls vary in size from a few centimetres to a few metres in diameter. The average diameter of reef flat microatolls is 50 cm. The tallness of the skeletons is directly related to the depth of water in which they grow. Naturally in one moat of standing water the microatolls will all grow to the same level and as the foundation is usually regular they have approximately the same thickness. Quite commonly, microatolls in one moat have a similar diameter. This suggests that conditions were suitable for coral planulae settlement in the moat for only a short time. Perhaps the moat floor became covered with sediment too fine to be conducive to planulae attachment.

(f) Coalescent growth

When polyps of the same species meet, they cause the structure to coalesce and normally to grow in the resultant direction (figure 6). The old structure decays and is shortly lowered such that the line of contact is bevelled down to the level of the microatoll plane. With time a flat pavement of coalesced microatolls develops (figure 7) revealing few or no living polyps. At the margins of reef flats many areas, several metres across, that although full of small holes were essentially flat, proved to be the surface of one or several coalesced microatolls that had suffered extensive bioerosion and encrustation (figure 13).

DESCRIPTION OF PLATE 3

FIGURE 15. Microatoll colonies of *Goniastrea* in open water on the windward side of Three Isles, Great Barrier Reef. Hammer is 30 cm long.

FIGURE 16. Underwater photograph of a tall-sided *Porites* microatoll in open water on the leeward side of Low Isles, Great Barrier Reef. The coral is 80 cm tall and the top dead surface is coated with a thin layer of a soft colonial anemone that is just exposed at low water springs.

FIGURE 17. Mini-atolls, 'algal cup reefs'. South shore, Bermuda. Photographed from 10 m altitude.

FIGURE 18. Coalescing mini-atolls. North Lagoon, Bermuda. Photographed from 200 m altitude.

FIGURE 19. Coalescing mini-atolls (showing former concentric growth pattern on dead centres). Kaneohe Bay, Oahu, Hawaii. Photographed from 150 m altitude.

FIGURE 20. Coalescing mini-atolls (some showing a secondary rim development; cf. the secondary rim development in microatolls). Central Region, Great Barrier Reef. Photographed from 250 m altitude.

3. DISTRIBUTION AND MODES OF OCCURRENCE

Each shallow water location on and around the reef flats of the patch reefs is characterized by a predominance of certain species and growth forms of microatolls.

(a) Open water

At the peripheral margin of patch reefs, some corals are exposed at extreme low tides and develop microatoll form. The common massive coral on the windward slopes of the outer margin of the reef flat is *Goniastrea*. This coral develops an irregular or cracked surface and usually grows up to about 75 cm diameter. The windward side of reefs normally suffer heavy surf and consequently water is periodically lapping onto the exposed corals and this could be responsible for the uneven surface (figure 15). On the leeward side the massive colonies of *Porites* grow up to the level of lowest low water and develop steep sided microatolls ('bommies') commonly 1–5 m in diameter (figure 16). The microatolls in open water commonly grow in close association with other corals (commonly branched) to build a framework of skeletons *in situ*. There is a marked change in the growth form of branched corals as they approach the maximum elevation at which they can grow. The density of branching is greater near to the low water level and the branches become flattened and stubby (figure 14).

(b) Ponded water on reef flats

(i) Windward side

Ponded water occurs in moats on the windward side of the reef flats. These moats lie between successive sets of loose or cemented shingle ramparts and are generally about 5–50 m wide (figure 22). The depth of water in moats varies according to the height of the lowest point, or sill, of the pile of sediment ponding the water above the moat floor and also according to the permeability of the ponding sediment. Most moats are between 5 and 40 cm deep at low tide.

As cemented ramparts are generally higher, more permanent and less permeable than loose ramparts, they pond the highest moats (figure 23) and consequently contain the highest living corals (see later). The microatolls contained in the more permanent moats are normally large, tabular colonies. As the water level stays fairly constant throughout the period of low water at each tide the microatolls here tend to develop a simple surface.

The migration to and fro of loose ramparts or spits that retain water brings about periodic changes in the level of pond water. These changes are reflected in the convoluted surface morphology of microatolls.

It is not uncommon for an established moat to suddenly become drained by a breach in the rampart, stranding and killing off all the contained microatolls (figure 24). In certain cases a breach in the rampart may merely lower the lowest level of the moat water and bring about the development of a lower secondary microatoll plane on the colonies. Such a case was observed at the seaward side of Third Island of the Three Isles reef (figure 21). As this reef was mapped by Spender in 1929 (Spender 1930) and again by Stoddart in 1973, it was possible to compare the positions of the ramparts. Since 1929 the rampart has moved on to the reef by about 50 m. In so doing a gap has developed, draining the ponded water over a sill. However, before the rampart was breached, part of it joined the low cliff of rampart rock on the Third Island and maintained a part of the moat. The cross section (figure 21) is drawn to scale to show the relative heights of lowest water currently in the moat and in the ponded water on the reef flat side of the sill. The corals in the moat have either an unaltered hemispherical shape or else a simple microatoll top; those behind the sill have either a simple microatoll top or a secondary microatoll plane below an exposed part of the colony indicating a lowering of lowest water level. It was also noted that through the feather edge side of the rampart were exposed round-topped dead coral skeletons that were about half the diameter of those currently living in the moat. The surface of these dead round-topped corals is considerably higher than those living in open water and they most probably grew in the moat created by the rampart in the 1929 position and were first buried then later exhumed during its subsequent advance.

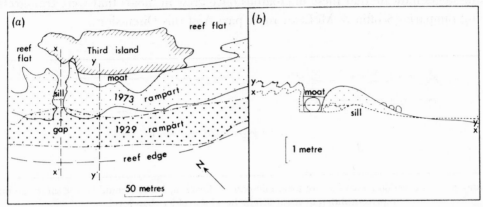

FIGURE 21. Plan (a) and cross section (b) of the area neighbouring Third Island on Three Isles Reef, Great Barrier Reef. The plan shows the 1929 and 1973 positions of the loose shingle rampart, indicating the positions of the present gap and moat. The cross section shows a profile through the moat superimposed on a profile along the gap over the sill.

(ii) *Leeward side*

The bulk of the draining water from the central part of the reef flat flows to leeward, the lower side of the reef flat. The water either drains freely or exits over a shallow sill at the flanks of the leeward sand cay and consequently it is rather shallow throughout low tide, supporting only thin microatolls. It is in these leeward moats that the branching coral microatolls are the most abundant. This probably relates to their greater ability to cope with the large quantities of bottom-moving sediments. Here also, asymmetric growth develops in the unidirectional currents as described earlier (figure 12). Leeward sand migration (driven by wind-induced surface currents) is common in these moats and sometimes sediment is piled up on the leeward sides of corals after strong winds. The leeward tails of sediment are ephemeral and disappear after a few days. Presumably currents, along with fish and other cavity dwellers, clear the sand.

(iii) *Central reef flat*

In the central part of the reef flat water is ponded at low tide in broad shallow pools. In these pools large thin microatolls grow and commonly coalesce to fill the pool with a flat coral pavement. Massive forms predominate and *Porites lutea* is the most common species. As the microatolls grow broad yet thin and unattached to a firm substrate they are readily broken by wave action during high tide. Both simple and convoluted upper surfaces are common.

4. Fossil microatolls

In-place fossil microatolls, up to 6000 years old, were found either exposed on the reef flats (figure 25) or in a matrix of cemented shingle in cliff exposures of limestone on the reef flats of the 'low wooded island' type of patch reefs (figure 26). The growth structure, species composition and distribution of these fossil microatolls are similar to those of the present forms living in moats and pools on reef flats.

The disposition of the fossil microatolls and the nature of the matrix and overlying sediments of the limestones indicate that these microatolls once grew in moats that were transgressed by advancing ramparts (Scoffin & McLean 1978, part A of this Discussion).

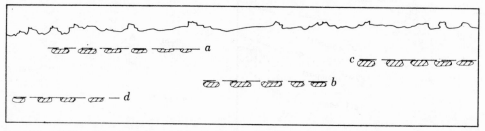

FIGURE 28. Cliff section view of sets of fossil microatolls. Levels *a, b* and *c* could be the same age but represent different moat levels. Level *d* is older than level *a*.

As the tops of living microatolls are all at the same height for one enclosed body of water, it follows that a group of fossil microatolls whose surfaces are all at the same height most possibly represent one earlier moat. Consequently layers of fossil microatolls in different strata may well be contemporaneous and just represent different heights of water in different (but neighbouring) moats (figure 28). The groups of microatolls in figure 28 at levels *a, b, c*, each represent a separate moat and they could have been (though not necessarily) contemporaneous. Field evidence for chronological difference in levels would be the existence of one group of microatolls directly under another, *a* and *d* in figure 28, not just lower and to the side.

5. Heights and ages of Northern Province microatolls

(a) A note on datum

All levelled profiles have been reduced to low water datum, which is the mean height of lower low waters at springs. Individual profiles have been related to low water datum by reference to predicted tides at Cairns: for most of the area covered by these surveys, the predicted tides do not differ significantly in amplitude or timing from those at Cairns and corrections to the Cairns curves only need to be made north of Cape Melville.

It is important to realize that there are several sources of error in these reductions and we cannot evaluate their magnitude. First, there may be actual differences in tidal curves between Cairns, the secondary stations for which predictions are available (these include several of the islands studied, notably Green Island, Hope Islands, Howick Island, Low Isles, Low Wooded Island) and the islands on which profiles were measured. Secondly, local tidal levels may have been distorted by meteorological effects: the trade winds blew strongly through most of the Expedition and it would be surprising if on some days at least tidal levels were not distorted by up to 0.3 m. Thirdly, and perhaps most importantly, all reductions to datum were made by observing a still water level at a known time on a particular profile and relating this to a predicted tide curve. Determination of a still water level is often difficult because of rough conditions, especially at middle and high waters. Relation of the still water level to the tidal curve is also more difficult on irregular low-amplitude neap tides than on springs: most of the surveys, however, were carried out on springs.

In the absence of tidal records at each of the sites surveyed, however, the height data represent the best estimate of elevations related to low water datum and enable different sections to be directly compared. The errors mentioned above are considerably reduced when the relation between one level and another is determined on one reef. All heights are given in metres. For comparison, at Cairns, mean high water springs are at 2.3 m, mean low water neaps at 1.2 m and mean low water springs at 0.5 m; these figures probably apply to all islands north to Cape Melville.

TABLE 1

location	elevation/m
Low Isles	0.55
Three Isles	0.25, 0.59, 0.88, 0.94
Low Wooded Island	0.63
Two Isles	0.26, 0.35, 0.44
Houghton Island	-0.07, -0.16, -0.13
Ingram	-0.47
Leggatt	0.09, 0.15

TABLE 2

location	elevation/m
Three Isles	0.76, 0.76, 0.85, 0.98, 0.99, 1.06, 1.09
Low Wooded Island	1.17
Two Isles	0.42
Watson	0.90
Leggatt	0.76

(b) Highest living open-water corals

The outer parts of reef flats of isolated patch reefs, seaward of the shingle ramparts, characteristically slope regularly seawards with inclinations of about 0° 40′ (about 1 : 80) and widths of 60–80 m (but sometimes over 100 m). Inclination increases rapidly about the level of low water springs and corals growing on these outer lower slopes are immersed at low spring tides. These represent the highest living open-water corals and their elevations are presumably directly related to tidal cycle and duration of immersion at different depths.

Sample elevations for living corals are given in table 1 (mean low water springs 0.5 m).

We conclude that mean low water springs represents an effective upper limit to coral growth in free-draining locations (figure 29).

(c) Highest living reef-flat microatolls

The highest uncemented rampart measured was at Watson Island and had an elevation of 3.12 m above datum. The highest cemented rampart was 3.42 m at Houghton Island. The uncemented ramparts pond water to a lower level than the cemented ramparts and this difference is reflected in the elevations of the microatolls living in the moats (figure 23).

Sample elevations of microatolls in moats of uncemented ramparts are shown in table 2.

Highest microatoll heights in moats of cemented ramparts at Watson are 1.53, 1.53, 1.54 and 1.55 m.

The highest living reef-flat microatoll (in fact the highest living coral) recorded by us on the Northern Province of the Great Barrier Reef was 1.55 m, i.e. approximately the level of mean high water neaps (1.6 m above datum). Even though it is theoretically possible for sea water to be ponded above high water neaps, it is doubtful that corals could grow in water that was not renewed every day. Thus it is concluded that the effective upper limit of growth for reef-flat microatolls is high water neaps (figure 29). Therefore, for this area, if we take present mean low water springs (0.5 m) as the effective upper limit of growing coral in open-water, free-draining localities then any fossil coral formed less than 1.1 m above this level could not be taken as evidence of any necessary change in sea level since it grew. This is so unless the fossil coral could be satisfactorily shown to be an open-water variety and not a reef-flat form.

(d) Corals suitable for sea level history determinations

When attempting to determine the earlier (low water) levels of open sea water, care should be taken in selecting those fossil corals for dating and levelling whose structures suggest that they grew up to sea level in open water rather than those characteristic of shallow pools on reef flats. Some criteria for distinguishing reef-flat from open-water forms are listed below.

(i) At low tide reef-flat pools are rarely deeper than 50 cm, for the amplitude of naturally formed ramparts of coral shingle rarely exceeds 50 cm for the *entire* perimeter of the pool. Therefore fossil microatolls having walls taller than 50 cm are likely not to have grown on reef flats.

(ii) The formation of a new rampart may seal an outflow channel and suddenly cause an increase in the level of pool water, and the new conditions continue for sufficient time to affect the coral structure; or conversely, a breaching of a rampart may rapidly lower pool level. Such changes are not uncommon on reef flats and will be reflected in the terraced structure of reef-flat microatolls. Ephemeral changes in water level of this sort do not occur in the open water and terracing of the upper surface of microatolls does not occur in this environment.

(iii) The dead surfaces of reef-flat microatolls are normally flat as a result of the stillness of the protected ponded water, whereas in the open water the waves lap onto the corals and more irregular upper dead surfaces are usual.

(iv) Reef-flat microatolls rarely form part of a contiguous coral framework. They are normally seated on cemented shingle and, with time, become enveloped in shingle and sand. The open-water microatolls grow to heights well above sediment level and are commonly part of a contiguous coral framework *in situ* which may include a variety of branching corals. It is also noted in open water that branching corals develop shorter stubby branches at the level of low water and the upper portion of the corals becomes much denser with branches (figure 14).

(e) High fossil microatolls

All the fossil microatolls that were collected for dating and whose elevations were measured were of the reef-flat type and were exposed either as isolated in-place corals on reef flats and among mangroves (figure 25) or in cliff sections where they were cemented in growth position in a matrix of coral shingle (figure 26). Sample heights of the highest of the exposed isolated types are given in table 3.

Carbon-14 dates for samples from each site are shown in table 4.

TABLE 3

location	height/m
Houghton Island	1.06, 1.16, 1.32, 1.35
Leggatt	1.08, 1.09, 1.10, 1.10,
	1.11, 1.12, 1.14

(roughly the level of present mean low water neaps: 1.2 m)

TABLE 4

sample location and code	age B.P./a
Houghton Island (ANU-1287)	5850 ± 170
Leggatt Island (ANU-1286)	5800 ± 130

(determinations by H. Polach for R. McLean)

Sample heights of the highest of those cemented in shingle ramparts are as follows given in table 5 (measurements by D. Hopley marked *).

TABLE 5

location	height/m
East Pethebridge Island	0.98,* 1.44*
Turtle I	1.93, 1.99
Nymph Island	1.16,* 1.65,* 1.66*
	1.76*, 1.80,* 2.10* 2.30*

Carbon-14 dates for high microatolls in cemented shingle are shown in table 6.

TABLE 6

sample location and code	age B.P./a
Turtle I (ANU-1478)	4420 ± 90
Nymph Island (ANU-1285)	3700 ± 90
Houghton Island (ANU-1595)	3250 ± 80
Three Isles (ANU-1380)	3750 ± 110

(determinations by H. Polach for R. McLean)

The only exposed fossil microatolls of reef-flat type that occur out of place with respect to present growth levels are those found above 1.6 m (present mean high water neaps). This is to say that several of the corals found exposed in different sections of cemented shingle which date about 4000 a B.P. must have grown under conditions of higher mean high water neaps than presently prevail. The highest fossil microatoll found (at Nymph) was 2.3 m above datum so mean high water neaps of about 4000 years ago here was at least 0.7 m higher than that of today. A summary of the highest levels of living and fossil microatolls in relation to tidal and moat levels is given in figure 29.

A fall of 0.7 m in the level of mean high water neaps over the last 4000 years can be accounted for by one or a combination of two processes: either sea level as a whole has lowered 0.7 m relative to the reef (i.e. eustatic or tectonic change) or sea level relative to the reef has not altered but the tidal range has changed (such an event over such a time would be accounted for most simply by a slight change in the configuration or depth of the neighbouring ocean basin). If 4000 a B.P. sea level, as a whole relative to the reef, was 0.7 m higher than it is today (but tidal ranges were similar to today's) then the highest levels of fossil open-water and reef-flat corals of this age would both now be raised 0.7 m above present highest levels; but if 4000 a B.P. sea level was as today's but tidal ranges differed such that mean high water neaps was 0.7 m higher than it is today, then the level of mean low water springs (height of open-water corals), would be *lower* than today, for when a tidal range increases the low levels are depressed while the high levels are elevated. So if the raising of the 4000 a B.P. microatolls relates only to a change in tidal range it may be expected that no evidence of fossil open-water corals higher than today's would be found.

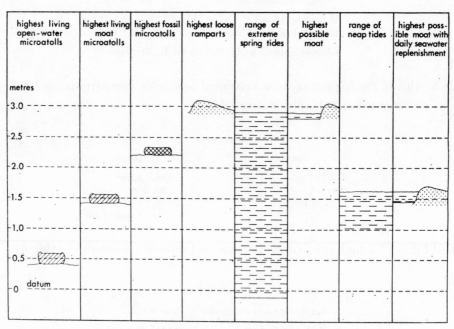

FIGURE 29. Highest levels of living and fossil microatolls shown in relation
to tidal ranges and moat levels.

A thorough search of all the islands in this region between Cairns and Cape Melville revealed no exposures of cliffs or cemented platforms of in-place open-water corals. It is not impossible that such features once existed but have subsequently been eroded, though in the light of the well preserved cemented ramparts and in-place exposed microatolls as old as 6000 a B.P. it is thought unlikely that such cliffs or platforms were ever exposed. However, a 0.7 m lowering of sea level would not perhaps be expected to expose open-water reefs as cliffs and platforms similar to those produced by the intertidal deposits of ramparts that build up to 3 m elevation; rather, the open-water structures would expose to produce low broad flattish areas with very gentle seaward slopes that would soon be covered with intertidal deposits of coral shingle. If a raised open-water reef exists it should have the same relative disposition to the fossil ramparts as the present open-water reef has to present ramparts. Though no in-place fossil branching coral framework was found it is possible that some of the very large dead microatolls that occur at the leeward flanks of some reef flats (figure 27) represent examples of raised open-water microatolls ('bommies'). Unfortunately, during the Expedition no cores were taken of these corals to check their thicknesses to confirm their open-water growth form, nor were dating samples taken to indicate their age, though the elevation of the surface of one (figure 27) was measured to be 0.3 m above low water springs, i.e. 0.8 m above datum. At present all that can be stated with any degree of certainty is that between 4420 and 3250 years ago the level of mean high water neaps was at least 0.7 m higher than it is today. Whether or not the level of mean low water springs has fallen by the same amount over this time is speculation.

6. Summary of conclusions

Microatolls are coral colonies whose dead, flat upper surfaces result from a restriction in upward growth by the proximity of the air/water interface. Long fluctuations in the water level are reflected in the surface morphology and internal structure of microatolls. As the basal surfaces of many microatolls are governed by the sand/water interface the vertical thicknesses of the colonies represent the depth of the water in which they grew. The reefs of the Northern Province of the Great Barrier Reef support at least 43 species (23 different genera) of corals having microatoll growth form. Forms living at the edge of reefs in open water (up to mean low water springs, i.e. 0.5 m above datum) can be distinguished from forms living in rampart-bounded ponds and moats (up to mean high water neaps, i.e. 1.6 m above datum) on reef flats. Because of this distinction it is theoretically possible, with the aid of exposed (datable) fossil microatolls of open-water and reef-flat types, to determine not only sea level history but also tidal history of raised coral reefs. The highest fossil microatolls found in this region were 4000 years old and their elevation indicates that mean high water neaps was, at the time of their growth, at least 0.7 m higher than it is today.

7. Discussion

(a) Ancient microatolls

A microatoll shape indicates a proximity to water surface during growth and therefore would be a useful indicator of extremely shallow water deposition if found in ancient limestones. Although massive colonial organisms have been building reefs since Lower Palaeozoic times, it is surprising that there are no reports of pre-Pleistocene atoll-shaped skeletons. Possible explanations for this lack are given below.

(i) Microatolls are found growing in abundance only on reef flats that have areas of ponded water during low tide. Special conditions have to be satisfied for such reef flats to develop and these conditions may not have been met in the past. Today on the Great Barrier Reef patch reefs, reef flats are formed above low water where steady strong waves pile up branching coral sticks into ramparts around the seaward margins. If the ramparts are stabilized by cement or mangrove growth, moats are established and maintained. A wide tidal range is necessary to ensure diurnal water replenishment in the high moats. Where wave energy is very high and the reef, though shallow, is narrow (e.g. outer barrier of the Great Barrier Reef) the shingle is carried right across the reef and dumped in deeper water on the leeward side.

(ii) Microatolls grow only at the summit of reef development and are therefore the first skeletons to be destroyed on exposure of the reef.

(iii) They have not been recognized as microatolls in cross section.

(b) *Patch reefs of atoll shape*

A structure that grows at a uniform rate in all directions from a locus on a flat substrate will build a hemispherical shape. If this structure is confined to lateral development once water level is reached an atoll shape is produced. This structure can be a coral skeleton or a patch reef that developed from one initial coral. The chief difference between the processes operating in the production of the atoll-shaped coral and atoll-shaped patch reef is that on the reef numerous builders and reef dwellers all contribute to the supply of loose sediment which can, once sea surface is reached, choke the central areas as it cannot easily escape over the lip of actively growing corals at the margin. The central areas of patch reefs may then subside to a level of sand by a more rapid rate of bioerosion than growth. Examples of 'mini-atoll' patch reefs from Bermuda, Hawaii and Australia are shown in figures 17–20. The patch reefs coalesce (figures 18 and 19) in a similar manner to the coalescing of microatolls, except that in reefs the influence of flanking loose sediment is more important. One example of patch reefs from the Great Barrier Reef (figure 20) has the appearance of a secondary split in the coral growth of the lip producing a form similar to microatolls that have enjoyed a period of renewed growth to a higher water level, giving a secondary microatoll plane. Whether or not it is possible to interpret recent sea level changes from the surface morphology of patch reefs just as we can determine pond level changes by using the surface and internal configurations of microatolls remains to be seen.

REFERENCES (Scoffin & Stoddart)

Agassiz, A. 1895 *Bull. Mus. comp. Zool. Harvard Coll.* **26**, 209–281.
Dana, J. D. 1872 *Corals and coral islands.* (398 pages.) New York: Dodd and Mean.
Darwin, C. R. 1842 *The structure and distribution of coral reefs.* (214 pages.) London: Smith, Elder and Co.
Garrett, P., Smith, D. C., Wilson, A. O. & Patriquin, D. 1971 *J. Geol.* **79**, 647–668.
Guppy, H. B. 1886 *Proc. R. Soc. Edinb.* **13**, 857–904.
Hoskin, C. M. 1963 *Recent carbonate sedimentation on Alacran Reef, Yucatan, Mexico.* National Academy of Sciences – National Research Council, Pub. 1089, pp. 1–160.
Kuenen, P. H. 1933 *Snellius – Exped. east. Part Neth. E.-Indies.* **5** (2), 1–126.
Kornicker, L. S. & Boyd, D. W. 1962 *Bull. Am. Ass. Petrol. Geol.* **46**, 640–673.
Krempf, A. 1927 *Mém. Trav. Serv. Océanogr. Pêche Indochine*, **2**, 1–33.
MacNeil, F. S. 1954 *Am. J. Sci.* **252**, 385–401.
Newell, N. D. & Rigby, J. K. 1957 *Soc. Econ. Palaeont. Miner. Sp. Pub.* **5**, 15.72.
Scheer, G. 1972 *Proc. Symp. Corals Coral Reefs 1969*, pp. 87–120. Marine Biological Association of India.
Scoffin, T. P. & McLean, R. F. 1978 *Phil. Trans. R. Soc. Lond.* A **291**, 119–138 (part A of this Discussion).
Semper, C. 1880 *Die natürlichen Existenzbedingungen der Thiere.* Leipzig. Two volumes.
Semper, C. 1899 *The natural conditions of existence as they affect animal life.* (Fifth edition, 472 pages.) London: Kegan Paul, Trench, Trübner and Co.
Spender, M. A. 1930 *Geogrl J.* **76**, 193–214 and 273–297.
Stoddart, D. R. & Scoffin, T. P. 1978 *Atoll Res. Bull.* (In the press.)
Wood-Jones, F. 1912 *Coral and atolls: a history and description of the Keeling-Cocos Islands, with an account of their fauna and flora, and a discussion of the method of development and transformation of coral structures in general.* (392 pages.) London: L. Reeve and Co.

Appendix: Determination of a collection of coral microatoll specimens from the northern Great Barrier Reef

By B. R. Rosen

British Museum (Natural History),
Cromwell Road, London SW7 5BD, U.K.

1. Introduction

The following list is based on the coral collection made by T. P. Scoffin and D. R. Stoddart in the course of their investigation of microatoll growth in the Northern Great Barrier Reef during the Great Barrier Reef Expedition in 1973. The specimens are all taken from microatolls of the reef flats between Cairns and Cape Melville and are intended to be representative of the range of reef flat species observed to grow in this form, within the localities visited by the Expedition.

TABLE A1. NUMBERS OF MICROATOLL SPECIMENS COLLECTED FROM EACH LOCALITY,
WITH SPECIES AND GENERA DATA

locality	position lat. S	position long. E	no. of specimens collected	no. of scleractinian genera and subgenera collected	no. of non-scleractinian genera collected	no. of scleractinian species collected	no. of non-scleractinian species collected
East Hope I.	15° 00′	145° 27′	1	0	1	0	1
Larks Pass Reef	15° 08′	145° 43′	1	0	1	0	1
Nymph I. lagoon	14° 39′	145° 15′	1	1	0	1	0
Three Is	15° 07′	145° 25′	3	3	1	3	1
Turtle IV I.	14° 43′	145° 12′	38	18	1	28	1
Two Is, east side	15° 01′	145° 26′	1	1	0	1	0
Two Is, west side	15° 01′	145° 27′	12	9	0	11	0
Watson I.	14° 28′	144° 54′	13	9	1	11	1
West Hope I.	15° 45′	145° 26′	2	2	0	2	0
totals for this collection			**72**	**22**	**3**	**40**	**3**

In preparing this list, I have accepted the collectors' concept of what constitutes a microatoll, but it is useful to draw attention to the fact that, in addition to the usual massive forms, branching and foliaceous corals are also included. The diagnostic feature for microatolls formed by non-massive species is that an imaginary surface defined by the outermost growth limits of their heads should have the same form as a massive microatoll, i.e. a living lateral margin with a dead or partially dead upper surface. From this, and the species listed here, it would appear that a very wide range of reef flat corals are potential microatoll builders. Probably it is only the laminar encrusting species and the free-living fungiids that are to be excluded. For the same reason, the list cannot be regarded as a check list of all microatoll builders, as one might predict that additional species are to be found with this growth form, especially among the poritids and faviids.

In the check list I have indicated the basis of the determinations under the heading 'Reference' for each specimen. I have given a published work or works and in instances where there is relevant material in the collections of the British Museum (Natural History) I have also cited particular specimens. In this way, I have intended to convey whose species concept I have followed for each determination. I believe this to be especially helpful in the great problematic genera like *Acropora* and *Porites*. As I have not given full synonymies, it follows that concepts of authors not cited are not necessarily excluded. The classification used is that by Wells (1956). The specimen numbers throughout are register numbers of the Department of Zoology, British Museum (Natural History), where the microatoll specimens are now placed. They consist entirely of dried material.

Geographical positions of the localities given for the specimens are listed in table A1, together with the number of specimens, genera and species collected at each one. The highly selective nature of the collection makes zoogeographic comment inappropriate. None of the coral genera listed is new to the Great Barrier Reef region. All but *Goniastrea* are listed by Wells (1955), this genus having evidently been accidentally omitted. Wells listed it in his earlier table of Indo-Pacific distribution (1954). Indication of new records at species level is of doubtful value at the moment, because of the uncertain status of so many coral species names. The excellent work begun by Veron & Pichon (1976) will eventually provide a proper picture of species distribution for the whole region. Such remarks as have been made here are restricted to those genera which have been treated so far in their currently published results of this project (above). It may be inferred from the present list that unless there are remarks to the contrary, species in the genera *Psammocora* and *Pocillopora* are already known from the Cairns–Cape Melville region.

I should like to thank Ms J. G. Darrell for her assistance in this work, and Dr Michel Pichon (James Cook University of North Queensland) for checking the manuscript.

2. Determinations

class Anthozoa Ehrenberg, 1834
subclass Zoantharia de Blainville, 1830
order Scleractinia Bourne, 1900
suborder Astrocoeniina Vaughan & Wells, 1943
family Thamnasteriidae Vaughan & Wells, 1943
genus *Psammocora* Dana, 1846

Psammocora contigua (Esper)
　Reference: Veron & Pichon (1976)
　Material: Turtle IV I. (1976.3.1.183)
Psammocora digitata (Edwards & Haime)
　Reference: Veron & Pichon (1976) and synonymy.
　Remarks: Veron & Pichon regard the subgenera of *Psammocora* as recognized by previous authors as indistinguishable. This species has previously been placed in *Psammocora* (*Stephanaria*) as the synonyms *P. togianensis* and *P. exaesa*. Not previously recorded between Cairns and Cape Melville.
　Material: Turtle IV I. (1976.3.1.65)

family Pocilloporidae Gray, 1842
genus *Pocillopora* Lamarck, 1816

Pocillopora damicornis (Linnaeus)
　Reference: Veron & Pichon (1976)
　Material: Turtle IV I. (1976.3.1.44); Two Is, west side (1976.3.1.20, 1976.3.1.30).

family Acroporidae Verrill, 1902
genus *Acropora* Oken, 1815

Acropora corymbosa (Lamarck)
   Reference: Brook's (1893) *Madrepora corymbosa* specimen (1845.18.12.11)
   Material: Turtle IV I. (1976.3.1.55)

Acropora cuneata (Dana)
   Reference: Wells (1954)
   Material: Turtle IV I. (1976.3.1.46)

Acropora cf. *cymbicyathus* (Brook)
   Reference: Wells (1954)
   Remarks: The present specimen differs from Wells's material probably in consequence of its habitat. There are too few well developed branches to be certain of complete identity. Brook's type is not available for examination at the B.M.(N.H.) because his species is in fact a nom.nov. for Ortmann's *Madrepora cerealis* Dana, said to be in the Strasburg Museum.
   Material: Turtle IV I. (1976.3.1.43)

Acropora palifera (Lamarck)
   Reference: Wells (1954)
   Material: Turtle IV I. (1976.3.1.59); Two Is, west side (1976.3.1.23).

Acropora cf. *rosaria* (Dana)
   References: Wells's (1954) and Crossland's (1952) specimens (1934.5.14.40 (= G.B.R.E. No. 368), 1934.5.14.357 (= G.B.R.E. No. 108)).
   Remarks: The growth form of the present specimen hinders more certain determination.
   Material: Turtle IV I. (1976.3.1.54)

Acropora squamosa (Brook)
   Reference: Crossland's (1952) specimen (1934.5.14.361 (= G.B.R.E. No. 107)) but not necessarily other material of Crossland and Brook (1893).
   Material: Two Is, west side (1976.3.1.22)

Acropora valida (Dana)
   Reference: Wells (1954) and Brook's (1893) specimens of *Madrepora valida* (e.g. 1893.4.7.118).
   Material: Three Is (1976.3.1.13).

genus *Montipora* de Blainville, 1830

Montipora hispida (Dana)
   Reference: Bernard (1897)
   Material: Turtle IV I. (1976.3.1.47)

suborder Fungiina Duncan, 1884
superfamily Agariciicae Gray, 1847
family Agariciidae Gray, 1847
genus *Pavona* Lamarck, 1801
subgenus *Pavona* Lamarck, 1801

Pavona (*Pavona*) *danai* (Edwards & Haime)
   Reference: Vaughan (1918)
   Material: Turtle IV I. (1976.3.1.62)

Pavona (Pavona) explanulata Lamarck
  Reference: Vaughan (1918)
  Remarks: See also *P. duerdeni* Vaughan and *P. diffluens* Lamarck.
  Material: Turtle IV I. (1976.3.1.67)

subgenus *Polyastra* Ehrenberg, 1834

Pavona (Polyastra) obtusata (Quelch)
  References: Quelch's (1886) type (1886.12.9.162) and second specimen (1886.12.9.163), and Wells (1936).
  Material: Turtle IV I. (1976.3.1.32)

genus *Coeloseris* Vaughan, 1918

Coeloseris mayeri Vaughan
  Reference: Vaughan (1918)
  Material: Watson I. (1976.3.1.6, 1976.3.1.11)

superfamily Poriticae Gray, 1842
family Poritidae Gray, 1842
genus *Goniopora* de Blainville, 1830

Goniopora cf. *tenuidens* (Quelch)
  References: Quelch's (1886) type (1886.12.9.304) and second specimen (1886.12.9.308), Vaughan (1918), and Bernard's *Goniopora* 'Great Barrier Reef (12)4' (1892.12.1.542).
  Remarks: Present specimens differ from the above material in having radial wall elements which are thicker than septal elements. The corallites appear like neat 'punctures' in the corallum surface, a character not really seen in the reference specimens. Vaughan (1918) suggested that Bernard's 'Great Barrier Reef (12) 4' and '(12) 5' probably belong with Quelch's species. The first of these does; the second does not. Crossland's (1952) *G. tenuidens* (1934.5.14.498) is certainly different from the reference material, and at present should be excluded from Quelch's species. There is a great variation in corallite depth in these corals. Very shallow corallites are developed on what were the downward facing surfaces of 1976.3.1.63, as orientated in life position. This may be a regular character in this species, or even the genus.
  Material: Nymph I., lagoon (1976.3.1.18); Turtle IV I. (1976.3.1.64).

genus *Porites* Link, 1807
subgenus *Porites* Link, 1807

Porites (Porites) andrewsi Vaughan
  Reference: Vaughan (1918)
  Material: Two Is, west side (1976.3.1.21)
Porites (Porites) compressa Dana
  Reference: Vaughan (1907)
  Material: Turtle IV I. (1976.3.1.39, 1976.3.1.51); Two Is, east side (1976.3.1.24).
Porites (Porites) lutea Edwards & Haime
  Reference: Vaughan (1918)
  Material: Turtle IV I. (1976.3.1.61, 1976.3.1.67); West Hope I. (1976.3.1.1).

Porites (Porites) mayeri Vaughan
  Reference: Vaughan (1918)
  Material: Watson I. (1976.3.1.4)
Porites (Porites) nigrescens Dana
  Reference: Vaughan (1918)
  Material: Two Is, west side (1976.3.1.25)

subgenus *Synaraea* Verrill, 1864

Porites (Synaraea) iwayamaensis Eguchi
  Reference: Wells (1954)
  Material: Turtle IV I. (1976.3.1.40, 1976.3.1.53)

suborder Faviina Vaughan & Wells, 1943
superfamily Faviicae Gregory, 1900
family Faviidae Gregory, 1900
subfamily Faviinae Gregory, 1900
genus *Favia* Oken, 1815

Favia favus (Forskål)
  Reference: Rosen (1968)
  Material: Turtle IV I. (1976.3.1.50, 1976.3.1.58, 1976.3.1.66)
Favia pallida (Dana)
  Reference: Vaughan (1918)
  Material: Turtle IV I. (1976.3.1.35, 1976.3.1.45); Two Is, west side (1976.3.1.16); Watson I.
(1976.3.1.7, 1976.3.1.70).

genus *Favites* Link, 1807

Favites abdita (Ellis & Solander)
  Reference: Wijsman-Best (1972)
  Material: Three Is (1976.3.1.26); Turtle IV I. (1976.3.1.34, 1976.3.1.36, 1976.3.1.57);
Two Is, west side (1976.3.1.19); Watson I. (1976.3.1.3, 1976.3.1.7).
Favites acuticollis (Ortmann)
  Reference: Wijsman-Best (1972)
  Material: Turtle IV I. (1976.3.1.37)

genus *Oulophyllia* Edwards & Haime, 1848

Oulophyllia crispa (Lamarck)
  Reference: Wijsman-Best (1972)
  Material: Watson I. (1976.3.1.69)

genus *Goniastrea* Edwards & Haime, 1848

Goniastrea australensis (Edwards & Haime)
  Reference: Wijsman-Best (1972)
  Material: Watson I. (1976.3.1.5)
Goniastrea palauensis (Yabe & Sugiyama)
  Reference: Wijsman-Best (1972)
  Material: Turtle IV I. (1976.3.1.61)
Goniastrea retiformis (Lamarck)
  Reference: Wijsman-Best (1972)

Material: Three Is (1976.3.1.15); Turtle IV I. (1976.3.1.31, 1976.3.1.49); Watson I. (1976.3.1.10, 1976.3.1.68).

genus *Platygyra* Ehrenberg, 1834

Platygyra daedalea (Ellis & Solander)
  Reference: Wijsman-Best (1972)
  Material: Turtle IV I. (1976.3.1.33, 1976.3.1.41, 1976.3.1.48, 1976.3.1.56); Watson I. (1976.3.1.71); West Hope I. (1976.3.1.2).
Platygyra lamellina (Ehrenberg)
  Reference: Wijsman-Best (1972)
  Material: Two Is, west side (1976.3.1.29)
Platygyra sinensis (Edwards & Haime)
  Reference: Wijsman-Best (1972)
  Material: Watson I. (1976.3.1.9)

genus *Leptoria* Edwards & Haime, 1848

Leptoria phrygia (Ellis & Solander)
  Reference: Wijsman-Best (1972)
  Material: Turtle IV I. (1976.3.1.42)

genus *Hydnophora* Fischer de Waldheim, 1807

Hydnophora cf. *microconos* (Lamarck)
  Reference: Wijsman-Best (1972)
  Remarks: The character of the corallites is consistent with *H. microconos*, but the growth form is encrusting as in *H. exesa*.
  Material: Two Is, west side (1976.3.1.27)

subfamily Montastreinae Vaughan & Wells, 1943
genus *Echinopora* Lamarck, 1816

Echinopora lamellosa (Esper)
  Reference: Matthai (1914)
  Material: Turtle IV I. (1976.3.1.60)

family Oculinidae Gray, 1847
subfamily Galaxeinae Vaughan & Wells, 1943
genus *Galaxea* Oken, 1815

Galaxea fascicularis (Linnaeus)
  Reference: Matthai (1914)
  Material: Turtle IV I. (1976.3.1.63)

family Mussidae Ortmann, 1890
genus *Lobophyllia* de Blainville, 1830
subgenus *Lobophyllia* de Blainville, 1830

Lobophyllia corymbosa (Forskål)
  Reference: Matthai (1928)
  Material: Two Is, west side (1976.3.1.17)

genus *Symphyllia* Edwards & Haime, 1848

Symphyllia nobilis (Dana)
  Reference: Wells (1954)
  Material: Turtle IV I. (1976.3.1.52); Two Is, west side (1976.3.1.28), Watson I. (1976.3.1.72).

suborder Dendrophylliina Vaughan & Wells, 1943
family Dendrophylliidae Gray, 1847
genus *Turbinaria* Oken, 1815

Turbinaria cf. *mesenterina* (Lamarck)
  References: Bernard's (1896) specimen (1876.5.5.40), Yabe & Sugiyama (1941).
  Remarks: Corallites of the present specimen are slightly smaller than in Bernard's specimen, and the coenosteum is finer and denser. Bernard's remark that *T. mesenterina* has 'very inconspicuous' septa is not correct. The above specimen of his has now been cleaned of dried soft tissue.
  Material: Turtle IV I. (1976.3.1.38).

subclass Alcyonaria de Blainville, 1830
order Stolonifera Hickson, 1883
family Tubiporidae Edwards & Haime, 1857
genus *Tubipora* Linnaeus, 1758

Tubipora musica Linnaeus
  Material: Larks Pass Reef (1976.3.1.14)

Order Coenothecalia Bourne, 1900
family Helioporidae Moseley, 1876
genus *Heliopora* de Blainville, 1834

Heliopora coerulea (Pallas)
  Reference: Wells (1954)
  Material: East Hope I. (1976.3.1.12)

class Hydrozoa Huxley, 1856
order Milleporina Hickson, 1899
family Milleporidae de Blainville, 1834
genus *Millepora* Linnaeus, 1758

Millepora platyphylla Hemprich & Ehrenberg
  Reference: Wells (1954)
  Material: Three Is (1976.3.1.15); Turtle IV I. (1976.3.1.67); Watson I. (1976.3.1.72).

References

Bernard, H. M. 1896 *Catalogue of the madreporarian corals in the British Museum (Natural History)*. Volume 2. The genus *Turbinaria*, the genus *Astraeopora*. (106 pages.) London: Trustees of the British Museum (Natural History).
Bernard, H. M. 1897 *Catalogue of the madreporarian corals in the British Museum (Natural History)*. Volume 3. The genus *Montipora*, the genus *Anacropora*. (192 pages.) London: Trustees of the British Museum (Natural History).
Bernard, H. M. 1903 *Catalogue of the madreporarian corals in the British Museum (Natural History)*. Volume 4. The family Poritidae. 1. The genus *Goniopora*. (206 pages.) London: Trustees of the British Museum (Natural History).

Brook, G. 1893 *Catalogue of the madreporarian corals in the British Museum (Natural History).* Volume 1. The genus *Madrepora.* (212 pages.) London: Trustees of the British Museum (Natural History).

Crossland, C. 1952 *Scient. Rep. Gt Barrier Reef Exped. 1928–29* **6**, 85–257.

Matthai, G. 1914 *Trans. Linn. Soc. Lond. Zool.* (2) **17**, 1–140.

Matthai, G. 1928 *Catalogue of the madreporarian corals in the British Museum (Natural History).* Volume 7. A monograph of the Recent meandroid Astraeidae. (288 pages.) London: Trustees of the British Museum (Natural History).

Quelch, J. J. 1886 *Rep. scient. Results Voy.* Challenger *Zool.* **16** (46), 1–203.

Rosen, B. R. 1968 *Bull. Br. Mus. nat. Hist. (Zool.).* **16**, 323–352.

Vaughan, T. W. 1907 *Bull. U.S. natn. Mus.* **59** (9), 1–427.

Vaughan, T. W. 1918 *Pap. Dep. mar. Biol. Carnegie Instn Wash.* **9**, 51–234.

Veron, J. E. N. & Pichon, M. 1976 *Aust. Inst. mar. Sci. Monogr. Ser.* **1**, 1–86.

Wells, J. W. 1936 *Ann. Mag. nat. Hist.* (10) **18**, 549–552.

Wells, J. W. 1954 *Prof. Pap. U.S. geol. Surv.* **260** (I), 385–486.

Wells, J. W. 1955 *Rep. Gt Barrier Reef Comm.* **6** (2), 1–9.

Wells, J. W. 1956 In *Treatise on invertebrate paleontology* (F), Coelenterata (ed. R. C. Moore), pp. F328–F444. Lawrence, Kansas: Geological Society of America and University of Kansas Press.

Wijsman-Best, M. 1972 *Bijdr. Dierk.* **42**, 1–90.

Yabe, H. & Sugiyama, T. 1941 *Sci. Rep. Tohoku Univ.* (2) Spec. Vol. 2, pp. 67–91.

Phil. Trans. R. Soc. Lond. B. **284**, 123–127 (1978)   [123]
Printed in Great Britain

Evolution of the far northern barrier reefs

By J. E. N. Veron†

Department of Biological Sciences, James Cook University of North Queensland,
P.O. Box 999, Townsville, Queensland, Australia 4810

The basement and much of the volume of the northern barrier reefs are Miocene in age. Formation of the ancestral northern barrier reefs was greatly affected by late Miocene crustal instability associated with formation of the Papuan Basin. Present physiographic features are primarily the product of Quaternary eustatic changes. Present morphologies of the deltaic and far northern dissected reefs are discussed in relation to what is known of the geological history of the area.

INTRODUCTION

In this paper, morphologic features of the northern barrier reefs are combined with geological data to suggest principal developments in the evolutionary history of the reef.

The inevitable complexity of this subject, combined with lack of information on almost all its aspects, has prevented, until recently, the proposal of any well substantiated model of reef development within the Great Barrier Reef (G.B.R.) Province. Early writers (Gardiner 1898; Hedley 1926; Hedley & Taylor 1908; Davis 1917; Andrews 1922; Richards 1922; Bryan 1928) mostly commented on or discussed various aspects of Darwin's model of reef evolution, or (Crossland 1931; Gardiner 1932) discussed Stephenson, Stephenson, Tandy & Spender's (1931) description of the surface features of Yonge Reef.

Fairbridge (1950) was the first to give an integrated account of the principal problems involved in the understanding of the evolution of the G.B.R. The subject was briefly reviewed by Jones (1966) and Jones & Endean (1967), then discussed in depth by Maxwell (1968, 1969). Recent reviews of the geology and geomorphology of Queensland's continental shelf (Bird 1971; Maxwell 1973; Heidecker 1973) provide a broad background for the interpretation of more detailed, localized studies and for comparisons between the G.B.R. and reef systems of other parts of the world.

The subject of this paper is restricted to the far northern barrier reefs as described in earlier papers (Veron & Hudson 1978; Veron 1978; both in this volume), to which studies the reader is directed without further reference.

EARLY GEOLOGICAL HISTORY

Recent petroleum exploration in the Gulf of Papua has given some insight into the nature oᵢ the reef basement and into its early geological history (Tanner 1969; unpublished well reports of Pasca No. 1 (Phillips Australia Oil Co.) and Anchor Cay No. 1 (Tenneco–Signal Oil Co.)).

The basement of the northern Great Barrier Reef and the associated open-sea platform reefs is lower Miocene in age and is underlain by Mesozoic sedimentary rocks (e.g. Anchor Cay) and highly deformed Palaeozoic metamorphics such as are exposed along much of the Queensland coast.

† Present address: Australian Institute of Marine Science, P.O. Box 1104, Townsville, Australia 4810.

The substructure of the modern reef is Miocene in age and probably much of the reef volume is Miocene (Tanner 1969). During the early Miocene, much of the Papuan Basin appears to have been divided into basins and shelf areas by a system of more or less continuous barrier or elongate reef zones. Those areas behind each barrier were built up by carbonate sedimentation with intermittent patch reefs developing in preferential bathymetric positions in similar manner to the reefs of today.

During late Miocene there began a period of crustal instability and emergence of the fringes of the Papuan Basin. The rapidly rising Niugini Highlands overwhelmed the rapidly sinking Papuan Basin with a flood of clastics. This was accompanied by volcanism, most common in the eastern portion of the basin near the deepest downwarp; but is evidenced in the southwestern portion of the basin where several present-day islands are composed of Recent basalts. Anchor Cay No. 1 well shows that there was more than a single out-pouring, as in places reef corals are interspersed between successive flows.

The ancestral G.B.R. extended northward across the Gulf of Papua and for an unknown distance across the present location of the island of Niugini. However, the above-mentioned late Miocene lowering of the Papuan Basin floor terminated reef growth at approximately 9° S lat. Northward of this line there developed a sharply defined eastward dipping flexure from a 'hinge line'. Stratigraphic intervals increase eastward (basinward) from the hinge line. Some ancestral reefs, e.g. the 'Pasca No. 1 reef', were overwhelmed during the late Miocene; others further to the south were able to keep pace with the sinking basement and appear today as the open-sea lagoonal platform reefs (Portlock Reef, Eastern Fields and Boot Reef) which in places rise abruptly from depths of 1000 m or more.

Barrier reef development continued through Pliocene time to the present. Thus, Anchor Cay well penetrated a reef section of Pliocene age and the reef's continuance is locally evident on seismic sections.

Quaternary geological history

By the close of the Tertiary, major physiographic zones of the northern G.B.R. were probably similar to those of today. The Queensland Trench was separate from the Southern Shelf Embayment, the Southern Marginal Shelf had been linked with that of the Northern Region and the Coral Sea Platform had been isolated (Maxwell 1969). Present surface features are primarily a consequence of Pleistocene and Holocene sea level fluctuations.

Although reliable interpretation of the eustatic record during the Tertiary is difficult, Maxwell (1969) concluded that there is substantial evidence to indicate sea levels that were much higher than the present. The lower sea levels of the Quaternary have led to the planation and erosion of older reefs and to the formation of the strand-lines primarily apparent in the Central and Southern Regions.

The important consequence of these fluctuations for reefs is that periods of prolific reef growth during transgressions alternate with periods of severe erosion and consequent sedimentation during regressions.

Whatever the effects of early Quaternary sea level fluctuation, it is clear that sea levels of the past 120 000 years are largely responsible for the present reef morphologies. It is widely agreed that during this period, eustatic changes oscillated between -100 m and $+10$ m. There appears to have been a major still-stand of about -50 m centred about 40 000 a B.P., after which the sea level decreased to approximately -100 m, 20 000 a B.P. However, the all-important detail of subsequent (Holocene) eustatic changes remains obscured behind a series of conflicting studies, both 'local' and world-wide.

During the last major (Wisconsin) glaciation, the sea must have withdrawn from most of the G.B.R. province, certainly from the Northern Region. The present northern barrier reefs would have formed a high limestone wall separating the ocean from the continental shelf, now an emerged lagoon-floor plain traversed by rivers and covered with steep-sided, flat-topped limestone hills in the positions of modern reefs as well as the continental emergences that now form the continental islands. The seaward face of the outer limestone wall would have been subject to rapid subaerial erosion and would probably have eroded into a line of karstic ridges, such as Chevalier (1973) envisaged for barrier reefs of New Caledonia.

Subsequent to the Wisconsin glaciation the sea rose to its present level – either at a more or less uniform rate as suggested by several authors (see Hopley 1974) or more rapidly to above present levels followed by fluctuations of various descriptions as suggested by others, including Fairbridge (1961) and Mörner (1969, 1971).

DISCUSSION

The lack of information on the age and nature of the surface and subsurface (presumably Quaternary) layers of the reef, combined with the lack of an adequate picture of Holocene sea levels, makes any further account of the reef's history extremely speculative. However, this study does indicate some of the factors which have led to the differentiation of the present barrier reef morphologies. Two aspects are considered.

1. Evolution of present morphologies

(a) The northern limit of the G.B.R. and the northern dissected barrier reefs

This study indicates that the northern limits of the barrier reefs at about $9\frac{1}{2}°$ S as well as the morphology of the reefs between $9\frac{1}{2}$ and $10°$ S have been determined by tectonic subsidences as described by Tanner (1969) rather than by sedimentation and/or low salinity from the rivers of the Gulf of Papua as suggested by Fairbridge (1973 b).

Present evidence for this view is as follows: (1) Coral cover of the far northern barrier reef is as great or is greater than that observed in the deltaic reefs and most of the ribbon reefs studied. (2) Coral abundance of the Murray Islands 28 km SW of the northern reef limit compares favourably with those of more easterly and southerly islands, and shows no evidence of adverse environmental conditions (Mayer 1918; Vaughan 1918; Veron, personal observations). (3) Freshwater from the Fly river has a strong easterly set and would be more likely to affect the barrier reefs of southern Niugini (Whitehouse 1973) than barrier reefs 110 km to the south. (4) Aerial photographs indicate increasing depth of water cover northward from the northernmost reef studied. (5) The two reefs studied mostly had steep sides to depths of 17–37 m and were surrounded by sand and rubble; there was no encroaching silt layer.

(b) Deltaic reefs

Although aerial photographs of the deltaic reefs and the more northern dissected reefs indicate that those two types have little or nothing in common, Veron has shown that they occupy similar bathymetric situations and have similar seaward faces with subcontinuous ridges crossing the mouths of their channels. Principal differences are (1) the nature of the channels, and (2) the formation of the deltaic complex at the back of the deltaic reefs.

Clearly the seaward channels of the deltaic reefs are a product of the flood tide (when water can only cross the reef line via the channels). Their steep sides are poorly illuminated and have little coral cover except near the surface and consequently would appear to be in the process of erosion. The whole of the reef back of the deltaic reefs, especially the westward projecting 'arms' of the channels, appear to be in the process of rapid development, no doubt as a result of the twice daily injection of oceanic water.

The deltaic system therefore appears to be in the process of change and development. The ancestral reef of the dissected and deltaic types was perhaps similar, with the present morphologies being associated with different changes in relative sea levels, the former from tectonic subsidence as indicated above, the latter from a relative decrease in sea level of primarily eustatic origin occurring during much more recent times (see §2 below).

(c) Ribbon reefs

Such a diversity of form is included in the 536 km length of ribbon reefs that few generalizations can be made about them here. However, it should be noted that (1) they are all situated at the very edge of the Queensland Trench and have very steep sloping outer faces descending to great depth and (2) that the reefs are separated by channels and passes of irregular depth and size which indicate a substantial structural separation between one reef and the next.

The origin of the ribbon reefs is probably closely associated with the origins of the Queensland Trench. Gardner (1970) has proposed that the Tasman geosyncline from Niugini (including the east Australian continental shelf) and the Papuan geosyncline (from Niugini and including New Caledonia and New Zealand) (David 1950) were a single feature with Niugini lying adjacent to northern Queensland. He proposed that, between late Eocene and late Oligocene, Niugini rotated anticlockwise to its present position, forming as it did so many of the physiographic features of the Coral Sea, including the Queensland Trench.

Whatever its origin, the absence of any continental slope between the Queensland Trench and the ribbon reefs suggests that some form of reef system has always lined its western rim. Otherwise a continental slope and a narrower shelf would surely have developed.

The detached reefs, including the Great Detached Reef, were probably once part of the main line of barrier reef. Likewise the present main reef line is probably situated westward of earlier lines.

2. Effects of Holocene sea levels on barrier reef outer reef flats

One of the most distinctive aspects of the ribbon and deltaic reefs is the outer reef flat which mostly consists of hard, very flat limestone devoid of any coral cover and, for extensive areas, having little or no cover of larger algae. Reef flats of this nature are not found with inner reefs; they are clearly associated with exposure to strong wave action. This study tentatively suggests that they are the wave-resistant result of planation by fall in relative sea level occurring after the last (Holocene?) emergence. This suggestion is not original; both Newell (1961) and Fairbridge (1973a) have made similar observations on other reefs.

Alternative explanations include: (a) that sufficient unobserved calcifying organisms are present in the rock matrix to balance erosion forces, (b) that the reef flats formed during the present sea level still-stand but that the calcifying organisms involved are no longer present, and

(c) that the reef flats formed during some pre-Holocene still-stand at present sea level and are sufficiently resistant to wave and aerial erosion to have maintained their present form during Holocene sea level fluctuations.

A study of the age and nature of the surface few metres of reef flat should readily clarify the above speculation. At present it is only possible to note that there is no evidence of subaerial erosion on the reef flats. This indicates that their present surfaces were not exposed during the last major regression and have only been exposed during present or above-present sea levels.

The author wishes to thank Professor W. G. H. Maxwell for his comments.

REFERENCES (Veron)

Andrews, E. C. 1922 *J. Proc. R. Soc. New South Wales* **56**, 10–38.

Bird, E. C. F. 1971 *Quaternaria* **14**, 275–283.

Bryan, W. H. 1928 *Rep. Gt Barrier Reef Comm.* **4**, 58–69.

Chevalier, J. P. 1973 In *Biology and geology of coral reefs* (eds O. A. Jones & R. Endean), vol. 1 (Geology 1) pp. 143–167. New York: Academic Press.

Crossland, C. 1931 *Geogrl J.* **77**, 395–396.

David, T. W. E. 1950 *The geology of the Commonwealth of Australia*. Vols 1–3. London: Edward Arnold.

Davis, W. M. 1917 *Am. J. Sci.* **194**, 339–350.

Fairbridge, R. W. 1950 *J. Geol.* **58**, 330–401.

Fairbridge, R. W. 1961 *Phys. Chem. Earth*, **5**, 99–185.

Fairbridge, R. W. 1973a In *New Guinea Barrier Reefs* (ed. W. Manser), Univ. Papua New Guinea, Geology Department, Occasional Paper 1, pp. 129–146.

Fairbridge, R. W. 1973b In *New Guinea Barrier Reefs* (ed. W. Manser), Univ. Papua New Guinea, Geology Department, Occasional Paper 1, pp 163–194.

Gardiner, J. S. 1898 *Proc. Int. Congr. Zool.* 1898, 119–124.

Gardiner, J. S. 1932 *Nature, Lond.* **129**, 748–749.

Gardner, J. V. 1970 *Bull. geol. Soc. Am.* **81**, 2599–2614.

Hedley, C. & Taylor T. G. 1908 *Rept. Aust. Ass. Adv. Sci.*, **1907**, 397–413.

Hedley, C. 1926 *Am. J. Sci.* **11**, 187–193.

Heidecker, E. 1973 In *Biology and geology of coral reefs* (eds. O. A. Jones & R. Endean), vol. 1 (Geology 1), pp.273–298. New York: Academic Press.

Hopley, D. 1974 *Proc. 2nd Int. Coral Reef Symp.* **2**, 551–562.

Jones, O. A. 1966 *Aust. nat. Hist.* **15**, 245–249.

Jones, O. A. & Endean, R. 1967 *Sci. J.* **3** (11), 44–51.

Maxwell, W. G. H. 1968 *Atlas of the Great Barrier Reef*. (258 pages.) Amsterdam, London and New York: Elsevier.

Maxwell, W. G. H. 1969 In *Stratigraphy and palaeontology, essays in honour of Dorothy Hill* (ed. K. S. W. Campbell), pp. 353–374. Canberra: Australian National University Press.

Maxwell, W. G. H. 1973 In *Biology and geology of coral reefs* (eds O. A. Jones & R. Endean), vol. 1 (Geology 1), pp. 233–272. New York: Academic Press.

Mayer, A. G. 1918 *Pap. Dep. mar. Biol. Carnegie Instn Wash.* **9**, 1–48.

Mörner, N. 1969 *Geol. Mijn.* **48**, 389–399.

Mörner, N. 1971 *Quaternaria* **14**, 65–83.

Newell, N. D. 1961 *Z. Geomorph. Supplbd.* **3**, 87–106.

Richards, H. C. 1922 *Qd geog. J.* **36–37**, 42–54.

Stephenson, T. A., Stephenson, A., Tandy, G. & Spender, M. A. 1931 *Scient. Rep. Gt Barrier Reef Exped. 1928–29* **3**, 17–112.

Tanner, J. J. 1969 *E.C.A.F.E.* **1969**, 1–5.

Vaughan, T. W. 1918 *Pap. Dep. mar. Biol. Carnegie Instn Wash.* **9**, 49–234.

Whitehouse, F. W. 1973 In *Biology and geology or coral reefs* (eds O. A. Jones & R. Endean), vol. 1 (Geology 1) pp. 169–186. New York: Academic Press.

Phil. Trans. R. Soc. Lond. B. **284**, 129–139 (1978)   [129]
Printed in Great Britain

The Recent development of the reefs in the Northern Province of the Great Barrier Reef

By T. P. Scoffin†, D. R. Stoddart‡, R. F. McLean§ and P. G. Flood‖

†*Grant Institute of Geology, University of Edinburgh, West Mains Road, Edinburgh EH9 3JW, U.K.*

‡*Department of Geography, University of Cambridge, Downing Place, Cambridge, U.K.*

§*Department of Biogeography and Geomorphology, Research School of Pacific Studies, Australian National University, Canberra, Australia* 2600

‖*Department of Geology and Mineralogy, University of Queensland, St. Lucia, Brisbane, Australia* 4067

[Plates 1–3]

Geophysical and well data suggest that the Recent reef growth on the exposed northern Great Barrier Reefs forms only a thin veneer on earlier reef limestones. Reef framework growth up to sea level is principally from local prominences which supply sediment of unstable coral branches to neighbouring valleys. Once a large area of a reef reaches sea level, the SE winds play an important rôle in further development by causing the major lateral growth to be to leeward and characteristic intertidal deposits to accumulate. Leeward sand cays form during prevailing conditions on reefs across the shelf and may be eroded during storms but windward rampart deposits form during storms and are eroded during prevailing conditions on those reefs exposed to heavy surf. The ramparts on the inner-shelf reefs are stabilized by mangroves and cements resulting in a progressive elevation of the surfaces of reefs towards the mainland.

1. Pre-Holocene

In this province the earliest unambiguously described subsurface reef rock is very early Pleistocene, though Miocene and Pliocene carbonates may well be of reef detritus (Hill 1974). During the Pleistocene, polar ice sheets expanded and retreated such that world sea level fell and rose with an amplitude of 100 m or more for at least 2 Ma (Stoddart 1973). During times of high sea level, reefs, if not too deeply drowned, would grow up to the sea surface and then expand laterally; during times of low sea level any exposed reef limestone would be eroded at the perimeter by marine erosion and at central elevated parts by freshwater solution. If exposure was for a long period the reef surface would be truncated at about sea level. Later growth would start from this surface and a discontinuity would exist between the two reef deposits. Geophysical evidence (Orme, Webb, Kelland & Sargent 1978, part A of this Discussion) indicates the existence of several discontinuity surfaces beneath the northern Great Barrier Reefs. The samples recovered from the borehole at Bewick (Thom, Orme & Polach 1978, part A of this Discussion) indicate that the last major advance of the sea covered a discontinuity surface on the reef at 4 m depth below present low water of spring tides (l.w.s.t.) at about 6300 a B.P. (Polach, McLean, Caldwell & Thom 1978, part A of this Discussion). Presumably very shortly after this the sea first reached its present level during the last major transgression, for the oldest in-place intertidal corals on the reef flats of nearby reefs are 5850 years old (Polach *et al.* 1978).

In this region the living reefs project from a shelf seabed that is about 30 m deep, so borehole evidence suggests that a maximum of only about one-eighth of the reefs' vertical thickness resulted from the growth during the last transgression of the sea. Consequently the major form of the reefs had developed long before the present sequence of continuous reef-top growth commenced.

2. Holocene

2.1. *Introduction*

In the Northern Province of the Great Barrier Reef, reefs near to sea level grow in two distinctly separate environments. One is at the shelf edge where long 'ribbon' reefs grow on the crest of a steep slope facing the open Pacific Ocean, the other is in the shelter of the barrier where individual reefs grow in roughly equidimensional planimetric proportions from a shelf floor of 20–40 m depth. The logistics of the expedition necessitated that most study should be given to the inner-shelf reefs and the results presented here relate principally to these. Our conclusions are drawn from a range of observations made during one visit to the region. These observations include aerial surveys, subsea surveys by Scuba diving, land surveys including drilling, and geophysical surveys.

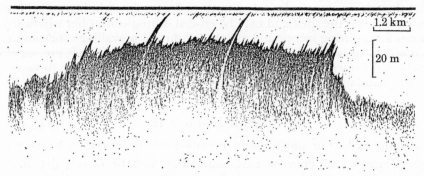

FIGURE 1. Echogram of a reef that does not reach sea level showing the steep coral covered projections; 16 km ENE of Three Isles.

Description of plate 1

FIGURE 2. Underwater photograph of a coral prominence (or bommie) showing the steep side with projecting branching corals. Leeward margin of Green Island; 3 m depth.

FIGURE 3. Underwater photograph of fallen branches of *Turbinaria* regrowing in a new orientation from a valley floor. Turtle III 'Blue Hole'; 3 m depth.

FIGURE 4. Underwater photograph of the upper limit of open-water reef growth. Massive *Porites* corals develop microatoll form. Leeward flanks of Low Isles; 1 m depth.

FIGURE 5. Oblique aerial photograph of the leeward sand split on E Pethebridge Reef. Note the refraction of the waves. Altitude: 200 m.

FIGURE 6. Oblique aerial photograph of Sinclair–Morris Reef showing leeward sand cay with sand spit, partial surround of beach-rock and vegetation cover. The windward side has a smooth arcuate margin and thin mangrove cover. Altitude: 200 m.

FIGURE 7. Aerial photograph of the windward margin of Howick Reef showing the narrow extent of windward growth from the granite island. Altitude: 200 m.

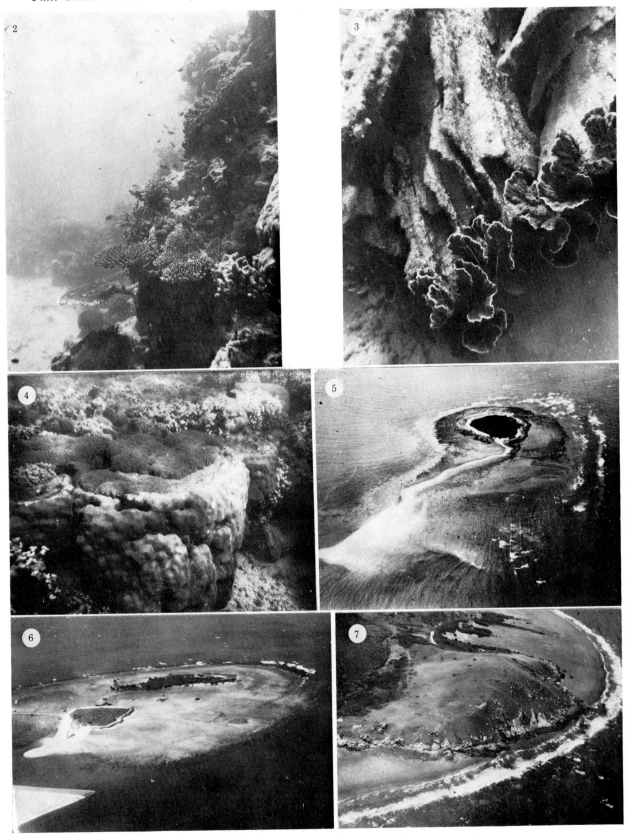

FIGURES 2–7. For description see opposite.

Phil. Trans. R. Soc. Lond. B, volume 284

Scoffin et al., plate 2

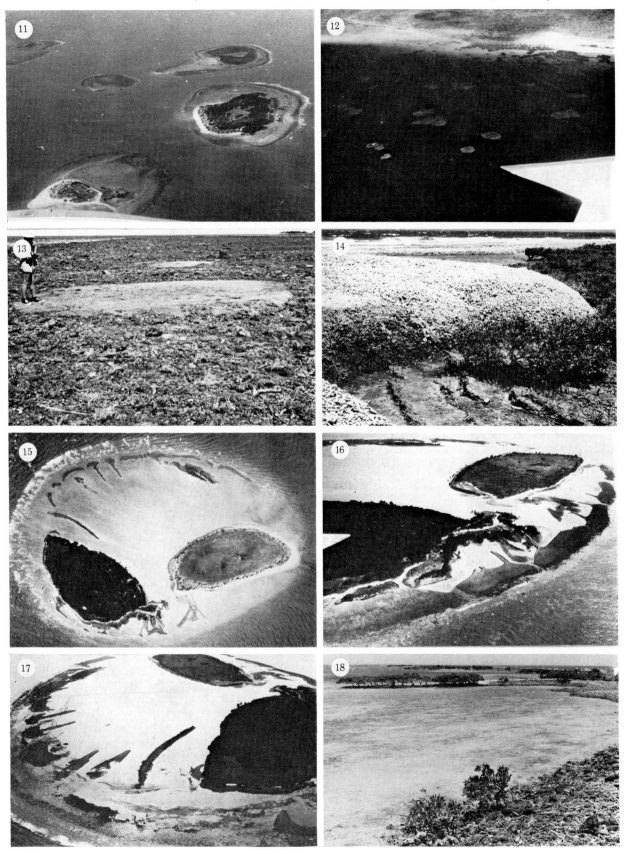

FIGURES 11–18. For description see opposite.

FIGURES 19–26. For description see opposite.

2.2. *Reef development up to low water spring tide level*

Echogram profiles (figure 1) and visual observations during Scuba diving on shallow reefs show that present growth develops an irregular surface of knobs and depressions. These observations and the evidence from the bores in Low Isles reported by Marshall & Orr (1931) indicate that reefs do not grow from depth up to sea level as even broad solid structures but rather that growth is localized at scattered prominences. These projections are built of massive and branching corals (figure 2, plate 1) but it is the greater stability and resistance to erosion that permits the massive corals to remain in place while many of the platy and branching corals eventually break from the framework and fall into the valleys. Some of the toppled branches continue growth from valley floors though renewed growth may be at an angle to former growth (figure 3, plate 1) but most add to the accumulation of rubble and sand in the depressions. All factors favour the growth of prominences:

(*a*) shallow corals are known to grow more rapidly than deep corals;

(*b*) the higher surfaces escape excessive sedimentation;

(*c*) upward pointing growth is stable, whereas outward growth more readily breaks off. Consequently, once a few projections are established they could supply sediment to neighbouring valleys making growth there more difficult.

Eventually the framework structure reaches l.w.s.t. level and upward growth stops. Characteristic growth forms of massive and branching corals are developed at this level (Scoffin & Stoddart 1978, this volume) and the frame expands laterally only (figure 4, plate 1). The valleys become occluded and eventually grown over or filled up with sand. This flat surface of patchy framework and sand is then later veneered by intertidal deposits.

2.2.1. *Wind influence*

The trade winds from the SE drive waves, which are commonly 2–3 m in amplitude, across the shelf. The steady force of waves on shallow reefs causes the development of characteristic

---

DESCRIPTION OF PLATE 3

FIGURE 19. Oblique aerial photograph of the windward margin of Mid Reef showing concentric banding. Altitude: 200 m.

FIGURE 20. Platy coral shingle (each about 15 cm diam.) on the windward margin of the reef flat showing varying degrees of corrosion resulting from the boring activity of microscopic filamentous algae. Hampton Reef.

FIGURE 21. A freshly broken shell of a giant clam revealing large lined borings by *Lithophaga* bivalves (about 1.5 cm diam.) and numerous small borings by *Cliona* sponges. Meguera Reef.

FIGURE 22. A coral boulder (50 cm diam.) corroded by the borings of numerous *Tridacna crocea* shells. Low Isles.

FIGURE 23. A 1 m high cliff of rampart-rock that is notched at the mean low water neap tide level revealing in place microatolls of a former reef flat. Watson Reef.

FIGURE 24. A flat platform of cemented shingle that ponds water to leeward, to the left of the photograph, up to a level of 1.6 m above datum, at low tide. The loose shingle, to the right, ponds water to 0.8 m above datum at low tide. The open sea level at low water spring tide is 0.5 m above datum. Corals currently grow in each body of water up to the level at low tide.

FIGURE 25. Bassett-edge relief of a partly eroded tongue of cemented shingle (the long tongue shown in figure 17). The level of planation of this rock (reef flat level) is down to a former level of reef flat. Three Isles.

FIGURE 26. Cliff of cemented shingle showing present erosion is down below the foundation level of the *in situ* microatoll preserved in the rock. Hammer 30 cm long.

windward and leeward coral assemblages. On the windward margin branching corals of *Acropora* species are dominant; on the flanks and leeward margins, though branching corals are common, massive corals, especially *Porites*, abound. The different growth forms of the windward and leeward corals are reflected in the type of coarse sediment around the reef rim (Flood & Scoffin 1978, part A of this Discussion).

The waves play an important rôle in the transport of sediment on shallow reefs. The swell meets the windward front at 90° and is bifurcated. The two wave sets are refracted as they pass the flanks of the reef and they impinge at an acute angle to the reef margin. This results in longshore drift of sand around reef sides such that on the leeward margins spits develop, or where the two opposing wave sets meet a deposit of sand accumulates (figu res 5 and 6, plate 1).

FIGURE 8. Rose diagram showing the direction of the long axes of 67 shallow inner-shelf reefs of the northern province. The orientation of Recent longitudinal dunes on the mainland at Cape Flattery is shown by the open arrow.

The wave refraction and consequent peripheral sand transport are thought to be the major factors responsible for the regular smooth arcuate form of the windward margins of so many inner-shelf reefs (figure 6). Most of those reefs without this regular arcuate windward margin are seen not yet to have grown close enough to sea level for the waves to have full effect. The dominant orientation of the longest dimension of the shallow inner-shelf reefs of this region is SE–NW (figures 5 and 8); thus it would appear that the trade winds play an important part in governing reef configuration. Winds may influence either the form of the reef foundation (for example in building longitudinal dunes as occur on the nearby mainland at Cape Flattery) or the lateral development of reefs during growth, or both. We have no data that allow us to propose the factors responsible for the foundation shape but we can see that there is a tendency for the present lateral development of reefs to be parallel to the wind.

Framework growth occurs all around the reef perimeter but even though the windward corals are more abundant and possibly grow faster than the leeward corals, the leeward sediment transport is at a much more rapid rate than windward framework expansion. The evidence for this is from aerial photographs, echogram profiles and sediment analyses.

Aerial photographs of the windward margins of the high (continental) islands (such as Howick, Noble and South Island, Lizard Group) show a very limited lateral development of reef structure to windward of the land compared with to leeward (figure 7, plate 1). This observation is further supported by evidence from echograms of the windward and leeward margins of the reefs. These are schematically shown in figure 9. The windward slopes are steep (averaging 50° between 4 and 30 m depth), commonly interrupted by terraces and in some cases surrounded by a narrow trough. The leeward slopes are more gradual (averaging 15° between 4 and 30 m) and lack terraces and a trough, though with scattered coral prominences (figure 10). The windward terraces occur principally at depths of 3–5 m, 10–13 m and rarely at 18 m. The trough, which is only a few metres wide, extends to one or two metres deeper than the surrounding soft seabed which is at about 30 m. The terraces are presumably the remnants of earlier erosion during periods of lower sea level. The origin of the windward troughs is unknown for they were not seen visually, only on echograms. They could result from scour related to currents from the SE, for they die away around reef flanks; if so, then they must be still active for such locations are normally the resting places of reef-front talus. The absence of terraces on the leeward slopes suggests that burial here is more rapid than on the windward slope.

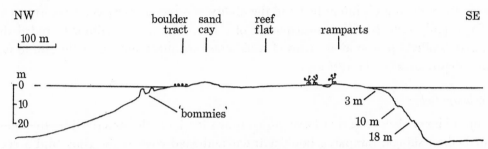

FIGURE 9. Schematic windward–leeward profile of an inner-shelf reef based on echograms from Low Isles, Three Isles and the Howick Group.

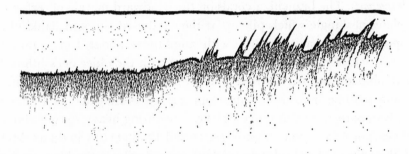

FIGURE 10. Echogram profile of the leeward margin of Michaelmas Reef (the reef flat is to the right) showing gradual sandy slope with steep coral covered projections or bommies. (Many bommies reach the surface (see figure 12) but were avoided during profiling.)

Dyed sand tracer experiments on Ingram–Beanley reef indicated a net leeward migration of sand across the reef flat (Flood & Scoffin 1978), supporting the conclusions of Marshall & Orr (1931) who conducted sediment trap experiments on Low Isles. Doubtless, wind-induced currents carry sand-sized and finer grains to leeward across reef flats during low water but also tidal currents drain to leeward across most of the central and leeward parts of the reef flat during ebb on account of the shallow slopes to leeward of reef flats. During flood tide the incapacity of the tidal current to carry sediment up-slope against still-draining water restricts the return of sediment. Both the physiographic features of reef marginal areas and the sediment composition (Flood, Orme & Scoffin 1978, part A of this Discussion) indicate the building of a leeward cone of sediment from a reef-top source. These submarine deposits plus the intertidal sediment tails and spits extending from innershelf reefs (figures 5, and 11, plate 2) cause the reef axes to parallel the wind direction and the main lateral development to be to leeward. The high coral patches (locally termed 'bommies') to leeward (figures 2, 10, and 12, plate 2) are progressively buried and incorporated into the reef (figure 13, plate 2).

It is difficult to assess the rate of this leeward extension of reefs. The observation of 30 000-year-old reef sediment only 4 m below l.w.s.t. level on the leeward edge of the leeward sand cay on Bewick Reef (Polach *et al.* 1978) suggests that here little leeward extension beyond the elevated foundation has occurred over Holocene time, though the Stapleton reef core data (Polach *et al.* 1978) where 11 m below l.w.s.t. level a fossil coral dates at only 5260 years old, suggest more rapid leeward extension.

2.3. *Reef development above low water spring tide level*

Once the reef has reached sea level the physical gradients are markedly sharpened. Wave erosion of reef-front corals is intensified and the ability of waves to carry coarse sediment across the reef is rapidly reduced. The consequence of this is that coarse sediment eroded during heavy waves is piled up around the rims of reefs. Finer sediment spills over the rim to fill any remaining depressions on the reef top.

2.3.1. *Sediment bodies*

The tops of inner-shelf reefs that have grown to sea level are characterized by the following sediment bodies: shingle ramparts, boulder tracts and sand cays at the rims, and a reef flat sand blanket.

2.3.1.1. *Shingle ramparts.* The branching corals on the windward margin of reefs readily break to form rods or plates of shingle which pile up at the windward rim and flanks of reefs as large asymmetric ripples, known as ramparts, with amplitudes about 1 m (figure 14, plate 2). Commonly several sets of ramparts occur on one reef with the older ones lithified intertidally by cements precipitated from sea water (Scoffin & McLean 1978, part A of this Discussion). The ramparts may pile one on to another developing ridges or they may be separated by narrow shallow moats. Ramparts form and move significantly during heavy storms whose frequency is on the scale of years or tens of years (figures 15, 16 and 17, plate 2) (Stoddart, McLean, Scoffin & Gibbs 1978, this volume; Fairbridge & Teichert 1948; Scoffin & Stoddart 1978, this volume).

2.3.1.2. *Boulder tracts.* At the leeward margin of reefs, massive coral colonies abound and provide a source of sediment. Discrete coral boulders are piled up during storms at the leeward rims of reefs, into linear boulder tracts. It is thought that boulders do not move significantly once deposited on the reef top for the tracts are discrete features, successive boulder tracts being separated by a moat. Also boulders are commonly partly surrounded by matrix and cement. The pattern of boring, intertidal notching and encrustation on boulders is consistent with one position since deposition. If a boulder tract is immobile after formation and if it can be assumed that little lateral erosion of fossil boulder tracts has occurred, then the distance between two boulder tracts on one reef could represent the lateral expansion of a reef over the period between the two periods of deposition. The scant dating data available on boulders suggest that this expansion is of the order of 1.5 cm/a. This is not large considering that the sediment supplied to the leeward margin is derived from all around the reef and the reef top.

2.3.1.3. *Sand cays.* Sand cays develop on the leeward margins of reefs at the confluence of the two opposing sets of refracted waves. Sands from the beaches are built by storm waves and wind action into cays above high water spring tide level. The cays consist of well sorted, rounded sand grains derived from the reef top and margin and consist principally of coral fragments, benthonic Foraminifera, calcareous algae and molluscs. Locally pumice and soil horizons occur. Two main cay terraces have been observed on the sand cays at 3.5 and 5 m elevations (McLean & Stoddart 1978, part A of this Discussion) and the loose cay sands range in age from 2190 ± 70 a B.P. to 4380 ± 80 a B.P. of 18 samples ^{14}C dated (Polach *et al.* 1978). The sand cays are stabilized by freshwater vegetation and by intertidal lithification by fibrous aragonite cements into beach-rock (figure 6). Though cays do show slight seasonal variation in form (Flood 1974; Spender 1930; Steers 1937; Stoddart, McLean & Hopley 1978, this volume) it is thought that their positions have not altered significantly since their inception as relic beach-rock is not found far removed from present cays on reef flats.

2.3.1.4. *Reef-flat sand blanket.* All inner-shelf reefs have a large part of the intertidal reef top – normally the central and leeward areas – veneered by sand. Probing with a metal rod showed this loose sand to be less than a metre in thickness. The surface sand is mobile in all the exposed areas of the reef top but where the reef top has a dense cover of mangroves, sand and mud sediments are trapped and immobilized. The sandy substrates on reef tops support a varied in-fauna (Gibbs 1978, this volume).

2.3.2. *Reef-top biology*

The sand bodies that build on the reefs and their cemented equivalents, can locally influence the ecology of the intertidal reef surface. Ramparts and boulder tracts can pond water into moats which remain flooded throughout low tide. Marine organisms including corals live and grow in the moats. Characteristic of the moats is the microatoll form of coral which because of its limited upward growth develops a dead, flat top and living lateral margins. Moats are normally less than 50 cm deep depending upon the height and permeability of the damming ramparts, consequently moat microatolls are broad thin colonies with high breadth : thickness ratios (open-water microatolls have low breadth : thickness ratios; see Scoffin & Stoddart 1978, this volume). The maximum elevation to which microatolls can live is high water neap tides (i.e. 1.6 m above datum). Branching corals, dominantly *Montipora*, *Porites* and *Acropora*, are also common in ponded water on reef flats and are occasionally found growing on the seaward slopes of ramparts where moat water seeps through during low tide.

There is a strong correlation between rampart distribution and the occurrence of mangroves. On the local level mangroves are seen to preferentially colonize the less exposed parts of shingle ramparts (figure 18, plate 2). On a broader scale the limits of mangrove distribution across the shelf away from the mainland correspond closely with the limits of rampart distribution (Scoffin & McLean 1978, part A of this Discussion). The distribution of mangroves on reefs indicates that they require a stable substrate in the intertidal position for colonization. Pioneer growth is on loose ramparts (e.g. E Hope), later growth spreads to the now sheltered substrates of the windward side of the reef flat (Two Isles, Three Isles, Pipon, Low Isles, Watson, W Hope), and eventually the flat is overgrown across to the leeward sand cay (Bewick, Howick, Houghton, Coquet, Newton, Nymph, Turtles).

The other common marine vegetation present in the permanently flooded areas of the reef flat consists of marine phanerogams (notably *Thalassia*) and algae. *Thalassia* and some of the soft algae require loose sandy substrates of at least a few centimetres thickness, other algae (e.g. *Sargassum*, *Caulerpa*) prefer firm rocky substrates. This edaphic control is partly responsible for the concentric zonation of vegetation across the margins of the reef flats, since soft and hard substrates (probably related to former shingle deposits) occur in roughly concentric bands near the rim (figure 19, plate 3).

The reef-top carbonate-producing organisms can be subdivided into two groups: first, those that are either unattached or else readily disintegrate on death and secondly, those that are anchored to a firm substrate. The dominant members of the first group include benthonic Foraminifera (with *Calcarina*, *Baculogypsina* and *Marginopora* being the most abundant) and branching calcareous algae (chiefly *Halimeda*) and molluscs.

The organisms cemented to the substrate include corals, crustose coralline algae and some molluscs (notably oysters). Analyses of reef flat and sand cay sand samples (Flood & Scoffin 1978) reveal that coral fragments are the most abundant components indicating the effective breakdown of coral colonies on the reef margin and reef flat. The reef-produced sand grains are eventually carried off the reef top, principally to lee, but do not spread very far from their source (Flood *et al.* 1978, part A of this Discussion).

On the windward margin of reefs, wave action persistently pulverizes coral branches and on the reef flat numerous organisms either graze the surface of, or bore into, carbonate grains and, with time, cause disintegration to smaller particles. The dominant grazers are molluscs and echinoids and the major boring organisms are algae (figure 20, plate 3), sponges (figure 21, plate 3), worms and bivalves: *Lithophaga* (figure 21) and *Tridacna crocea* (figure 22, plate 3). The zone of most effective bioerosion is in the intertidal position between high water neaps and low water neaps. Any stable carbonate projections such as coral boulders, dead microatolls and large dead clams, are truncated down to the level of sand accumulation on the reef flat, that is about low water neap tides. Cliffs of cemented ramparts are undercut by bioerosion, producing a slight notch at the level of the present reef flat (figure 23, plate 3); this intertidal notch is an intermediate stage in the planation of the rampart rock.

Thus, in summary, the reef top is a surface of both aggradation and degradation, where the processes of growth and erosion compete. On shallow reefs, prevailing physical forces interact with the growing reef to produce intertidal bodies of sediment and in turn these bodies support characteristic faunal and floral assemblages. Ultimately a large portion of the reef top may become supratidal if the depositional processes dominate the erosional.

2.4. *Recent sea level and reef-top history*

There are presently exposed on the inner-shelf reef flats in-place intertidal microatolls that date at 6080 ± 90 a B.P., 4870 ± 70 a B.P., 3700 ± 90 a B.P., 2370 ± 70 a B.P., 800 ± 60 a B.P. and modern. From this it is deduced that the first time during the last marine advance that the sea reached its present level was about 6000 years ago but that since then the tidal range from l.w.s.t. to h.w.n.t. (i.e. the limits of intertidal microatoll growth) has always overlapped to some degree with the present tidal range from l.w.s.t. to h.w.n.t. (currently this is 1.1 m).

The highest in-place fossil microatolls are found in rampart rocks at 0.7 m above present high water neaps (Scoffin & Stoddart 1978, this volume) and they date about 3500 a B.P. (Polach *et al.* 1978). This indicates that 3500 years ago, high water neap tide level was at least 0.7 m higher than it is at present. However, this difference in level could result from a change in tidal range, or in mean sea level, or both. If there has been a change in mean sea level then other sea level indicators besides reef-flat microatolls should be preserved as fossils at high levels:

(*a*) Open-water reef framework (present maximum elevation l.w.s.t. 0.5 m). No exposures were found of fossil open-water reef framework, though this absence is not conclusive proof of no change in l.w.s.t. level, for a 0.7 m drop in sea level would place former open-water framework only at the level of the present reef flat. A few shallow excavations were made below the reef flat rocky and sandy substrates to maximum depths of 50 cm but they revealed only intertidal deposits. It is unlikely that high open-water reef framework did exist on the reef flats and has been subsequently eroded as so many other high features of this period have been preserved, including microatolls, ramparts and boulders. Presently large *Porites* colonies (locally termed 'bommies') grow on the flanks and leeward slopes of reefs (figures 10 and 12) up to levels of 0.5 m above datum. Exposures of large dead *Porites* colonies at reef rims (figure 13) with their surfaces at about 0.8 m above datum possibly represent fossil open-water 'bommies' though regrettably cores were not taken to confirm the low breadth/thickness proportions of these corals to indicate their open-water origin, neither were samples collected for ^{14}C dating.

(*b*) Limestones. Our observations suggest that beach-rock forms up to extreme high water spring tide level (2.9 m). Therefore a recent fall in sea level should expose beach-rock higher than 2.9 m on leeward sand cays whose loose sand dates at 3500 a B.P. or more. Though high beach-rocks are rare, there is evidence from Stapleton and Houghton of raised beach-rock about 0.4 m above present e.h.s.t.

Of special interest in this area of the Great Barrier Reef are the high platforms of cemented shingle (figure 24, plate 3). The deposits are basically lithified ramparts and the flat surface is one of deposition, not erosion (Scoffin & McLean 1978). Flat depositional surfaces form today at various intertidal levels and consequently it is not possible to interpret the high platform levels (maximum 3.5 m) by this criterion. If we use lithification levels and it is assumed that the maximum elevation of lithification of ramparts is the same as beach-rocks, i.e. e.h.s.t. (though a cautionary note should be added that rampart rocks all have a muddy matrix and it is not impossible that under the influence of capillary forces lithification could take place in the supratidal zone), then the surfaces of high platforms dated about 3500 a B.P. represent a former e.h.s.t. of 0.6 m higher than today.

(*c*) Erosional features. Present levels of reef flat erosion are about low water neap tides. Where high rampart rock is eroded flat surfaces of older shingle deposits are exhumed and their level corresponds to the dominant level of the present reef flat (figures 17 and 25, plate 3). No high erosional features are evident on the inner shelf reefs though the positions of the high microatolls in cliff sections (figure 26, plate 3) point to the seaward erosion after 3500 a B.P. of first those shingle deposits which formed the foundation on which these microatolls were seated, and secondly, the higher deposits which formed the barrier that ponded the water in which the microatolls grew.

It is the blending of these erosional features with masking depositional features that makes the interpretation of the recent history of the inner-shelf reef flats so complicated. Nevertheless, though complex, the features observed do present a consistent picture related to a recent change in mean sea level (McLean, Stoddart, Hopley & Polach 1978, part A of this Discussion). The available dating and levelling evidence point to the following conclusions:

About 6000 years ago, sea level reached its present level, and rose a minimum of 0.7 m (and a probable maximum of 1.1 m) about 3500 years ago. Subsequently, it fell to present levels perhaps by about 2000 a B.P.

2.5. *A summary of the variation in surface features of exposed reefs across the shelf*

2.5.1. *Outer barrier*

The reefs of the outer barrier have relatively little surface relief though some have a shallow lagoon. Only one reef in this region has a sand cay. Large scattered boulders principally on the windward flanks are common but there are no leeward boulder tracts. The algal rim, though with little elevation, is well developed on the ocean side of the reefs. Living corals occur over a large portion of the reef top, and very low sheets of coral rubble are common.

2.5.2. *Outer shelf*

Less than 50 % of the exposed reefs on the outer shelf have a supratidal sand cay and most of these cays have only a pioneer cover of vegetation and some beach-rock. Many of the reefs are wide with very low relief, the bulk of the reef tops are sand covered, though some reefs have shallow lagoons with living corals. Exposed ramparts are absent though the reef margins show a concentric pattern. Scattered boulders at the edges of reefs are common.

2.5.3. *Inner shelf*

The majority of the exposed reefs on the inner shelf have a leeward sand cay and these cays commonly have quite mature vegetation. On most reef tops the entire surface is elevated above l.w.s.t. and has notable relief (up to 3 m). Loose and cemented ramparts are exposed on most of the reefs, and leeward boulder tracts are normally present. There is a general trend for the extent of mangrove cover of the reefs to increase towards the mainland. Corals abound in moats but otherwise the reef tops have few living colonies.

Reefs with shingle ramparts, mangroves, sand cays and vegetation were termed 'low-wooded islands' by Steers (1929) and 'island-reefs' by Spender (1930). The appearance of the low-wooded islands on the shelf is fairly abrupt though a reef such as East Hope with some of its surface still well below low tide level and without well established ramparts could be considered to be intermediate between outer-shelf type and inner-shelf type.

Spender (1930) proposed a sequential development of surface features for reefs in this region which he noted developed across the shelf from edge to mainland. The reef classes of Spender progressively develop first a leeward sand cay, then ramparts, and finally mangroves spreading from the ramparts. At least part of this sequence has been verified by recent surveys of mangrove cover (Stoddart *et al.* 1978, this volume). Spender (1930) suggested that the elevation of the reef flat was crucial in determining which surface features were developed. Though the reefs near the shelf edge have generally lower reef flats than those towards the mainland, the difference is small (normally less than 50 cm). It seems more probable that it is the ease with which sediment can be removed from the reef top (this property relates to both reef-flat elevation, including presence of a lagoon, and prevailing physical conditions) which is vital in the building of reefs above low water level. Once the whole reef top is at the intertidal level, ramparts can build and, providing mangroves and cements are available, they can be stabilized. The effect of an initial stabilized body of reef top sediment is cumulative, for when the level of the reef rim is slightly raised it reduces the power of waves to spread new sediment, so that additional deposits can bank up on reef margins. Eventually the force required to remove the compounded sediments is beyond the scope of the environment. It is probably this exponential effect that causes the relatively sudden appearance of low-wooded islands on the shelf.

The main deposits that build reefs high above l.w.s.t. consist of coarse coral shingle and these are principally deposited at reef rims during storms, but it is left to the normal prevailing waves to disperse these sediments after storms. Consequently, shelf edge reefs are more easily freed of storm deposits than inner shelf reefs. The landward increase in elevation of reef surface features across the shelf can thus be explained entirely by the decrease in the prevailing wave energy from the barrier edge to the mainland. Leeward sand cays on the other hand are building and not eroding during normal prevailing conditions. They are therefore found on many reefs across the shelf that have at least some of their surface at sea level.

REFERENCES (Scoffin *et al.*)

Fairbridge, R. W. & Teichert, C. 1948 *Geogrl J.* **11**, 67–88.

Flood, P. G. 1974 *Proc. 2nd Int. Coral Reef Symp., 1973*, vol. 2, pp. 387–394. Brisbane: Great Barrier Reef Committee.

Flood, P. G. & Scoffin, T. P. 1978 *Phil. Trans. R. Soc. Lond.* A **291**, 55–71 (part A of this Discussion).

Flood, P. G., Orme, G. R. & Scoffin, T. P. 1978 *Phil. Trans. R. Soc. Lond.* A **291**, 73–83 (part A of this Discussion).

Gibbs, P. E. 1978 *Phil. Trans. R. Soc. Lond.* B **284**, 81–97 (this volume).

Hill, D. 1974 *Proc. 2nd Int. Coral Reef Symp., 1973*, vol. 2, pp. 723–731. Brisbane: Great Barrier Reef Committee.

Marshall, S. M. & Orr, A. P. 1931 *Scient. Rep. Gt Barrier Reef Exped. 1928–29* **1**, 93–133.

McLean, R. F. & Stoddart, D. R. 1978 *Phil. Trans. R. Soc. Lond.* A **291**, 101–117 (part A of this Discussion).

McLean, R. F., Stoddart, D. R., Hopley, D. & Polach, H. A. 1978 *Phil. Trans. R. Soc. Lond.* A **291**, 167–186 (part A of this Discussion).

Orme, G. R., Webb, J. P., Kelland, N. C. & Sargent, G. E. G. 1978 *Phil. Trans. R. Soc. Lond.* A **291**, 23–35 (part A of this Discussion).

Polach, H. A., McLean, R. F., Caldwell, J. R. & Thom, B. G. 1978 *Phil. Trans. R. Soc. Lond.* A **291**, 139–158 (part A of this Discussion).

Scoffin, T. P. & Stoddart, D. R. 1978 *Phil. Trans. R. Soc. Lond.* B **284**, 99–122 (this volume).

Scoffin, T. P. & McLean, R. F. 1978 *Phil. Trans. R. Soc. Lond.* A **291**, 119–138 (part A of this Discussion).

Spender, M. A. 1930 *Geogrl J.* **76**, 193–214 and 273–297.

Steers, J. A. 1929 *Geogrl J.* **74**, 232–257 and 341–370.

Steers, J. A. 1937 *Geogrl J.* **89**, 1–28 and 119–146.

Stoddart, D. R. 1973 *Geography* **58**, 313–323.

Stoddart, D. R., McLean, R. F. & Hopley, D. 1977 *Phil. Trans. R. Soc. Lond.* B **284**, 39–61 (this volume).

Stoddart, D. R., McLean, R. F., Scoffin, T. P. & Gibbs, P. E. 1977 *Phil. Trans. R. Soc. Lond.* B **284**, 63–80 (this volume).

Thom, B. G., Orme, G. R. & Polach, H. A. 1978 *Phil. Trans. R. Soc. Lond.* A **291**, 37–54 (part A of this Discussion).

Phil. Trans. R. Soc. Lond. B. **284**, 141–147 (1978)   [141]
Printed in Great Britain

Archaeology and the Great Barrier Reef

By J. M. Beaton

*Department of Prehistory, Research School of Pacific Studies, Australian National University,
Canberra, Australia 2600*

The Great Barrier Reef and its associated islands form one of Australia's most neglected archaeological provinces. Only now can some of the most obvious parameters be described. Islands which evolved in Holocene times have been incorporated into local coastal Aboriginal economy yet also show evidence of having been visited by Torres Strait peoples. Islands which were inland hills during lower sea level Pleistocene times could have had a near continuous use by Aboriginal Australians since the continent was first colonized by them. Even this early approximation of the area's prehistory suggests complex and varied adaptation to this peculiar habitat. This reconnaissance shows that there are highly varied archaeological remains. Other evidence, some of which comes from living Aborigines, attests to the importance of this complex habitat even to most recent times. The earliest human use of the area is not known, but much can be inferred.

Arguments can be put forward to suggest that the Great Barrier Reef area was occupied as early as any other part of Australia. One argument reasons from the established early dates for man in Australia. Lampert (1975, p. 197) has claimed as much as 40000–50000 years ago for the first colonization of Australia. Whatever the foundation date might be, the continent was certainly well colonized by 26000 a B.P. (Bowler, Thorne & Polach 1972). This date (ANU-375B) comes from about latitude 34° S and since colonization is presumed to have been from south Asia, this date should be considered as an estimate for the widespread occupation of the continent. Other Pleistocene dates for man in the south of the continent come from Koonalda Cave, Devil's Lair, in the southwest, and elsewhere (Jones 1973). Recently, Bowdler (1975, p. 24) has reported on the human occupation of what is now Hunter Island, Tasmania, at about 23000 a B.P. (ANU-1498).

Closer to the Barrier Reef, human occupation has been described at Kenniff Cave in the area of the Great Dividing Range in southeast central Queensland (Lat. 26° S) at about 19000 a B.P. (Mulvaney 1975). Rosenfeld has more recently dated sediments overlying engraved rock art at Laura (Lat. 16° S) as being about 13000 a B.P. (ANU-1441) (personal communication). Also near Laura, Wright (1971) has estimated a terminal Pleistocene date of about 10000 years for a rockshelter site. C. White and J. P. White respectively have dated pre-22000 a B.P. edge-ground axes from Arnhem Land (White 1967) and waisted blades from New Guinea (White 1970).

These early dates are not presented to suggest that man was evenly or densely distributed in these times, in fact quite the opposite is the more likely case (Bowdler 1976). The dates do imply that man had at least attempted to establish himself throughout Australia. That he should have somehow overlooked the northeast coast does not seem possible. Moreover, Golson's suggestion (1971) that the northern and coastal plant communities would be more suitable than interior plant communities only enhances the probability of an early occupation of this coast. We might reasonably even underline this argument if we accept the recent environmental reconstructions which show a significantly drier than present Australia (Nix & Kalma 1972; Kershaw & Nix 1975) in the late Pleistocene.

Human groups would then have occupied an open-forest covered shelf that becomes increasingly closed from 20 000 to 8000 a B.P., if the reconstruction suggested by Nix & Kalma is correct (1972, pp. 85–89). Whatever might have been the plant community composition and distribution on the continental shelf, some change due to burning by Aboriginal man could be expected from the general Australian case (Jones 1969) and for northeast Queensland in particular (Kershaw 1975, p. 187).

Jennings (1972, p. 37–37) has reasoned from glacio-eustatic time–depth curves and other data that the low land-bridge between Cape York and New Guinea became a straits between 5000 and 11 000 a B.P. If we take 8000 years ago as an estimated date for the submersion of the continental shelf south of the Torres Strait, we are invited to speculate that man has existed as many as 30 000 years on the continental shelf before its latest evolution as a reef habitat. One unfortunate consequence of the inundation of the reef platform is that most of, if not all of, the archaeological remains deposited in the early prehistory of the area would now be under water. Some unduly hopeful prehistorians have long awaited the opportunity to explore Pleistocene coastal archaeological sites via underwater archaeology. Moore (1967, p. 422; 1971, p. 9) voiced these hopes for the archaeology of the Torres Strait. But coastal archaeological sites which are usually found on foreshores and deposited on unconsolidated sand dunes, often deflated by winds and eroded by rains, have only the action of surf and tide to look forward to as the sea envelops them. No archaeologist should have *a priori* expectations that the fragile elements that constitute an archaeological deposit could survive this exposure *in situ*. Although the Barrier Reef province may not be as rigorous an intertidal environment as, say, that of the Bass Straits, neither submerged shelf should preserve significant archaeological deposits. Even if my predictions are incorrect and there are significant archaeological deposits preserved in benthic sediments, as they apparently have been elsewhere (Selwyn 1965; Emery & Edwards 1966), present archaeological technology and priorities argue strongly against underwater exploration and excavation. For a picture of man's use of the continental and reef islands of the Great Barrier Reef we must look to Holocene deposits as we now find them.

Two kinds of islands are considered below: reef islands, which are recent formations and continental islands, which were mainland land forms in the Pleistocene but are now separated from their parent land by straits.

LIZARD ISLAND: A CONTINENTAL ISLAND

Lizard Island was the largest island (*ca.* 6 km²) visited. The island's transverse ridge (elevation 359 m), forming the northwest half of the island, slopes abruptly to the southwest. The southwest half of Lizard island is comparatively flat with occasional rock outcrops and vegetated dunes. Since time was limited, the survey specialized on the central saddle portion of the island and south-facing dunes.

The largest site discovered on Lizard Island is a midden deposit bordering the recent airstrip. The overall site dimensions are roughly 100 m north–south by 75 m east–west. A profile, cut by tractor preparation of the airstrip, showed a deposit of about 50 cm of midden. There is enough ash and decomposed organic matter to give the soils a light brown colour contrasting with the Recent white sands and the cemented red Pleistocene sands. Faunal remains included a clam-type mollusc, probably *Tapes turgida*, the giant clams, *Tridacna* spp., the oysters, *Saxostrea* and *Crassostrea* and others. No bone or turtle carapace was seen. The absence of bone, the fact that the site sits on a Pleistocene base and the chalky appearance of much of the shell gives one the impression, however speculative, that the site is considerably older than other island sites which are situated close to the littoral zone. Many quartz flakes with obvious edge wear can be seen on the surface. No projectile points were seen. One ground stone artefact, much like that described for north Queensland by Roth (1904, pl. 10, fig. 66), eroded from the bank cut for the airstrip. The dense porphyritic cobble had been battered at one end and fractured. The other end of the cobble had been bifacially polished which resulted in an adze or axe-like ground edge. The tool measured $7.0 \times 5.5 \times 2.5$ cm.

A series of dune sites were found facing the seasonally windward southeast. These sites are almost certainly the same sites seen by the *Endeavour's* Captain, James Cook and naturalist Joseph Banks on 12 August 1770 (Banks 1962, p. 103; cf. Cook 1768–1771, p. 229):

> Distant as this Isle was from the main, the Indians had been thus in their poor embarcations, sure sign that some part of the year must have very settled fine weather; we saw 7 or 8 frames of their huts and vast piles of shells the fish of which I suppose had been their food. All the houses were built upon the tops of eminences exposed entirely to the SE, contrary to those of the main which are commonly placed under the shelter of some bushes or hillside to break off the wind.

The dune sites continue as occasional shell scatters nearly to the top of the narrow peninsula that forms the southernmost tip of the island. Dune sites differ from the airstrip site in that no dark midden soils are observable, at least on the surface, and the majority of surface shell is *Trochus niloticus*.

Other areas of Lizard Island which were not surveyed include the extreme southwest perimeter, the high rocky granitic formations and the eastern coast.

Howick Island: a continental island

Howick Island has a complex geography, the three 'high' formations (elevations 30, 47 and 56 m) of the southeastern corner being separated from the vegetated sand cay to the northwest by 4000 m of dense mangrove. The vegetated sand cay was surveyed and found to be littered with *Trochus* shell. No dark soil midden deposits were apparent on the sand cay but one dugout canoe and one modern grave was recorded. The canoe was originally collected by a previous phase of the expedition on the northwest side of Ingram Island, 12 km northwest of Howick Island and moved to the Howick sand cay.

The modern grave is bordered with drift timbers and beach-rock and decorated with *Trochus* and *Triton* shell. The timbers are arranged as a rectangle, 2×1 m, and a crucifix, also of drift timber, decorates one end of the grave plot.

'Drift' canoes from the Torres Strait and New Guinea are not uncommon flotsam on Barrier Reef islands. The Ingram–Howick canoe is decorated with a 5 cm wide 'strip' in 2 cm relief which runs nearly the entire length of the 5 m canoe parallel to and 10 cm below the gunwale. Lashing marks on the gunwale imply parallel boom pairs for fixing at least one outrigger. The actual number of outriggers cannot be specified as only one gunwale remains.

The three elevated portions of Howick Island were not surveyed. One attempt at a landing on the windward side of Howick was made, but wind and surf conditions discouraged the effort. The elevated peaks of Howick probably have archaeological remains. MacGillivray (1882, p. 110), on the voyage of the H.M.S. *Rattlesnake*, observed a 'party of natives' on the highest of the hillocks. MacGillivray landed on the Howick sand cay and noticed remains of turtle and fish (p. 111). Turtle bone and carapace are still present along with dugong bones on the sand cay.

Noble Island: a continental island

One site, a rock shelter at *ca.* 5 m elevation on the southeast side of Noble Island, was found. The site is chiefly a broad (*ca.* 20 m) apron deposit spread at *ca.* 180° from the rock shelter proper which is 15 m wide (at the mouth) and relatively shallow (2 m). No rock art was seen on the friable shelter roof or walls. The deposit, a dark midden, may be 1 to 2 m deep. Stone tools seen on the surface were wholly quartz flake tools but vegetation obscured most of the surface from scrutiny. Fragments of burnt wood, not seen in most of the other island sites, were exposed on the surface. *Trochus* was the most abundant shell. Other sites may occur on Noble Island, especially the west side of the island which is formed by heavily vegetated dunes.

Pipon Island: a reef island

Pipon Island, 5 km off Cape Melville, is a low island vegetated almost entirely by mangrove. The only elevated portion of Pipon Island has been the site of a lighthouse for many years and earthworks at the lighthouse site may have obscured any archaeological deposits. Dugong bones, and some turtle carapace, were found, but no artefacts, other than those of recent European origin, were found.

Bewick Island: a reef island

Bewick Island is a low island covered by mangrove flora, *Rhizophora stylosa* and *Avicennia marina*. There is an elevated portion, roughly 400 m² of vegetated coral sand dune, on the northwest side of the island. The dunes are elevated to a maximum of about 4 m above mean sea level. The dunes, especially those portions immediately fronting the littoral, are scattered with shell, turtle bone and carapace, dugong bone and exotic stone materials. Because the reef islands are formed by coral growth, the stone material found on Bewick must derive from mainland or continental island sources. Since no excavation was performed and no natural soil profiles existed, no estimate of the persistence of archaeological materials throughout the vertical dune strata could be made.

One modern grave on Bewick Island was recorded. The grave is outlined in a rectangular fashion, much like the grave on Howick Island, except that the outline material is wholly beach-rock. The grave is decorated with *Trochus* and *Hippopus* shell, glass fishing floats and a hexagonal bottle. Two large *Pinctada margaritifera* decorate each end of the grave. No cross was apparent on the grave.

Ingram Island: a reef island

On the highest part of Ingram, about 8 m above m.s.l., on the windward side of the island, McLean sampled the sediments and discovered continental rocks where none should reasonably have been. Two quartz rocks, weighing 55 and 15 g, and two quartz bearing sedimentary rocks, weighing 39 and 6 g, were collected. While none is an ideal typological specimen of a flake tool, their association in a dark grey sedimentary layer, which contrasts strikingly with the natural off-white colour of the natural coral sand sediments, argues very strongly for their transportation from elsewhere, probably the mainland. *Tridacna*, the largest bodied shellfish food source available, and *Melo* shells were associated with this strata as well. *Melo* shell ornaments, usually pendants, were traded well inland from sources on Cape York (Tindale 1975, pp. 26, 87, 199; Roth 1910–13, p. 18). *Melo* shell has also been used for utilitarian purposes, e.g. cook pots, water containers and a kind of knife (Tindale 1975, p. 85). The square *Melo* shell fragment from Ingram measures 6 × 6 cm and does not have any abraded or sharpened edge.

Nymph Island: a reef island

Like Ingram, Nymph Island was noted by McLean to have a dark grey coloured surface deposit contrasting in colour with the coral sand sediments found beneath it. As on Ingram, the deposit was found on the highest part of the island (4.2 m), but unlike Ingram, the deposit was found on the lee side of the cay. Four continental rocks of different kinds were collected in his strata sampling. They were a quartz-bearing quartzite (24 g), a silcrete flake (11 g) and two oxidized red decomposing sandstone fragments (269 and 22 g). *Melo* and *Tridacna* both occur in the deposit as they do on Ingram. The *Melo* fragment, triangular in shape, has one smooth abraded edge in contrast to its other two margins, which are roughly broken as are all the edges of the example from Ingram. This *Melo* fragment measures about 8 cm on each of its three sides.

The Turtle Islands: reef islands

From McLean's soil pit 1, 4.5 m above m.s.l., a sample has been found to contain six continental rocks of various materials. In the relevant soil sample, taken from just below the surface, two fractured quartzite rocks (80 and 45 g), one concave quartzite spall (28 g) suggestive of heat induced fracture, two mudstone rocks (260 and 69 g) and one breccia (486 g) have been identified. The 80 g fractured quartzite rock is a fraction of a cobble and retained part of the original cobble cortex. On the cortex surface a battered area, very suggestive of a hammer function, has been noted. None of the other rocks or rock fragments showed unequivocal evidence of use wear. McLean's soil sample, which included the continental rocks, is dark in colour and ashy in appearance. These two attributes make the sample appear very unlike the light coloured natural sand beneath the dark upper strata. The colour and texture of the upper strata are consistent with the kind of rich organic deposits found as types of archaeological shell middens, particularly those that are not eroded or deflated. The characteristics of the rocks present in this sample are puzzling in that for the most part they are not suggestive of quality stone material suitable for flake tool manufacture; they are suited to either hammer or grinding functions, yet none show evidence of having been so used. While it is clear that these rocks were carried to the islands and incorporated in human midden deposits, it is not clear why.

A note of caution to soils scientists should be added here. Aboriginal man on these islands (sand cays) is known to have dug deep and probably broad pits to collect low-saline water. Where there is no good indication of horizontal integrity of the sedimentary strata and radiometric tests do not show significant age differences where they might well be expected, then man must be considered as a source of reworking of the sediments.

Summary and discussion

Both kinds of islands, those created by rising sea levels, and evolving sand cays which have developed after the sea level has been relatively stabilized, show evidence of human use. The continental islands may have ancient archaeological sites which were deposited when their basal land forms were well removed from intertidal zones. The sites on the sand cays, however, due to the recent evolution of the cays themselves must be as young or younger than three or four thousand years.

The stone axe and flake tool components of the airstrip site on Lizard Island are in keeping with those known from interior sites in Queensland. The stone material associated with the sand cays is quite different and does not fit the interior assemblage model nor does it fit my expectation that *any* stone carried to the islands from another stone source should be well used.

Of the 13 continental rocks collected by McLean on sand cay islands only one appears to have obvious use characteristics. Even though this is quite in keeping with the 10–20 % use-worn tools I recognize from total stone inventories from excavated sites in interior Queensland, the sample is certainly too small to assess the Aboriginal use of stone resources on the Barrier Reef. Similarly, these limited observations provide only a bare beginning for an archaeological study of the Barrier Reef. We do at least know that the islands are a part of the Aboriginal economy in prehistory. McLean's samples, though useful, are 'incidentals'. My own observation, from visiting Lizard, Bewick, Noble, Pipon and part of Howick, only underscores the fact that archaeological resources do indeed exist in quantity on the Barrier Reef islands and cays.

The kinds of questions we need to direct future sampling towards are: what are the differences in the archaeological prehistories of the continental and reef islands, and what is the explanation for those differences? We have at least now the beginnings of observations upon which to base further sampling and research. Clearly, there are two kinds of sites on Lizard Island alone. I suspect that one represents a pre-island condition and that the other, the dune sites first discovered by Banks and Cook and rediscovered by me, represented a later seasonal shellfish gathering camp. I suspect too that the middens on the sand cays represent evidence of the exploitation of food resources by the coastal inhabitants. Certainly coastal hunter–gatherers would be attracted by the comparatively very high proportion of rich intertidal resources to coast-line that these islands offer.

I am indebted to Dr Roger McLean who recorded archaeological remains on Ingram, Bewick, Two Isles, The Turtles and Nymph Island. I remain accountable for the conclusions and interpretations in this paper. I thank Dr D. R. Stoddart, the Royal Society, the Universities of Queensland and the other Expedition participants for this opportunity. For permission to use unpublished material I thank Athol Chase, John Flinders, Bob Flinders, Robert Layton and Peter Sutton. The Queensland Department of Aboriginal and Island Affairs gave permission for the work and Kate Sutcliffe considerably expedited the permit. Rhys Jones was instrumental in organizing my part in the exercise. Professor J. Golson gave useful suggestions. The Department of Prehistory, Research School of Pacific Studies, A.N.U. provided the support.

REFERENCES (Beaton)

Banks, J. 1962 *The Endeavour journal, 1768–1771* (ed. J. C. Beaglehole), 2 vols. Sydney: Angus & Robertson.

Bowdler, S. 1975 *Aust. Archaeol. Newslett.* no. 3, Department of Prehistory, Research School of Pacific Studies, A.N.U., Canberra.

Bowdler, S. 1976 In *Sunda and Sahul, prehistoric studies in island southeast Asia, Melanesia and Australia* (eds J. Allen, J. Golson & R. Jones). London: Academic Press. (In the press.)

Bowler, J. M., Thorne, A. G. & Polach, H. A. 1972 *Nature, Lond.* **240**, 48–50.

Cook, J. 1768–1771 *Captain Cook's Journal.* London: Elliot Stock.

Emery, K. O. & Edwards, R. L. 1966 *Am. Antiq.* **31**, 733–737.

Flinders, Bob, Flinders, John, Chase, A. C., Layton, Robert & Sutton, Peter, undated. *Flinders Island Tapes:* unpublished transcripts, Australian Institute of Aboriginal Studies, Canberra.

Golson, J. 1971 In *Aboriginal man and environment in Australia* (eds J. Mulvaney & J. Golson), pp. 196–238. Australian National University.

Jennings, J. N. 1972 In *Bridge and barrier: the natural and cultural history of the Torres Straits* (ed. D. Walker). Publication B6/3, Department of Biogeography and Geomorphology, Research School of Pacific Studies, A.N.U., Canberra.

Jones, R. 1969 *Aust. nat. Hist.* **16**, 224–228.

Jones, R. 1973 *Nature, Lond.* **246**, 278–281.

Kershaw, A. P. 1975 In *Quaternary studies* (eds R. P. Suggate & M. M. Cresswall), pp. 181–187. Wellington: The Royal Society of New Zealand.

Kershaw, A. P. & Nix, H. A. 1975 *Australian Conference on Climate and Climate Change,* Monash University, Victoria, 7–12 December 1975.

Lampert, R. J. 1975 *Antiquity* **49**, 195, 197–206.

MacGillivray, J. 1882 *Narrative of the voyage of H.M.S. Rattlesnake.* London: J. & W. Boone.

Moore, D. R. 1967 *Aust. nat. Hist.* **15**, 418–422.

Moore, D. R. 1971 Paper presented, section 25, ANZAAS, Brisbane, roneo'd.

Mulvaney, D. J. 1975 *The prehistory of Australia.* Melbourne: Penguin Books.

Nix, H. A. & Kalma, J. D. 1972 In *Bridge and barrier: the natural and cultural history of the Torres Straits* (ed. D. Walker), Department of Biogeography and Geomorphology, Research School of Pacific Studies, A.N.U., Canberra.

Roth, W. E. 1904 *North Queensland Ethnography,* Bulletin No. 7, Brisbane.

Roth, W. E. 1910–13 North Queensland Ethnography, Bulletin No. 14, *Records of the Australian Museum,* vol. VIII.

Selwyn, B. 1965 Unpublished Ph.D. dissertation, Columbia University, New York (University microfilm order No. 65–13990, Ann Arbor).

Tindale, N. B. 1975 *Aboriginal tribes of Australia.* Canberra: A.N.U. Press.

White, C. 1967 *Antiquity* **41**, 149–152.

White, J. P. 1970 *Proc. Prehist. Soc.* **36**, 152–170.

Wright, R. V. S. 1971 In *Aboriginal man and environment* (eds D. J. Mulvaney & J. Golson). Canberra: Australian National University Press.

Phil. Trans. R. Soc. Lond. B. **284**, 149–159 (1978)   [149]

Printed in Great Britain

Evolution of reefs and islands, northern Great Barrier Reef: synthesis and interpretation

By D. R. Stoddart,† R. F. McLean,‡ T. P. Scoffin,§ B. G. Thom‡
and D. Hopley‖

†*Department of Geography, Cambridge University, Downing Place, Cambridge, U.K.*
‡*Department of Biogeography and Geomorphology, Australian National University, Canberra,
Australia* 2600
§*Grant Institute of Geology, University of Edinburgh, West Mains Road, Edinburgh EH9 3JW, U.K.*
‖*Department of Geography, James Cook University of North Queensland, P.O. Box* 999,
Townsville, Queensland, Australia 4810

This paper brings together the evidence from shallow coring, surface geomorphology, lithology of exposed rocks, superficial sediment accumulations, vegetation patterns, and the historical record derived from radiometric dating to suggest a sequence of reef and island development on the northern Great Barrier Reef in Holocene time. Reefs initially grew vertically as the sea rose rapidly from glacial low levels. This continued until vertical growth was limited by the air/sea interface as the rate of sea level rise slowed. Vertical growth was then replaced by reef flat formation at low intertidal levels, and by the lateral extension of reefs, especially to leeward. Superficial sediment accumulations on the reef flat define a series of changing habitats for further organic growth, and also record the sequence of Holocene events. Controls of the transition from vertical to horizontal reef growth will be discussed and some comments offered on latitudinal variation in reef form along the Great Barrier Reef.

Introduction

The work of the 1973 Royal Society and Universities of Queensland Expedition has thrown new light on the history and morphological development of reefs of the northern Great Barrier system in response to sea level change since the last glacial low stand of the sea. In this final discussion we wish to review some of the main conclusions from these new data, in the context of our present knowledge of reef development elsewhere in the world.

Before this work was undertaken, we inferred that as the continental ice sheets melted between 14 000 and 7000 a B.P., the level of the world ocean rose over several thousand years at a mean rate of 1 m/100 a and also that sea level continued to rise after the greater part of the ice had disappeared, partly because of 'isostatic decantation' of water from areas rebounding after ice load had been removed and partly because of differential adjustments of the crust to the new water load (Bloom 1971; Walcott 1972). The complex interplay of these factors led to continuing change of sea level after 6000 a B.P., and probably also to differences in the record of sea level at different localities, both on a continental and a local scale, thus casting doubt on the heuristic utility of the concept of a general eustatic sea level curve. Much of the controversy in the coral seas over the Holocene record of sea level change has concerned the interpretation of events during these last 6000 years, which is also the period during which the surface morphologies of existing reefs have developed.

SEA LEVEL AND REEF GROWTH

We also suspected that the growth potential of reefs was such that they could not keep pace with the transgressing sea during the major part of its last rise. Chave, Smith & Roy (1972), in a first approximation based on calcification rates of common organisms and their relative abundance on reefs, suggested a gross potential equivalent to a vertical accretion rate of 0.7 m/ 100 a, but a net rate of only 0.1 m/100 a. Subsequent estimates of carbonate production, based on analyses of sea water alkalinity, yield mean figures equivalent to accretion on reef flats of 0.3–0.5 m/100 a and in lagoons of 0.06–0.1 m/100 a (Smith & Kinsey 1976). Although some rates of reef growth derived from the recent geological record exceed these figures (see for example, Chappell & Polach 1976), the disparity is not great, and the data are consistent with the inference by Mesolella, Sealy & Matthews (1970) from the raised reefs of Barbados that during times of rapid sea level rise reefs are drowned, and that only during stable or slowly rising sea levels do reefs have the capacity to grow vertically from suitable substrates to the surface and then to expand laterally.

We knew further that, once emersed by a negative movement of sea level, reefs are relatively persistent structures, subject to superficial karstic modification and to diagenetic fabric changes but retaining their gross forms. Estimates based on the solubility of carbonate minerals suggested vertical erosion rates of 0.5–1.0 mm/100 a, but these have been revised upwards by Trudgill's direct measurements on Aldabra Atoll to 26 mm/100 a for subaerial and 50–400 mm/100 a for marine erosion (Trudgill 1976a, b). These rates are nevertheless extremely slow by comparison with the magnitudes of reef structures and the periods available for karst erosion during low glacial sea levels.

Hence it is not surprising that many workers have concluded that modern reefs thinly veneer older karst-eroded reef structures; indeed, in many parts of the world eroded remnants of pre-Recent reefs protrude through more recent accretionary sequences (Stoddart 1969, 1973). Subsurface exploration in the Marshall Islands first revealed evidence of stratigraphic discontinuities between pre-Recent and Recent limestones, marked by diagenetic changes and marked radiometric age disparities between adjacent units; and the Eniwetok bores showed that such horizons occurred several times in the reef column, corresponding to successive periods of emersion (Schlanger 1963). Similar records have been found by drilling in the Tuamotus, the Leeward Hawaiian Islands, New Caledonia and British Honduras, and demonstrate that many reefs have formed by incremental accretion of thin limestone sequences on periodically submerged but intermittently emerged and karst-eroded older reefs, a model first fully envisaged by Tayama (1952, p. 170) and recently beautifully demonstrated at Aldabra Atoll by Braithwaite, Taylor & Kennedy (1973). Purdy (1974a, b) went further by using shallow cores to interpret seismic profiles, and was able to show that phases of reef accretion are preferentially located on topographic highs, and hence that reef topography mimics that of underlying karst features.

We have similar evidence for the northern Great Barrier Reef. On the shelf itself the seismic record shows an extensive subsurface discontinuity at depths of 40–100 m, while the bore on Bewick Island passed from Holocene accretionary sediments into altered older limestones at only 5.6 m below datum (the comparable depth at Stapleton was 14.6 m); the material above this discontinuity is less than 6000–7000 years old, and that below is Pleistocene. These figures compare with minimum depths to this 'Thurber Discontinuity' in other reef areas of 13 m at

Eniwetok, 31 m at Funafuti, 37 m at Midway, 10 m in New Caledonia, 6–11 m at Mururoa in the Tuamotus, and 9–26 m in British Honduras, though at many reef localities the discontinuity rises above present sea level to outcrop as pinnacles of *feo*, *ironshore*, *makatea* or *champignon*. This broad comparability is perhaps surprising in view of the probable diversity of erosional environment in different parts of the tropics during the last low stand of the sea. Not enough is yet known of Pleistocene climates in the northern Great Barrier Reef area to make any precise statements about the character of karst erosion. There is some agreement that full glacial periods were arid, with rainfalls cut by one-half compared to the present, and that there was a moister period about 8000 a B.P., with rainfall increased by a similar amount (Webster & Streten 1972; Bowler *et al.* 1976).

Bewick is one of the most 'mature' of the low wooded island reefs, in the sense of having the greatest proportion of its reef top covered with Holocene supratidal sediments and mangroves. This advanced stage may result from the existence of an unusually high reef residual on which the Recent reefs have grown, whereas other reefs with deeper foundations and perhaps greater area, requiring greater volumes of carbonates to bring them to present sea level, have lagged in stage of development. The Bewick and Stapleton records may also be compared with that at Heron Island in the extreme south of the Barrier Reef, where, as interpreted in terms of topography, stratigraphy and seismic structure, the pre-Recent unconformity lies at a depth of 15 m (Davies 1974; Davies, Radke & Robison 1976; Flood 1976). Unfortunately no uppermost Pleistocene and Holocene records are available from any of the other Barrier Reef bores. There is a further point of difference in that the shelf surrounding Bewick is up to 30 m deep, and that in the Capricorn Group is at 35–40 m. Subtracting the known Holocene reef increment, we infer that up to 26 m in the north and 20–25 m in the south represents the thickness of pre-Wisconsin reef growth. Little is known about this sequence, but the record of accretion and erosion that it represents is undoubtedly complex.

Nowhere on the Great Barrier Reef is Pleistocene reef framework exposed above present sea level. The 'raised reef' described by Jukes (1847, vol. I, pp. 340–342) near Raine Island is a deposit of storm boulders, and there is no evidence at Raine Island itself, in spite of frequent references to the contrary, of any emergence. This absence of Pleistocene reefs is in marked contrast to the situation in Western Australia, much of the Indian Ocean, mobile Indonesia, parts of Polynesia and the Caribbean. The generally ubiquitous last interglacial reef limestones, aged 90000–130000 years, are not found, although individual corals of this age (112000–143000 a) have been reported from the mainland coast of eastern Australia at 28° and 34 °S, beyond the present limits of the Great Barrier Reef (Marshall & Thom 1976). The absence of outcropping Pleistocene reefs with the Barrier Reef province itself leads to the inference that subsidence rather than stability has dominated reef development in that area.

SEA LEVEL AND REEF-TOP MORPHOLOGY

Microatolls and the cessation of vertical growth

The sea reached its present level on the northern Great Barrier Reef by about 6000 a B.P., as shown by dated microatolls (Fisher 6310 ± 90; Low Wooded 6080 ± 90; Houghton 5850 ± 170; Leggatt 5800 ± 130). On several reefs extensive fields of microatolls grew on reef tops, many within the vertical range of modern corals, between 6300 and 4800 a B.P. This confirms the conclusion previously reached from other evidence that in Australia, sea level had reached its present level substantially earlier than appears to have been the case in Micronesia and Northern Hemisphere areas, where the record appears lagged by up to 2500 years (Thom & Chappell 1975).

The highest of these fossil microatolls reaches 0.7 m above the present maximum height at which modern moated microatolls form (1.55 m, or approximately h.w.n.), and is dated at 3700 ± 90 a B.P. Some at least of the fossil microatolls may not have been moated, but may have lived on open reef tops (the amount of sediment accumulated on the flats cannot have been great in the initial stages of flat formation), and we must infer either a marginally higher sea level or a slightly greater tidal range than at present at about 3500 a B.P. Hopley (1975) has described similar fossil microatolls on a fringing reef at Middle Island, near Bowen, at about 1.6 m (approximately m.s.l.), with ages of 5210–5290 a B.P.

The microatoll is a diagnostic feature of a critical transition in the mode of reef growth, from unconstrained vertical upgrowth when the reef top is still below sea level, to inhibited vertical and extended lateral growth when the top reaches sea level (or more precisely the level of low water neaps). We have no information from any Queensland reef of the internal facies distributions resulting from this transition, but the age scatter of microatolls from 6000 a B.P. to the present reveals that the transition is diachronous, as a result of differences in basement depth and topography, reef size, and reef environment and ecology. Many extensive surface reefs of apparently simple structure probably conceal complex growth histories. Marshall & Orr (1931, pp. 117–123) showed at Low Isles how much of the reef volume is sedimentary infill rather than growth framework, and there are several cases in the Northern Province of irregular reefs in the process of apparent coalescence to form larger and simpler reefs (Turtle II and East Hope are examples). Conversely it is possible, at least in the Princess Charlotte Bay area, that some large platform reefs are thin veneers on flat-lying Mesozoic sedimentary rocks, and have simpler histories.

It is instructive to compare the Barrier Reef record of reef growth with other areas. A fringing reef in Panama shows a period of rapid vertical growth beginning about 7000 a B.P. at -15 m, with reef flat formation about 2500 a B.P.; similarly a reef on Oahu started growing about 7000 a B.P. at -10 to -15 m, and its reef flat formed about 2000–3000 a B.P. (Macintyre & Glynn 1976; Easton & Olson 1976). In both cases the approach of the reef top to the sea surface led to the replacement of dominantly vertical growth by lateral growth, with consequent ecological and facies changes. During the last 7000 years, in both localities, the sea itself rose by 10–15 m. Thus reef growth began when the sea flooded the -15 m level, and was presumably successful because the rate of sea level rise had by then decelerated sufficiently for incipient reefs not to be drowned. At Eniwetok in the Marshall Islands, however, the formation of a reef flat with microatolls occurred between 4980 and 1885 a B.P., when active coral growth was replaced by rubble accumulation (Tracey & Ladd 1974; Buddemeier, Smith & Kinzie 1975) the Eniwetok record is closer in time to that of the Great Barrier Reef than that in Panama and in Oahu.

Rampart and platform formation

Shingle ramparts on the windward sides of inner reefs of the northern Great Barrier are storm-deposited materials of varying ages; records in this century demonstrate their variability in form and location. Platforms are their lithified equivalents and occupy comparable positions on the reefs. Steers (1937) showed the widespread distribution of 'upper' and 'lower' platforms, though the former are more restricted in extent than the latter; in accepting his distinction, we do not necessarily wish to imply in all cases either accordance, contemporaneity, or even constant relation between the two features so named. Steers believed that both platforms had similar origins as cemented ramparts, but that the upper owed its greater elevation to its formation during a former higher sea level in the Holocene.

The upper platform stands at 2.6–3.8 m (datum m.l.w.s.) and is often being dissected by erosion; radiometric ages on constituents range from 3050 ± 70 to 4420 ± 90 a B.P. and cluster in the interval 3300–3600 a B.P. Two dates on cement suggest that lithification took place about 1000 years after the formation of the clasts. The lower platform is usually separated from the upper, where both occur together, by an interval of 1–1.5 m. Its height range is 1.6–2.4 m and ages range from 380 ± 80 to 1460 ± 70 a B.P. More recent rampart deposits are unlithified, and probably overlap with younger lower platform deposits in age. The modern ramparts, in process of formation, and the lower platform are clearly related to present sea level: we could, therefore, infer, as did Steers, that the upper platform derived from ramparts built during a sea level up to 1 m higher than present, thus reinforcing the inference already made from the existence of microatolls of similar age (*ca.* 3500 a B.P.) up to 0.7 m above their modern counterparts. Two points of caution must be noted: (*a*) there is uncertainty about the level to which platform cementation can take place with respect to sea level, and (*b*) it is curious that if sea level did stand higher, no elevated reef framework, as distinct from moated microatolls, has been found on reef tops.

Neither ramparts nor platforms are unique to the northern Barrier Reef; indeed there is some danger that the use of a distinctive terminology implies an unwarranted degree of singularity. Steers (1940) drew attention to loosely comparable 'promenades' on the Pedro Cays, Jamaica, and comparison can also be made between the Barrier Reef platforms and the widespread rock ledges of windward coasts in the Carolines, Marshalls, central Polynesia, and the Tuamotus. These are, however, always single features, and being in microtidal situations they are less spectacular than the Queensland examples; their upper surfaces are at or slightly above the level of h.w.s. At Mururoa, Tuamotus, such a ledge rises to + 3 m and is dated at 3610 a B.P.; dates generally range from 1270 to 4350 a B.P., and most cluster between 2000 and 3000 a B.P. (as at Ebon, Jaluit, Ailinglapalap, Aitutaki and Mopelia: Curray, Shepard & Veeh 1970; Guilcher *et al.* 1969; Stoddart 1975). The age range taken as a whole is greater than that for the Queensland examples, nor do most of the ledges necessarily suggest formation during any sea level higher than the present (Curray *et al.* 1970; Newell & Bloom 1970). They are best interpreted as lithified storm-built ridges, which in the absence of any marked storm periodicity or sea level change need not show any marked clustering in heights or ages.

Sand cays

Sand cays are one of the most studied features of the coral reefs, but very little is known of their stratigraphy or history, and the few radiometric dates hitherto available have been either ambiguous or uninformative. We have shown that sand cays on the northern Great Barrier consist of an extensive high core surrounded in some cases by a discontinuous lower terrace. The high core effectively delineates the present gross topography of the cay. Its elevation varies from 5 to 7.3 m, and ages of constituent sediments range from 3020 ± 70 to 4380 ± 80 a B.P., averaging about 3500 a B.P. Hence the cays formed after 6000 a B.P., when the reef flats first appeared at the surface, but were essentially complete in shape and size by 3000 a B.P. The lower terrace, where present, stands at 3.5–4.5 m; constituent sediment ages range from 2190 ± 7 to 3280 ± 80 a B.P., with an average of approximately 2700 a B.P. However, the cartographic evidence of Steers's surveys and our own shows that on at least two islands where the sediment ages are greater than 2000 a B.P., the lower terrace as a topographic feature has formed in large part since 1936.

Nevertheless the clustering of dates for the two levels, supported by sedimentological, pedological and vegetational differences, and the remarkably constant height interval between them, requires explanation. The lack of cay sediments dating as younger than 2000 a B.P., if not caused by sampling bias, is notable: even the sediments of an unvegetated and ephemeral cay (Pickersgill) are 2330 ± 70 a old. The only younger sediments are gravels from unconsolidated storm-deposited banded shingle ridges, which on a number of islands show a scatter of ages from 510 ± 50 to 1550 ± 70 a; comparably scattered dates have been published by Hopley for islands near Townsville and Bowen (Hopley 1971, 1975).

Beach-rock

Beach-rock is extensive on sand cay shores; it has an aragonite cement and forms intertidally. On islands with both upper and lower terraces there are also two distinct generations of beach-rock: an older beach-rock on the shores of the higher part of the cay, and a younger, less consolidated rock on the shores of the lower terrace. On some islands only higher beach-rock is found, generally forming narrow horizontal shelves; on some smaller islands only the lower beach-rock occurs, forming, as on lower terrace beaches, characteristic inclined ledges. The higher rocks, which are well consolidated, have height ranges up to 2.5–3.0 m (m.h.w.s. 2.3 m), and their ages range from 2030 ± 70 to 2670 ± 70 a B.P. The lower rocks, between 1 and 2.3 m, are modern. The existence of these two types of beach-rock is thus consistent with the evidence of two aggradation levels on the cays and of two periods of rampart formation in the rampart rocks.

Rocks interpreted as old beach-rocks have been found on islands further south near the mainland coast between Bowen and Cairns, and especially in the Palm Islands. These have maximum elevations of 4.1–7.8 m, and ages from 3240 to 5250 a B.P. On at least two of these islands there is evidence of two levels of beach-rock, with a height difference of about 3.5 m (Hopley 1971). In the Bowen area three levels of beach-rock *sensu lato* have been reported in the height range 4.8–6.4 m, with ages in the range 3350–6020 a B.P. (Hopley 1975). The sea level record in these islands, however, appears to be different from that of Barrier Reef islands to both north and south, possibly for tectonic reasons (Hopley 1974, 1975).

Summary

Any interpretation of the development of reefs and islands on the northern Great Barrier Reef and any reconstruction of sea level history must therefore take account of four main groups of facts.

(1) Reef flats formed, as shown by the presence of microatolls, close to present sea level, on several reefs between 6000 and 5000 a B.P., between 12° 14′ S and 15° 06′ S. Not all reefs formed horizontal surfaces at sea level simultaneously, and some are still doing so, but this period represents the earliest during which substrates were available for reef islands to form on. Hopley (1977) has subsequently obtained similar dates for reef flat formation on outer ribbon reefs.

(2) Three groups of features formed in the time interval centred on the period 4000–3000 a B.P. Storm ramparts (now represented by upper platform rocks), often overlying fossil

microatolls, formed over the range 4420–2050 a B.P. and cluster during 3600–3300 a B.P. The core areas or upper terraces of sand cays formed by normal wave refraction rather than by storm activity were built from sediments dated 4380–3020 years, and clustered round 3500 a B.P. High beach-rock formed round the shores of these high-standing cays during the period 4380–2030 a B.P. Each of these features stands at higher elevations than comparable features associated with present sea level: in broad terms this height anomaly reaches 1 m for the ramparts represented by the upper platform, 1.5 m for the high cays, and 0.7 m for the high beach-rock; the corresponding anomaly for the highest fossil microatolls (dated 3700 a B.P.) is 0.7 m.

(3) Three groups of features, which appear to be related to present sea level, have substantially younger dates. Storm ramparts represented by the lower platform, with elevations within present tidal range, date between 1460 and 380 a B.P. The lower terrace of sand cays, with heights of 3.4–4.5 m, is formed of sediments dated between 3280 and 2190 a B.P. The storm-deposited banded shingle ridges formed between 1550 and 510 a B.P.

(4) Contemporary changes shown by cartographic evidence over the last 45 years include minor adjustments of cay shores, both erosional and aggradational; the often substantial relocation of rampart tongues and ridges during storms; and the episodic extension of mangroves on reef tops.

INTERPRETATION

Similar clusterings of dates for reef features have been found in other parts of the world, notably in the Marshall Islands (Buddemeier et al. 1975) and in Micronesia (Curray et al. 1970). The factor most frequently invoked to explain such data is Holocene sea level change: in the reef seas this was emphasized by Daly (1920) and powerfully reinforced by the influential work of Fairbridge (1961). It is, however, only one of several factors which need to be considered. These may be broadly grouped as *external* and *internal* controls of changing reef morphology.

External controls

Of these controls, sea level is the most obvious, but in a macrotidal situation such as that of north Queensland it is often difficult to relate particular reef features to specific tidal levels. Thus we have shown that corals can grow in moated situations to levels more than 1 m above those of open-water reef flats. Further ambiguities arise in considering the relations between lithified features such as platforms, the sedimentary accumulations from which they are formed, and cementation processes in relation to tidal levels. Thus the surfaces of upper platforms stand up to 1.5 m above still-water h.w.s. tide levels (2.3 m), but the highest astronomical tides reach 2.9 m, storms may carry local levels considerably above this, salt spray during high-tide rough weather may wet deposits higher still, and where interstices of deposits are mud-packed, capillarity may carry internal water levels in ramparts (and hence potential cementation levels) above those of external tides. We are, therefore, hesitant in too readily adopting a higher sea level explanation for the cluster of features dated 4000–3000 a B.P., for at least four further reasons:

(a) we have nowhere found constructional reef framework of this age exposed on the northern Great Barrier Reef;

(b) as Hopley (1974, 1975) has indicated, it is possible that tectonic movement associated with structural lineations on the mainland could have differentially affected reef levels along the Queensland shelf;

(*c*) on theoretical grounds Chappell (1974) has proposed that reef levels across the shelf may also have been affected by transverse deformation;

(*d*) any interpretation is at present based on assumed constancy of tidal range. However, if tidal range had contracted in the Holocene this would explain both the existence of high-standing fossil microatolls and the absence of high-standing reef framework, without recourse to changes in mean sea level.

Clustering of ages of sediments could also be explained by absolute variations in storminess in Holocene times, affecting the rates of mechanical destruction of corals and the process of lodgement of debris on the reef flat. We know that over the period 1909–69 decennial frequencies of cyclones in the area north of Cooktown varied by a factor of two. Further, in the same period, the area south of Cooktown had nearly 45 % more cyclones than that to the north (Coleman 1971), so that latitudinal shifts over time as well as changes in absolute frequency could lead to variations in sediment supply. The only unambiguous evidence for cyclone frequencies, however, would come from the ages of large storm-deposited reef blocks. We have dates on two such blocks on the reef flat at Ingram (4310 ± 100 a B.P. for a block 130 m from the reef edge, 640 ± 70 a B.P. for a block 80 m from the edge), and also for a block (2420 ± 70 a B.P.) in upper platform deposits at Howick, but these are quite inadequate for any conclusions to be drawn. Peaks of storminess could explain periods of rampart building and hence clusters of high corals in rampart-bounded moats; but they are less likely to explain periodicities in sand cay formation, since these result from normal swash and wind processes rather than storms.

There is no evidence in the reefs studied of secular changes in wind direction over the last 6000 years, of the kind identified in this century in Indonesia by Verstappen (1954).

Internal controls

The processes of reef and island formation involve not only responses to external controlling factors but also complex interrelations between processes, some of which are self-damping and others self-reinforcing. The existence of these negative and positive feedback linkages has rarely been explicitly considered in analyses of reef geomorphology, but they add a further complication to the assessment of reef responses to presumed changes in external factors. Three such linkages may be considered.

First, as Neumann (1972) suggested, since reef growth lags behind a rising sea level, there will be a period immediately after sea level has stabilized but before reef crests reach the surface when wave activity will not be damped by the coral baffle or excluded by coral barriers. This 'Holocene energy window' will be closed by the growth of the reefs themselves, leading to an apparent decrease in energy both across reef tops and on mainland coasts protected by reefs. This is a concept more relevant to reef-bordered continental coasts than to the tops of small patch reefs, and on the northern Barrier where reefs have reached the surface at different times over the last 6000 years it does not necessarily imply a finite time interval.

Second, we might infer that as greater proportions of a reef top are covered with sedimentary accumulations such as ramparts, cays and mangroves, the area available for active carbonate production is reduced, and that ultimately a steady-state situation is reached where a given amount of sediment formed during a high-production period is resorted and relocated on the reef and where any additional sediment supply is balanced by sediment export. This could account for the absence of young sediments (other than recent storm shingle ridges) and the clustering of most sediments in the period 4000–3000 a B.P. when the reef flats were still open and productive. Not only would sediment supply be affected by such a mechanism, but, by an extension of Neumann's principle, wave energy on reef tops would be diminished by the spread of islands and mangroves. Hence this might account, for example, not only for clustering of sediment ages but also for the formation of discrete upper and lower levels on cays. That this cannot be a complete explanation is shown by the fact that upper and lower terraces with high and low beach-rock are found on reefs which are almost completely open (such as Ingram and East Hope), and the lower terrace may be absent on reef flats almost completely covered by land (such as Bewick). Further, in assigning a dominant rôle in sediment production to corals on reef tops, we may well underestimate the contributions of other calcifying organisms, notably Foraminifera and algae, in pools and grass beds. If, however, sea level had fallen marginally below its present level since 3000 a B.P., the reef tops would have been emergent and calcification completely inhibited.

Both the above are examples of negative feedback linkages. The most obvious positive feedback process is the way that once sedimentation is initiated, particularly by storm deposition of rubble and shingle on windward reef edges, these deposits both form the nucleus of further deposition and also reduce energy conditions on the reef top so that mangroves can colonize. Reef-top levels are thus rapidly raised from the level of open-water coral growth (low water neaps, 1.2 m) to the maximum of mangrove accumulation just above high water springs (2.5 m). The mangroves themselves serve as an energy baffle leading to further accumulation of shingle ridges on their windward sides. These incremental growth processes, once initiated, act rapidly, and lead to the replacement of bare reef flats by the complex array of high depositional features associated with the low wooded islands. The crucial initiating step is the formation of storm ramparts, which, *contra* Spender (1930), is less a function of reef height (which is an effect rather than a cause of the changes) than of varying energy conditions across the shelf: on outer reefs coarse sediments are spread across the flats because wave energy is high, whereas on inner, more protected reefs the coarse material is simply lodged on the windward edge, where it triggers the sequence of developmental changes (Stoddart 1965). This initiating trigger may be considered a random event in that it is contingent both on existing reef morphology and on individual cyclone behaviour, and this, together with the rapidity of subsequent morphological changes, may explain some of the spectacular differences between the surface features of closely adjacent reefs. Since sand cays are built by normal rather than by storm wave processes, through refraction at the leeward ends of reefs, their formation is independent of the sequence of events leading to the development of low wooded islands, and they have indeed a much wider distribution.

CONCLUSION

These investigations by the Royal Society and Universities of Queensland Expedition to the northern Great Barrier Reef have revealed both the complexity of recent reef history and also the ambiguity of the many kinds of evidence available for reconstructing Holocene history. Indeed we believe that no useful purpose would be served by attempting to fit our data to pre-existing sea level curves, such as that most recently revised by Fairbridge (1976). Rather we would like to suggest three areas of further enquiry which might help to resolve some of the difficulties which we have found.

First, we require a detailed knowledge of the internal structure of a coral reef of the Great Barrier system, to determine its response to changing sea level. This knowledge can only be achieved by closely spaced drilling and radiometric dating; and ideally it is needed for outer ribbon reefs, islandless platform reefs and fringing reefs, as well as low wooded island reefs.

Secondly, we need to test the hypothesis of fluctuating storm frequency during the last 6000 years by radiometric dating of sequences of storm-deposited reef blocks, both on individual reefs and on different sectors of the Barrier.

Thirdly, we need to compare the data from the reefs with the record from the mainland coast of Queensland. Hopley (1970) began this work on aggradational sequences with his study of the Burdekin delta to the south, and it now needs to be extended to the beach ridge sequences of the lower Normanby round Princess Charlotte Bay and to the Pleistocene coastal dunes north of Cooktown. Steers (1929, 1937) also noted the possible correspondence between features on the reefs and wave-cut platforms on the mainland coast and high islands, and these require investigation as integral parts of the problem of the Great Barrier Reef.

Finally, it is salutary to recall that all the events which we now so painstakingly attempt to reconstruct were witnessed by man himself. People have lived in Queensland for at least 18000 years, and undoubtedly settled on the coastal shelf, possibly in rock shelters at the foot of limestone hills which now form the reefs, before it was flooded by a rapidly transgressing sea about 8000 years ago. There is much to be learned about the features of the shelf itself to supplement the ideas put forward by Maxwell (1968), and it is possible that the archaeological record may supplement the geological and geomorphological here as well as on the islands.

REFERENCES (Stoddart *et al.*)

Bloom, A. L. 1971 In *The late Cenozoic glacial ages* (ed. K. K. Turekian), pp. 355–379. New Haven: Yale University Press.

Bowler, J. M., Hope, G. S., Jennings, J. N., Singh, G. & Walker, D. 1976 *Quat. Res.* **6**, 359–394.

Braithwaite, C. J. R., Taylor, J. D. & Kennedy, W. J. 1973 *Phil. Trans. R. Soc. Lond.* B **266**, 307–340.

Buddemeier, R. W., Smith, S. V. & Kinzie, R. A. 1975 *Bull. geol. Soc. Am.* **86**, 1581–1584.

Chappell, J. 1974 *Quat. Res.* **4**, 429–440.

Chappell, J. & Polach, H. A. 1976 *Bull. geol. Soc. Am.* **87**, 235–240.

Chave, K. E., Smith, S. V. & Roy, K. J. 1972 *Mar. Geol.* **12**, 123–140.

Coleman, F. 1971 *Frequencies, tracks and intensities of tropical cyclones in the Australian region November 1909 to June 1916.* Canberra: Department of the Interior, Bureau of Meteorology. 42 pages.

Curray, J. R., Shepard, F. P. & Veeh, H. H. 1970 *Bull. geol. Soc. Am.* **81**, 1865–1880.

Daly, R. A. 1920 *Geol. Mag.* **57**, 247–261.

Davies, P. J. 1974 *Proc. 2nd Int. Coral Reef Symp.* **2**, 573–578.

Davies, P. J., Radke, B. M. & Robison, C. R. 1976 *BMR J. Aust. Geol. Geophys.* **1**, 231–240.

Easton, W. H. & Olson, E. A. 1976 *Bull. geol. Soc. Am.* **87**, 711–719.

Fairbridge, R. W. 1961 *Phys. Chem. Earth* **4**, 99–185.

Fairbridge, R. W. 1976 *Quat. Res.* **6**, 529–556.

Flood, P. G. 1976 *25th Int. Geol. Congr. Excursion Guide* 6AC, 15–20.

Guilcher, A., Berthois, L., Doumenge, F., Michel, A., Saint-Requier, A. & Arnold, R. 1969 *Mém. ORSTOM* **38**, 1–103.

Hopley, D. 1970 *James Cook Univ. Dept. Geog. Monogr. Ser.* **1**, 1–66.

Hopley, D. 1971 *Z. Geomorph.* N.F. **15**, 371–389.

Hopley, D. 1974 *Proc. 2nd Int. Coral Reef Symp.* **2**, 703–714.

Hopley, **D.** 1975 In *Geographical essays in honour of Gilbert J. Butland* (eds I. Douglas *et al.*), pp. 51–84. Armidale: University of New England.

Hopley, D. 1977 *Proc. 3rd Int. Coral Reef Symp.*

Jukes, J. B. 1847 *Narrative of the surveying voyage of H.M.S.* Fly. (Two volumes, 423 and 362 pages.) London: T. and W. Boone.

Macintyre, I. G. & Glynn, P. W. 1976 *Bull. Am. Ass. Petrol. Geol.* **60**, 1054–1072.

Marshall, S. M. & Orr, A. P. 1931 *Scient. Rep. Gt Barrier Reef Exped. 1928-29* **1**, 93-133.

Marshall, J. F. & Thom, B. G. 1976 *Nature, Lond.* **263**, 120–121.

Maxwell, W. G. H. 1968 *Atlas of the Great Barrier Reef.* Amsterdam: Elsevier. 258 pages.

Mesolella, K. J., Sealy, H. A. & Matthews, R. K. 1970 *Bull. Am. Ass. Petrol. Geol.* **54**, 1899–1917.

Neumann, A. C. 1972 *Abstr. 2nd natn. Conf. Am. Quat. Ass.*, pp. 41–44.

Newell, N. D. & Bloom, A. L. 1970 *Bull. geol. Soc. Am.* **81**, 1881–1894.

Purdy, E. G. 1974*a Spec. Publs Soc. econ. Paleontol. Mineral.* **18**, 9–76.

Purdy, E. G. 1974*b Bull. Am. Ass. Petrol. Geol.* **58**, 825–855.

Schlanger, S. O. 1963 *U.S. geol. Surv. prof. Pap.* 260-BB, 991–1066.

Smith, S. V. & Kinsey, D. W. 1976 *Science, N.Y.* **194**, 937–939.

Spender, M. A. 1930 *Geogrl J.* **76**, 194–214 and 273-297.

Steers, J. A. 1929 *Geogrl J.* **74**, 232–257 and 341–367.

Steers, J. A. 1937 *Geogrl J.* **89**, 1–28 and 119–139.

Steers, J. A. 1940 *Geogrl J.* **95**, 30–42.

Stoddart, D. R. 1965 *Trans. Inst. Br. Geog.* **36**, 131–147.

Stoddart, D. R. 1969 *Biol. Rev.* **44**, 433–498.

Stoddart, D. R. 1973 *Geography* **58**, 313–323.

Stoddart, D. R. 1975 *Atoll Res. Bull.* **190**, 31–57.

Tayama, R. 1952 *Bull. hydrogr. Dept, Tokyo* **11**, 1–292.

Thom, B. G. & Chappell, J. 1975 *Search* **6**, 90–93.

Tracey, J. I., Jr & Ladd, H. S. 1974 *Proc. 2nd Int. Coral Reef Symp.* **2**, 537–550.

Trudgill, S. T. 1976*a Z. Geomorph.* N.F. Suppl. **26**, 164–200.

Trudgill, S. T. 1976*b Z. Geomorph.* N.F. Suppl. **26**, 201–210.

Verstappen, H. T. 1954 *Am. J. Sci.* **252**, 428–435.

Walcott, R. I. 1972 *Quat. Res.* **2**, 1–14.

Webster, P. J. & Streten, N. A. 1972 In *Bridge and barrier: the natural and cultural history of Torres Strait* (ed. D. Walker), pp. 39–60. Canberra: Research School of Pacific Studies, Australian National University.

Phil. Trans. R. Soc. Lond. B. **284**, 161–162 (1978)   [161]
Printed in Great Britain

Concluding remarks

By J. A. Steers
Professor Emeritus of St Catharine's College, Cambridge, U.K.

The Great Barrier Reef Expedition of 1928–29 and the Royal Society and Universities of Queensland Expedition of 1973 are separated in time by only 44 years. In that period coral reef studies expanded greatly and the geomorphology of reefs and reef areas became a major theme of research. Up to 1928 it is almost true to say that no work had been done on coral *islands*, and the writings on reefs were almost entirely biological.

W. M. Davis's *The Coral Reef Problem* appeared in 1928. It was the first book that examined coral reefs from a geomorphological point of view. Davis made extensive use of deductive reasoning, but he did not examine, on the ground, the structure and formation of coral islands, although he travelled widely in coral regions and made great use of charts. He did, however, emphasize and discuss at length the importance of studying reefs in relation to the coast of the land they border. Thus he enlarged on the point, first made by Dana, that drowned valleys imply not only subsidence of the land they traverse but also of the foundation of the reefs in front of that land. This was a major contribution. Two other comments made by Davis are only of subsidiary importance – the unconformable contacts of reefs on the rocks on which they rest, and the disposal of detritus in lagoons, especially in subsiding areas. Davis's conclusion was that Darwin's theory of subsidence was the most convincing explanation of the way in which barrier reefs and atolls are formed. He also, rightly, acknowledged the force of Daly's glacial control theory, but fluctuations of sea level in the Quaternary were regarded as relatively minor incidents in comparison with the much greater subsidence that was necessary to explain deep drowned valleys and the evidence of the Funafuti bore. Incidentally this bore and the much shallower one on Michaelmas Reef were the only ones that were made before 1928. Davis also demonstrated the significance of cliffing in the marginal belts and the general absence of cliffs within reefs in the truly coral seas. In short, up to about 1928 discussions on the origin of coral reefs were wholly theoretical; despite the swing towards Darwin's hypothesis there were still supporters of the theories propounded by Murray, Agassiz, Gardiner and others.

The Expedition of 1928–29 was primarily biological, but its organizers sought help from the Royal Geographical Society. That Society asked me to join the Expedition as a geographer and invited Michael Spender to accompany me. We were to work independently of the main expedition which was based at Low Isles, where living accommodation and field laboratories had been built. (Sir) Maurice Yonge, the leader, and some others visited more distant parts of the reef, and Spender and Marchant (who joined the geographical party at Cooktown) both stayed at Low Isles after my return and, with the help of other members of the expedition, compiled two large scale and very detailed maps of Low Isles and Three Isles.

The point I want to emphasize is that when Spender and I began work on the reefs we had no definite idea of what there was to do, and how we were to do it! Discussions with geographers and biologists before we left England were optimistic rather than helpful, because no one interested in geomorphology had visited the Barrier. However, I arrived at Townsville, after very helpful discussions with Professor H. C. Richards and his colleagues at Brisbane. Since Spender travelled a little later than I did, I was able to make a short visit to the Bunker and

Capricorn Islands on the lighthouse ship *Cape Leeuwin*, and then was put ashore at Mackay and went on to Townsville by train. Spender arrived on 22 August, and we began our cruise on 29 August. We covered the coast and inner reefs between the Whitsunday group of islands and Flinders Island. We had to find our problems as we sailed along the coast. Many features impressed us: both the presence and absence of cliffs on parts of the mainland and high islands, the raised platforms on the mainland and rocky islands, dead and elevated fringing reefs, shingle and sparsely scattered sand cays, sand and shingle spits, and the low wooded islands with their shingle ramparts, 'promenades', mangroves, reef flat and sand cay. We stayed a short while at Low Isles – the most southerly of the low wooded islands – with the main party and were then able to have some useful discussions on reef problems.

Earlier writers had described reefs, and atolls figure largely in the literature. But little had been said about the islands as distinct from the reefs on which they rest. Along the Queensland coast the low wooded islands (or island reefs) are unique. Somewhat similar but by no means identical islands are found in other coral seas, and J. H. F. Umbgrove and P. H. Kuenen (of the *Snellius* Expedition) were studying those in the Bay of Batavia (Djakarta) at much the same time.

The work we were able to do in 1928 provoked interest in the study of reef islands off Queensland. Earlier geological work by E. C. Andrews, H. C. Richards, C. Hedley, G. A. V. Stanley, F. Jardine, W. H. Bryan and others had called attention to the general nature and structure of the mainland coast and high islands, but virtually no work had been done on the less spectacular, but by no means less interesting, coastal features and islands.

In 1936 I was asked to revisit the reefs and continue work begun in 1928. F. E. Kemp accompanied me as surveyor. We started from Brisbane in the schooner *Cambria* on 7 May and sailed for the Bunker and Capricorn Islands, low sand and shingle islands but quite unlike the low wooded islands. Thence we sailed as far north as Cape Direction and mapped many cays and low wooded islands.

Since then and especially since the end of the 1939–45 war, a considerable amount of geomorphological work has been done on the Queensland coast by R. W. Fairbridge, W. G. H. Maxwell, E. C. F. Bird, D. Hopley and others. There was, however, need for a much more detailed examination of the northern part of the reef area. In 1967 I was able to discuss this with friends in Brisbane and Townsville and found that they viewed the possibility of another expedition with interest. On my return to England I was able to discuss the suggestion with Sir Maurice Yonge and the Southern Zone Research Committee of the Royal Society. After the Pacific Science Congress in Sydney in 1971, Sir Maurice Yonge and I visited Brisbane and Townsville for further discussion. Within a few months funds had been found to launch the 1973 Expedition with Dr David Stoddart as leader.

Since 1928 many countries, particularly America and France, have sent expeditions or research workers to many coral seas. During and after the war several deep bores were put down on atolls, new techniques of dating and analysing deposits were invented, geophysical methods of obtaining sub-surface information are now in general use, and in the last two or three decades astonishing advances in our knowledge of the oceans and seas have been made. Our views on the structure of the globe have been revolutionized. The coral reef problem remains, and must be considered afresh. With this background – and with modern instruments and techniques – the Stoddart Expedition has been able to extend the pioneer investigations of 1928 and 1936.

INDEXES

As the contents page indicates, parts A and B of this book are paginated separately, each beginning at 1. Where both A and B page numbers occur in the same entry, they are separated by a semicolon.

AUTHOR INDEX

Page numbers in bold type are the starting pages of papers in this book.

INDEX OF PLACE NAMES

TAXONOMIC INDEX

GENERAL INDEX